GAIA

AN ATLAS
OF PLANET
MANAGEMENT

GAIA

AN ATLAS
OF PLANET
MANAGEMENT

REVISED AND UPDATED EDITION
GAIA BOOKS LIMITED

General editor
Dr NORMAN MYERS

ANCHOR BOOKS
DOUBLEDAY
NEW YORK LONDON TORONTO SYDNEY AUCKLAND

A GAIA ORIGINAL

Conceived by Joss Pearson

Written by Norman Myers
with Uma Ram Nath and Melvin Westlake
and Philip Parker, Timothy O'Riordan, and
Frank Barnaby (revised edition)

Direction	Joss Pearson
	Patrick Nugent

REVISED EDITION

Editorial	Philip Parker *Project Editor*
	Michelle Atkinson
Production	Susan Walby

FIRST EDITION

Editorial	Michele Staple *Project editor*
	Roslin Mair
	John Spayne
	David Black
	Erik Ness
	Cecilia Walters
Design	Chris Meehan
	David Cook
	Marnie Searchwell
Picture Research	Elly Beintema

An Anchor Book
Published by Doubleday, a division of
Bantam Doubleday Dell Publishing Group, Inc.,
666 Fifth Avenue, New York, New York 10103

Anchor Books, Doubleday and the portrayal of an
anchor are trademarks of Doubleday, a division of
Bantam Doubleday Dell Publishing Group, Inc.

Library of Congress Cataloging in Publication Data

Gaia. an atlas of planet management/general editor,
Norman Myers. – Rev. and updated ad.
  p.   cm.
  "Anchor books."
  Includes bibliographical references and index.
  ISBN 0-385-42626-7
  1. Environmental policy. 2. Human ecology. I. Myers,
Norman.
  9.E5M94   1983
  7—dc20   92-18126
                                    CIP

ISBN 0-385-42626-7

Filmset by Marlin Graphics Ltd, Sidcup, Kent

Reproduction by F E Burman Ltd, London and
Mandarin Offset, Hong Kong

Printed in Hong Kong

Anchor Books editions: 1984, 1993

10 9 8 7 6 5 4 3 2 1

Dedication
To the poor of the world, denied their
share of the world's rich resources.
And to Jim Lovelock, whose Gaia
Hypothesis first alerted us to the idea that
we might inhabit a "living" planet. Gaia,
in ancient Greece, was Goddess of
the Earth.

Preface to the revised edition

This is no ordinary atlas. It maps and analyses a living planet at a critical point in its evolution – as one species, our own, threatens to disrupt and exhaust its life-support systems. It charts the fate of the human family, its failures and divisions, its achievements, hopes, and fears of approaching catastrophe. And it proposes that we still have the chance to redirect our course towards a sustainable future, by acting with wisdom *today*, as the caretakers of *tomorrow's* world.

In almost a decade since its first publication **Gaia: An Atlas of Planet Management** has reached and inspired millions of people throughout the world, from schoolchildren to world leaders, and become *the* definitive guide to the crisis of the planet and humanity. Its challenge has been taken up by many. The environment movement has grown worldwide, governments have recognised the concept of sustainability, and the vision of Gaia, the living planet, has begun to awaken a new love for our earthly home. There has been an enormous rise in activity and concern by individual citizens, as consumers, as voters, as campaigners, as caretakers of the Earth.

In the same decade much has changed in the world. Some of the predictions of **Gaia: An Atlas of Planet Management** have proven true, or worse than true. It has been a decade of records – of successes as well as disasters. We have experienced the highest world temperatures ever measured, focusing our minds on global warming, and levels of ozone in the upper atmosphere fell to record lows. Yet we are making huge improvements in using energy more efficiently, and the family of nations have got together to ban CFCs in an unparalleled show of global co-operation. Famine and drought have struck many regions of the developing world. Yet the unprecedented relief efforts have been accompanied by the raising of world consciousness as to the underlying causes of hunger. And efforts to provide clean drinking water for all have helped hundreds of millions of people in the developing world. Our wanton destruction of species and habitats has escalated to record highs – from tropical forests to African elephants killed for their ivory. Yet we have made a landmark agreement to conserve the pristine environment of Antarctica from exploitation, and negotiated international treaties to begin to conserve biodiversity. Our economic troubles have led to world recession; poverty, debt, and hunger are affecting more people than ever before. But these very problems are sparking new economic approaches.

The worst news is that the twin pressures of human numbers and overconsumption of the Earth's resources are still increasing. Our numbers are now predicted to rise not to 8 billion, but to 11 or 12 billion. The gap between the haves and have-nots, between the resource-intensive North and low-income South, is still increasing.

Damage to the land, pollution, and destruction of species and habitats are proving intractable under this relentless pressure. But many nations are now making major efforts to conserve forests and land, improve human health and access to birth control, and conserve and recycle resources.

The best news has been the end of the Cold War, sudden and inspiring, and the rise of global concerns for democracy, internationalism, and peace. Global military expenditure reached a record high of $1000 billion a year, but the peak is passed and we now live in a world less fearful of nuclear conflict, and with a record number of countries enjoying liberal democracies. These changes have shown that a startling revolution in human perception is possible – and so given us hope. We may yet achieve an about-turn from our course to environmental collapse. We may yet leave for our children a world community that seeks to live without war and in balance with Gaia.

This completely revised edition of **Gaia: An Atlas of Planet Management** addresses these new directions and new issues in today's context of urgency. Threats have become reality; hopes have become plans and worldwide efforts for change. Thoroughly updated for the 1990s, with a range of new topics and illustrations, the Atlas now tackles the agenda to 2000 and beyond.

We are particularly honoured by the contributions of leading "thinkers and doers" from North, South, East, and West, for this revised edition. The chapters that follow are introduced by: a leading representative of the indigenous people; highly influential scientists and environmentalists from East and West; the Directors of two of our leading international organizations; and one of the foremost exponents of grass-roots action. These individuals represent the diversity of human thought and action we need to tap to secure our common future.

The book takes a clear and structured approach to this challenge. It is divided into seven topics — Land, Ocean, Elements, Evolution, Humankind, Civilization, and Management – and for each topic it organizes the mass of data, analyses, and opinions available into three perspectives: Potential as a sustainable resource; Crises; and Management alternatives. This structure allows the reader to examine any critical area of concern and weigh up: first, what it has to offer; second, why, where, and how things are obviously going wrong; and third, how we might set about doing things right.

More than a structure for a book, this analytical formula offers one possible approach to planet management – the art of bringing our human activities within the capacities of the planet. We hope that this new edition of **Gaia: An Atlas of Planet Management** will continue to spur the rising tide of change and global debate on our future prospects.

CONTENTS

Foreword

by Gerald Durrell

I am delighted that this most important book is coming out in a new, updated edition. Now, when at last the whole world seems conscious of the biological dangers that beset it, this is a most timely book for everyone concerned with the future of the planet. It explains our place on this Earth and the damage we are doing to ourselves. But it is not merely another "gloom and doom" book, for it shows us how we can mend our ways to our advantage.

This book is, in fact, a sort of blueprint for our survival – both for you who are reading this Foreword and I who am writing it. It is an attempt to show what a complex and magnificent world we have inherited, how it works and, most important of all, what bad stewards we often have been, and are still being, of this inheritance. It shows how we are plundering our planet in the most profligate and dangerous way, but it also shows what we can do to redress the situation.

Anyone reading and absorbing the message in this book surely must conclude that nearly all the ills that beset us, from starvation and disease to war itself, can be traced back inexorably to three root causes: overpopulation, political stupidity, and wasteful misuse of the planet's treasures, both finite resources, and renewable living wealth.

We are told that Adam and Eve were banished from the Garden of Eden into the world. We have – ever since we started walking upright-set about the task of banishing ourselves from our own Garden of Eden – our planet Earth. Perhaps banishment is the wrong word to use, for it assumes that there is somewhere else we can be banished to, and in our case, once we have ruined and used up this Eden there is no other, no second world hanging in the sky that we can all blithely move to, as if we were changing houses. This beautiful and endangered planet is the only one we have.

At the present rate of "progress", and unless something is done quickly, disaster stares us in the face. Erosion, desertification and pollution have become our lot. It is a weird form of suicide, for we are bleeding our planet to death. We are led by sabre-rattling politicians who are ignorant of biology, beset by sectarian groups noted for their narrow-mindedness and intolerance, surrounded by powerful commercial interests whose only interest in nature is often to rape it. We are misguided and misled, trotting to oblivion as obediently as the Gaderene Swine. I wonder what the attitude will be if, in a hundred years time, this book is read by our starving grandchildren – and they see that this decimation of their inheritance was recognized, cures for it were available, and still nothing was done?

Make no mistake about it, a cure is possible, as is pointed out in this book. But it requires a worldwide effort on everyone's part. Now that the threat of nuclear war is receding, perhaps instead of the billions that were being spent on the arms race, this money can be more profitably used to fight the environmental battles which are still in full swing and getting hotter. We are, all of us, regardless of race, colour or creed, facing the same biological problems, all of which directly or indirectly are of our own making. Think of some of the facts revealed here and then ask yourself whether we have not been blandly ignoring a threat which would make even nuclear war seem pale in comparison.

We are facing, over the next three decades, an average loss of one species per hour in tropical forests alone, and this is a very conservative estimate. These are species that for the most part we know little or nothing about, and could well be of enormous benefit to humanity. By felling tropical forests in the senseless way that we are doing, who knows what riches we are squandering? These tropical forests, once eliminated, can never be restored. They are gone, taking with them untold numbers of foods, drugs, and other products useful to humanity. With them go innumerable birds, mammals, reptiles, and insects, many of which could be of use to us, as the humble armadillo has helped to treat leprosy, and the owl monkey has assisted in treating malaria. We all know people who have attics or cupboards filled with an astonishing assortment of things, preserved because "they might come in useful". Well, the world is our attic, and we should preserve everything in it, for we do not know when it will "come in useful".

The world is still an incredibly rich storehouse (our *only* storehouse). It can (if we live with nature and not outside it) provide all of us with all we and future generations need. But we must learn to manage it intelligently. Most of nature – if it is not destroyed or corrupted by us – is a resource that is ever renewing itself. It offers us, if managed wisely, a never-ending largesse. It is our world, but we have yet to learn to treat it with respect and gratitude. Let's hope we do so before it is altogether too late, and we find ourselves breeding like a mass of greenfly on a cinder.

INTRODUCTION
The fragile miracle

The sphere of rock on which we live coalesced from the dust of ancient stars. Orbiting round the huge hydrogen furnace of the sun, bathed by radiant energy and the solar wind, the globe is white hot and molten beneath the crust: continents ride in a slow dance across its face, ocean floors spread. And between its dynamic surface and the vacuum of space, in a film as thin and vibrant as a spider's web, lies the fragile miracle we call the biosphere.

When the first astronauts circled the Earth in their tiny craft, millions of listeners heard them describe the beauty of this planet, "like a blue pearl in space", and were caught up in a moment of extraordinary human revelation. Since then, much has been written about "Spaceship Earth", on whose finite resources we all depend. And the more we explore the solar system, the more singular we understand our world to be. The atmospheric mix of gases, for instance, is entirely different not only from that of nearby planets but from what would be predicted by Earth's own chemistry. This "improbable" state of affairs appears to have arisen alongside the evolution of life, and persisted (with minor fluctuations) despite all possible accidental perturbations of cosmic travel, for perhaps two billion years. Life, by its very presence, is apparently creating, and maintaining, the special conditions necessary for its own survival.

It was a group of space scientists devising life-detection experiments for other planets who first stumbled on this phenomenon of the self-sustaining biosphere – and named it Gaia, the living planet. Since then, we have begun to learn much more about the planetary life-support systems which rule our lives – sadly, mainly by disturbing them.

Within this life realm, every organism is linked, however tenuously, to every other. Microbe, plant, and mammal, soil dweller and ocean swimmer, all are caught up in the cycling of energy and nutrients from sun, water, air, and earth. This global exchange system flows through various transport mechanisms, from ocean currents, to climate patterns and winds; from the travels of animals to the processes of feeding, growth, and decay. Information, too, flows through the bios-phere – reproduction transfers the store of genetic coding to new generations and creates new experi-ments; learning and communication occur between individuals. And throughout the life zone, change and diversity, specialization and intricate inter-dependence, are found at every level.

It is with this remarkable planet, and what we are doing to it, and to ourselves, that this book is concerned. UFOs apart, we are unlikely to find another Gaia – should we destroy the one we have.

The protective atmosphere
Like the feathers of a bird, the 9 or 10 layers of the atmosphere provide equable surface temperatures, shield life from the "rain" of cosmic particles, and block lethal ultraviolet radiation.

Exosphere

Ionosphere

O_2

Photosynthesis

Stratosphere
Troposphere

C

Planetary life-support systems
The living world, or biosphere, stretches around our planet in a film as thin as the dew on an apple. A hundred kilometres down beneath our feet, the globe is already white hot, at 3000°C. Thirty kilometres up above our heads, the air is too thin and cold for survival. In between, the green world flowers, richest around the tropical zones where the ice age glaciations have never reached. Here, in the tropical forests and the shallow sunlit seas and reefs, much of Earth's living wealth of species is concentrated.

The Earth's green cover is a prerequisite for the rest of life. Plants alone, through the alchemy of photosynthesis, can use sunlight energy, and convert it to the chemical energy animals need for survival. It was the emergence of photosynthesizing algae in the oceans which first released free oxygen into the atmosphere – a cataclysmic event for existing life forms, but a pre-condition for present-day existence. The ocean microflora still supply 70% of our oxygen, and this in turn maintains the protective ozone layer in the upper atmosphere. The oceans act as a "sink" for carbon dioxide from the air.

Plant cover provides the basis of all food chains, mediates water cycles, stabilizes microclimate, and protects the living soil – the foundation of the biosphere. Legions of soil micro-organisms, and of anaerobic microbes in the shallow muds of sea floor and swamp, work ceaselessly to recycle decaying matter back into the nutrient system.

3000°C
100 kilometres down

Temperature feedback

Earth's life-support systems are still little understood. Surface temperatures have remained suitable for life for several aeons, despite changes in solar flux. Feedback systems for temperature control are the levels of carbon dioxide and water vapour in the air – both affected by plant cover. Carbon dioxide acts as an insulator – the "greenhouse effect" (pp 110-11). Another is Earth's "albedo" – its shininess.

High albedo from light-covered areas cools the Earth. Most of Gaia's albedo value comes from cloud cover (influenced by vegetation), and from ice caps and oceans. Microflora in the oceans and plants on the land can darken or lighten these areas, thus altering their albedo. Atmospheric pollution raises carbon dioxide levels, while forest clearance and desertification raise albedo. We must act with care.

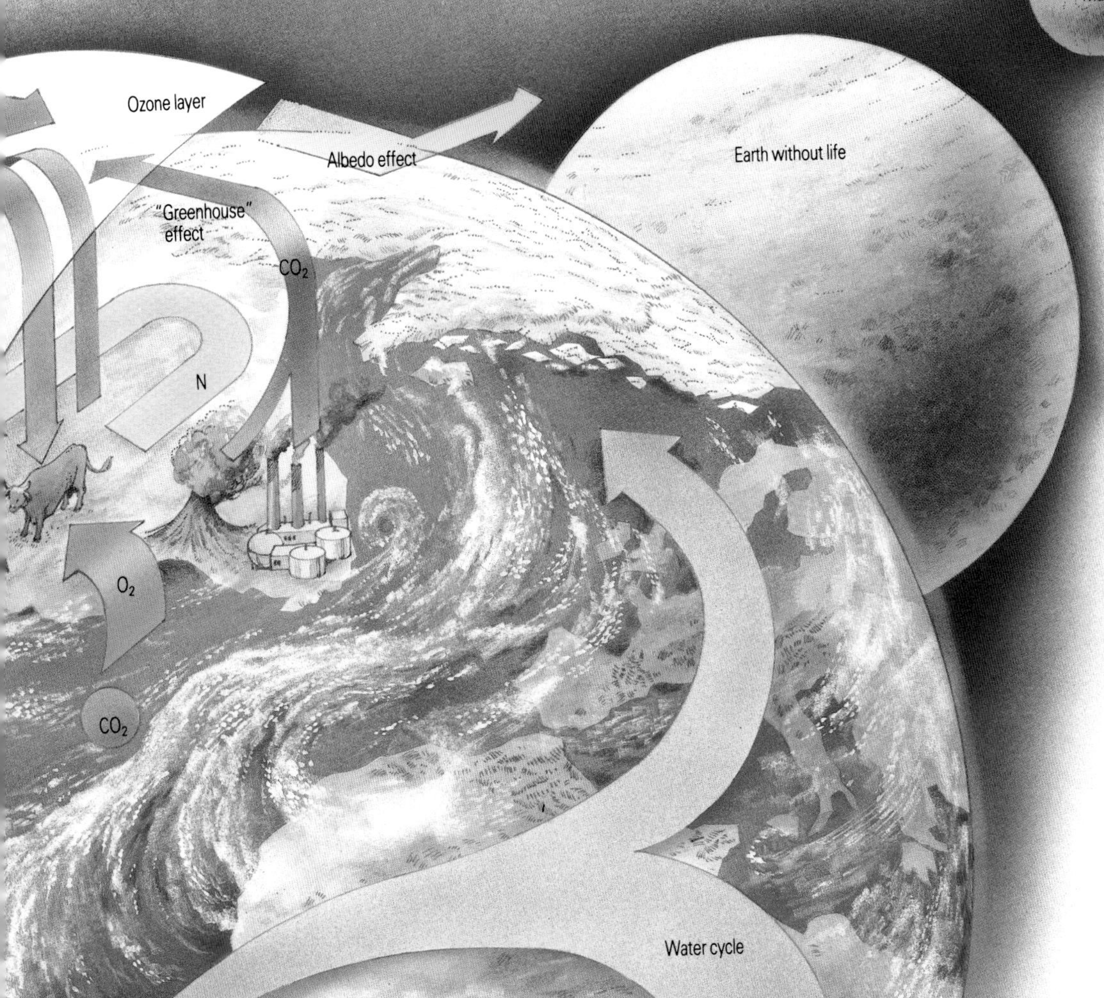

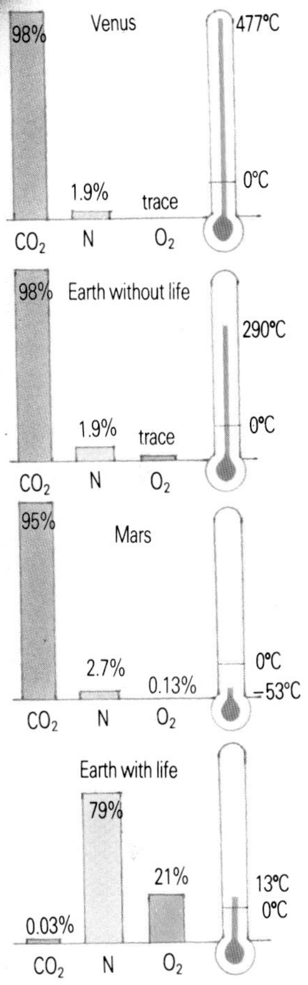

Venus

Mars

Ozone layer

Albedo effect

"Greenhouse" effect

CO_2

N

O_2

CO_2

Earth without life

Water cycle

Venus			477°C
98%			
	1.9%	trace	0°C
CO_2	N	O_2	

Earth without life			290°C
98%			
	1.9%	trace	0°C
CO_2	N	O_2	

Mars			0°C
95%			
	2.7%	0.13%	−53°C
CO_2	N	O_2	

Earth with life			13°C
	79%	21%	0°C
0.03%			
CO_2	N	O_2	

The unique planet

The most remarkable characteristic of living matter is that it is self-organizing. In contrast with the overall trend towards disorder or "entropy" evident in the universe, life *creates* order from the materials around it, exporting waste in the process. Thus life has the capacity to influence its environment.

When space scientists began devising life-detection experiments, one group suggested that a life-bearing planet might show an unexpected mix of gases in its atmosphere if life's chemistry were at work. When they looked at Earth in this light, their predictions were borne out with a vengeance. Earth's mix of gases, and temperature, were hugely different from what they predicted for a "non-living" Earth, as well as from neighbouring planets (above). The fact that these conditions appeared to have arisen and persisted alongside life led to the Gaia hypothesis – the proposal that the biosphere, like a living organism, operates its own "life-support" systems through natural feedback mechanisms.

Accelerating evolution

Though we wake to a largely humanized world of suburbs and cities, governments and wars, each of us carries within us the birth and death of stars, and the long flowering of Gaia.

Evolution is usually dated from the emergence of life, that "almost utterly improbable event with infinite opportunities of happening" (J. Lovelock). But this event itself was a stage in a process that has continued since time, as we know it, began – when, some 15 billion years ago, the Big Bang sent pure *energy* flooding out into a waking universe.

As this energy dispersed and the universe cooled, a patterning set in, and stable "energy structures" emerged as in a new order, *matter*. Over billions of years, the particles, atoms, and elements of matter formed and were processed and reprocessed in the heart of stars, until a higher order emerged, *life*.

Our probes into space have found life's chemical precursors widely distributed – indeed, space seems to be littered with the "spare parts of life" awaiting the right conditions for assembly. On our primeval planet, these conditions were found: the fierce energies of radioactivity and ultraviolet radiation, the abundant presence of hydrogen, methane, ammonia, and water. In Earth's oceans, the first strands of DNA, and then the self-replicating double helix, must have formed and broken countless times. But once the seed was set, the birth of the biosphere had begun.

Over nearly four aeons, the experiments, the increasing diversity and complexity continued until, as the life-support systems of our planet stabilized, a still higher order of complexity emerged – *intelligence* and *conscious awareness*.

Throughout its 15 billion years, the pace of the universe's development has been accelerating, each new wave of innovation building up to trigger the next, in a series of "leaps" up to further levels of diversification and change. Compress this unimaginable timescale into a single 24-hour day, and the Big Bang is over in less than a ten-billionth of a second. Stable atoms form in about four seconds; but not for several hours, until early dawn, do stars and galaxies form. Our own solar system must wait for early evening, around 6 p.m. Life on Earth begins around 8 p.m., the first vertebrates crawl on to land at about 10.30 at night. Dinosaurs roam from 11.35 p.m. until four minutes to midnight. Our ancestors first walk upright with ten seconds to go. The Industrial Revolution, and all our modern age, occupy less than the last thousandth of a second. Yet in this fraction of time, the face of this planet has changed almost as much as in all the aeons before.

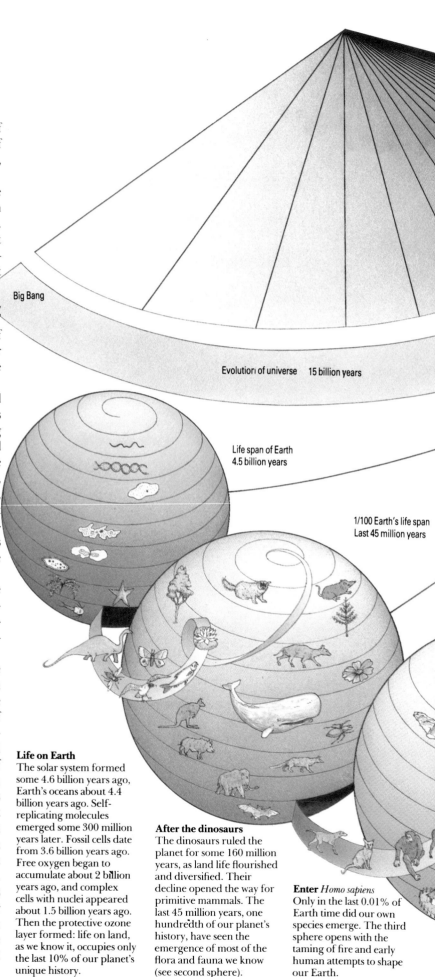

Big Bang

Evolution of universe   15 billion years

Life span of Earth
4.5 billion years

1/100 Earth's life span
Last 45 million years

Life on Earth
The solar system formed some 4.6 billion years ago, Earth's oceans about 4.4 billion years ago. Self-replicating molecules emerged some 300 million years later. Fossil cells date from 3.6 billion years ago. Free oxygen began to accumulate about 2 billion years ago, and complex cells with nuclei appeared about 1.5 billion years ago. Then the protective ozone layer formed: life on land, as we know it, occupies only the last 10% of our planet's unique history.

After the dinosaurs
The dinosaurs ruled the planet for some 160 million years, as land life flourished and diversified. Their decline opened the way for primitive mammals. The last 45 million years, one hundredth of our planet's history, have seen the emergence of most of the flora and fauna we know (see second sphere).

Enter *Homo sapiens*
Only in the last 0.01% of Earth time did our own species emerge. The third sphere opens with the taming of fire and early human attempts to shape our Earth.

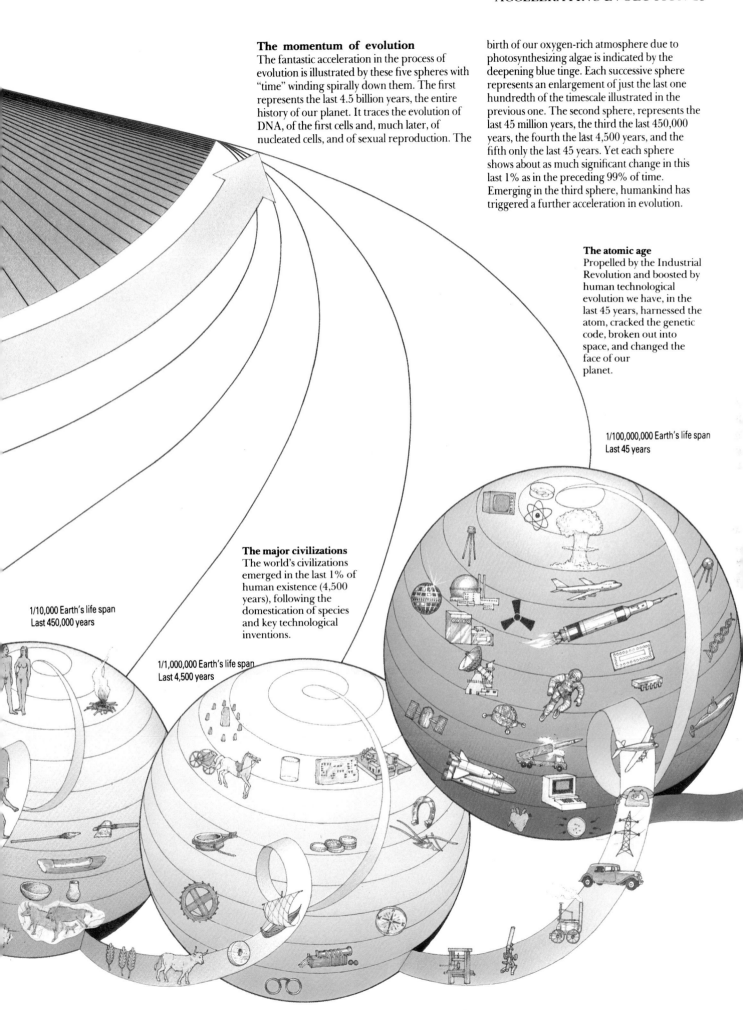

The momentum of evolution
The fantastic acceleration in the process of evolution is illustrated by these five spheres with "time" winding spirally down them. The first represents the last 4.5 billion years, the entire history of our planet. It traces the evolution of DNA, of the first cells and, much later, of nucleated cells, and of sexual reproduction. The birth of our oxygen-rich atmosphere due to photosynthesizing algae is indicated by the deepening blue tinge. Each successive sphere represents an enlargement of just the last one hundredth of the timescale illustrated in the previous one. The second sphere, represents the last 45 million years, the third the last 450,000 years, the fourth the last 4,500 years, and the fifth only the last 45 years. Yet each sphere shows about as much significant change in this last 1% as in the preceding 99% of time. Emerging in the third sphere, humankind has triggered a further acceleration in evolution.

The atomic age
Propelled by the Industrial Revolution and boosted by human technological evolution we have, in the last 45 years, harnessed the atom, cracked the genetic code, broken out into space, and changed the face of our planet.

1/100,000,000 Earth's life span
Last 45 years

The major civilizations
The world's civilizations emerged in the last 1% of human existence (4,500 years), following the domestication of species and key technological inventions.

1/10,000 Earth's life span
Last 450,000 years

1/1,000,000 Earth's life span
Last 4,500 years

Latecomers to evolution

Gaia went its creative way for several billion years, becoming steadily more diverse, complex, and fruitful. Then, in the last few seconds of life's "evolutionary day", *Homo sapiens* appeared – a creature that has wrought changes as great as several glaciations and other geological upheavals together, and has done it all within a flicker of the evolutionary eye. The evolution of *Homo sapiens* has produced a being that can think: a being that is aware, that can speculate about tomorrow.

Evolution has also equipped us to create our own form of planetary ecosystem. Whereas natural selection works through a trial-and-error process, undirected and unhurried, we can choose preferred forms of evolution, creating changes that might otherwise have taken millions of years to occur.

The greatest natural development through evolution in terms of energy conversion was the emergence of photosynthesis, two billion years ago. A mere 50,000 years ago we learned to harness fire, and thus to use the stored energy of plants in the form of wood. A few hundred years ago we moved on to exploit coal, then oil. Now, however, we are on the verge of widespread exploitation of the sun's energy through solar cells – potentially as marked an advance for Earth's course as that of photosynthesis itself. Similar breakthroughs include domestication of wild species and genetic engineering: quantum leaps to match the evolution of sexual reproduction.

Among the greatest advances of all is our ability to control disease, and thus to increase our numbers. Within the last 150 years, the human population has grown from around one billion in the 1830s, to two billion in the 1930s, to four billion in 1975, and to five billion in 1987, with a further increase to 6.25 billion projected by the end of the century. Herein we witness the phenomenon of exponential growth, a process that marks not only our increasing numbers, but also our consumption of energy and resources, our accumulating knowledge, and our expanding communications network.

Exponential growth is one of the most important concepts we shall encounter in this book. It is growth that is not simply additive (two plus two equals four and another two makes six); rather it is self-compounding (two multiplied by two equals four, multiplied by two equals eight). Very few people realize its implications for our future existence on Earth. If Africa, for example, maintains its present three percent growth rate until this time next century, its current 630 million people will increase to 9.5 billion.

The advance beyond our entrenched expectation for exponential growth in consumption will probably represent the greatest evolutionary leap of all.

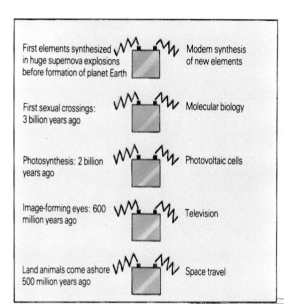

First elements synthesized in huge supernova explosions before formation of planet Earth — Modern synthesis of new elements

First sexual crossings: 3 billion years ago — Molecular biology

Photosynthesis: 2 billion years ago — Photovoltaic cells

Image-forming eyes: 600 million years ago — Television

Land animals come ashore 500 million years ago — Space travel

Evolution: revolution
We stand at a unique point in evolution, when planet-wide changes are proceeding faster than ever before. Four examples of accelerating growth are shown here, right. With the invention of the space rocket, our unique species is able to leave the biosphere and travel into space – a development as revolutionary as the step from water on to land 500 million years ago. Similar breakthroughs are shown in the box, left.

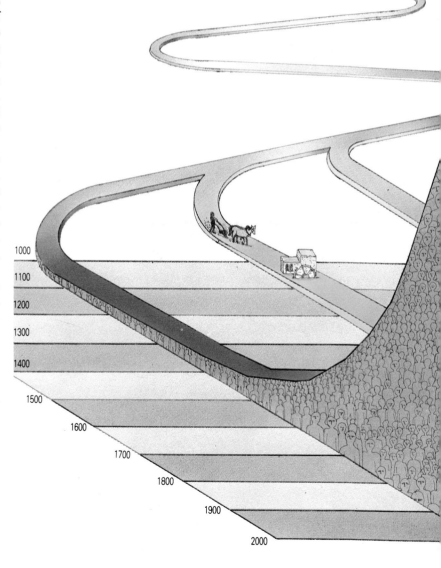

Population
(billions)

Energy
Consumption per capita
(watts)

Information
(number of scientific journals)

Mobility
(km per hour)

5

4

3

2

1

11,000

10,000

9,000

8,000

7,000

6,000

5,000

4,000

3,000

2,000

1,000,000

100,000

10,000

1,000

100

10

25,000

10

The long shadow

Today, the rise of human numbers casts a shadow over planet Earth. We have reached a total of five and a half billion people, and we are plainly failing to feed, house, educate, and employ many of these in basically acceptable fashion. Worse, the human community is projected to reach at least ten billion before the population explosion fizzles out into zero growth early in the 22nd century.

The problem does not lie only in a sheer outburst of human numbers. It lies also in an outburst of human consumerism. One billion over-affluent people enjoy lifestyles that impose a grossly disproportionate pressure on our planetary ecosystem. This consumerism is powered in turn by a sudden expansion in technological know-how, enabling us to use and misuse ever-greater stocks of natural resources – even to use them up. In fact, rather than a "population crisis" or a "resource crisis", we should speak of a single over-arching crisis: the crisis of humankind. The shadow stems from all of us, and it will darken all our lives.

On land, we plough up virgin areas, even though most of them are marginal at best. Soil, one of the most precious of all resources, is washed or blown away in billions of tonnes every year. To compound this tragedy, large tracts of productive cropland are paved over each year, or "developed". Deserts expand, or rather degraded lands are tacked on to them, at a rate threatening a third of all arable land in the next 75 years. Forests in the tropics are chopped down with a zest that will leave little by the middle of the next century. As the forests fall, species in their millions lose their habitats, many of them disappearing for ever.

In the oceans, we ravage one fishery after another. We cause dolphins, seals, and other marine mammals to follow the sad track of the great whales. We pollute the seas, just as we poison lakes and rivers in virtually every part of the world. We use the skies as a dustbin, and we desecrate our landscapes with growing piles of refuse, some of it toxic. In the atmosphere, we disrupt the carbon dioxide balance, triggering climatic dislocations that will upset agriculture worldwide.

Not surprisingly, this overtaxing of the Earth's ecosystem leads to breakdowns of other sorts. As more people seek greater amounts of declining resources, conflicts erupt: more people have been killed through military conflagrations since World War II than all the soldiers in that war. In fact, it is breakdown in our social systems, our economic structures, and our political mechanisms that generate the greatest threat of all. The shadow over planet Earth will never be deeper and darker than when it is lengthened by a mushroom cloud.

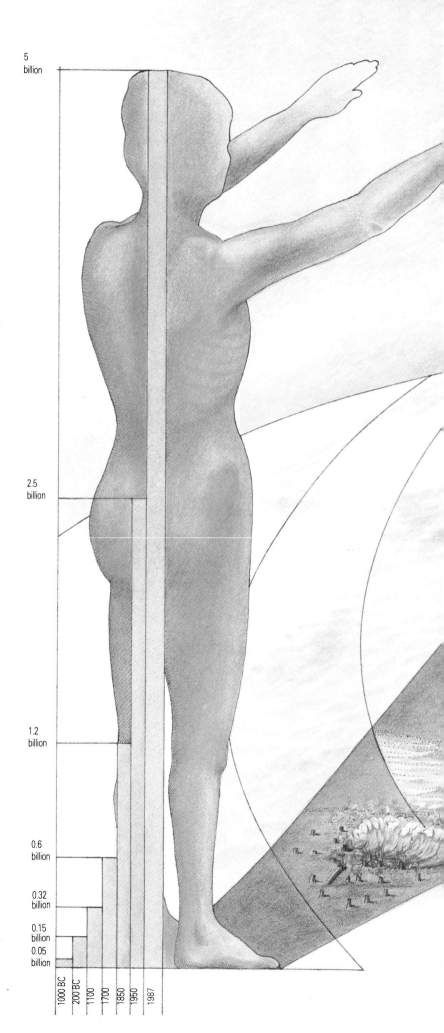

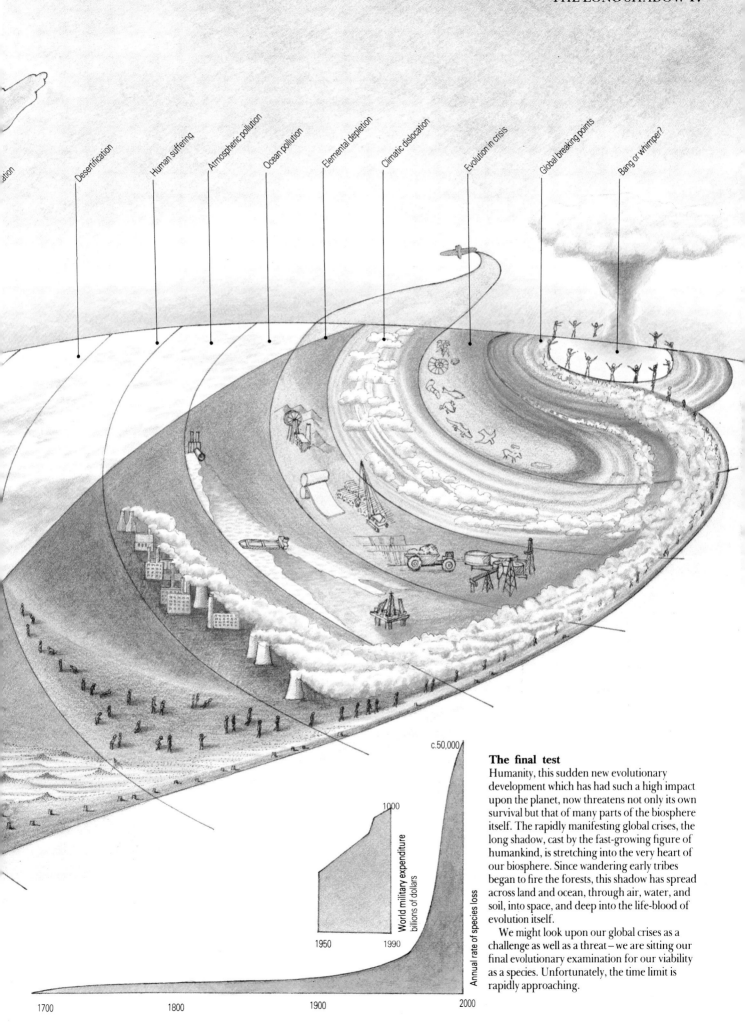

Desertification · Human suffering · Atmospheric pollution · Ocean pollution · Elemental depletion · Climatic dislocation · Evolution in crisis · Global breaking points · Bang or whimper?

World military expenditure
billions of dollars

c.50,000

1000

1950   1990

Annual rate of species loss

The final test

Humanity, this sudden new evolutionary development which has had such a high impact upon the planet, now threatens not only its own survival but that of many parts of the biosphere itself. The rapidly manifesting global crises, the long shadow, cast by the fast-growing figure of humankind, is stretching into the very heart of our biosphere. Since wandering early tribes began to fire the forests, this shadow has spread across land and ocean, through air, water, and soil, into space, and deep into the life-blood of evolution itself.

We might look upon our global crises as a challenge as well as a threat – we are sitting our final evolutionary examination for our viability as a species. Unfortunately, the time limit is rapidly approaching.

1700   1800   1900   2000

Crisis or challenge?

Humankind can be seen as either the climax of evolution's course, or as its greatest error. No other creature is a fraction so precocious. No other can think about the world, plan to make it better, and dream of the best possible. Yet no other reveals such capacity for perverse behaviour – for gross misuse of its habitat and for reckless proliferation of numbers, without thought for the consequences.

In certain senses, humanity is becoming a super-malignancy on the face of the planet, spreading with insidious effect and fomenting ultimate crisis in covert fashion. A cancer cell is unusually vital, since it replicates itself with remarkable vigour; it is also exceptionally stupid, since it ends by killing the host upon which it depends for its survival. But unlike the cancer cell, we are coming to realize the nature of what we are doing. Can we learn fast enough, act soon enough?

This is not the first time that the Earth's community has encountered crisis. Gaia has even benefited from periodic upheavals. Were it not for the dramatic demise of the dinosaurs, there would have been scant opportunity for mammals to become pre-eminent – with all that has meant for the supreme mammal, *Homo sapiens*. Out of crisis can come advance, provided the impetus of change does not "overshoot" into catastrophe. On past occasions, there have been thousands of years, even millions, for the corrective workings of the biosphere to adapt and adjust to new stresses. This time, there are just a few decades, far too short a period for Gaia to work its restorative course, unless it is done with the symbiotic support of humankind.

If we can match up to the crisis, Gaia may well move forward into an unprecedented period of development – development in its proper broad sense, embracing development of Earth's resources and of humanity's capacity for caring. If, however, we fail, *Homo sapiens* could eventually be discarded as an evolutionary blind-alley.

To achieve a breakthrough, we must learn a tough lesson. While it is often all right to adapt through small steps, improving an established course through "fine tuning", there are times when one must do an "about turn", and take more drastic corrective action. The tale of the French schoolchildren and their experimental frog is salutary. They took the frog and dropped it into a saucepan of boiling water, whereupon the frog skipped right out – instant rejection of an environment that proved distinctly unsuitable. But when the schoolchildren dropped the frog into a saucepan of cold water, and slowly heated it up, the frog swam round and round, adapting itself to the rising heat . . . until it quietly boiled to death.

Crises have both positive and negative characteristics. They can represent a threat to the status quo but at the same time can be seen as a symptom that something is wrong. They thus represent an opportunity to correct an imbalance and move on to a new level of organization. This is reflected in the Chinese character "ji" (here painted by a classical Chinese calligraphist) indicating both a crucial point, and an opportunity.

"Once a photograph of the Earth, taken from the outside is available ... a new idea as powerful as any other in history will be let loose".
FRED HOYLE (1948)

LAND

Introduced by Jorge Terena
Founder of the Nucleus for Indigenous Rights
and a representative of the Union of Indian Nations

My people, the Terena people, live in western Brazil near the Bolivian border, where the Amazon really starts. Many of the farmers who have moved into our territory use their land for cattle raising. It is because of this and the sugar cane plantations that most of our trees, our forests, and our land have gone. As children we would often go hunting and we always used to find deer, wild pigs, and tapirs in the forest nearby. Now they are very scarce. Even the fish have gone. We have started a project to replenish the land and try to bring the ecosystem into equilibrium again. It began at the Indian Research Centre in Goiania, but involves the regeneration of many different types of area: savannah, pre-Amazon, and Amazon. When the government demarcated our land, part of it was already degraded. We are trying to replenish this, although we cannot touch the farmers' land all around us. The farmers have left huge holes in the forests, but we are trying to fix these holes; if we don't, we will not survive.

We have to help the forest to regenerate because once the larger trees have been cut, the smaller ones stop growing; they never reach their full height. We have to plant the trees that used to be there, and maybe they will help each other as they grow. Once this is done the animals and birds come back because they depend on these trees and their fruits, just as we do. Also, in a deeper way, we need the environment spiritually; it is where our spirits live. If our spirits go away my people will walk, wandering, with no guide.

The Indian Research Centre is also working with the University of Goias so that we can learn techniques that will help us to speed up the regeneration of the forest. Although what non-indigenous people call "development" has destroyed our land and culture, we now need some of that culture's knowledge to undo the damage. Six students have studied biology at the university, in order to understand the ecosystems of the different regions that have been devastated by sophisticated agriculture. The students have graduated, and are completing a final course before returning to their villages to use appropriate technology with traditional methods to reintroduce native species of trees, fruits, and animals. Not many forest people can get to the Research Centre, so it is moving to them. Part of the Centre is moving into the forest to meet the needs of indigenous people and rubber tappers.

Non-indigenous people do not really understand our needs. They are only concerned about their own needs. I think they people have somehow lost their spiritual touch. I have been to the USA, England, and Italy, and everywhere I went I was surrounded by big buildings. As I looked at the people, I saw them walking around, really sad, not knowing where to go. I came to the conclusion that these people have lost contact with nature.

As I look at the people, the places, and the countries where I go, I see that, economically speaking, they are very powerful, and in a material way have everything they need. My people do not have the technology, the money, the buildings – but we do have our forests and we are happier. Many non-indigenous people do not understand that because of their greed they have lost contact with their spirituality and the spirit of the land. It is something that hurts me to see.

THE LAND POTENTIAL

Marvellous stuff, soil. Sterile and boring as it may appear, often mucky too, this thin layer covering the planet's land surface is the biosphere's foundation, our primary resource.

Soil is as lively as an army of migrating wildebeest and as fascinating, even beautiful, as a flock of flamingoes. Teeming with life of myriad forms, soil deserves to be classified as an ecosystem in itself – or rather, as many ecosystems. One hectare of good-quality soil in a temperate zone may contain at least 300 million small invertebrates – mites, millipedes, insects, worms, and other mini-creatures. As for micro-organisms, a mere 30 grams of soil may contain one million bacteria of just one type, as well as 100,000 yeast cells and 50,000 bits of fungus mycelium. Without these, the soil could not convert nitrogen, phosphorus, and sulphur to forms available to plants.

According to Harvard's Professor Edward O. Wilson, there is far more biological complexity in a handful of soil in Virginia than on the entire surface of Jupiter. Yet we invest more money in exploring the planets than in finding out how our basic life-support systems work here on Earth.

So next time you tread on earth (as opposed to concrete or asphalt), take a look at what lies at your feet. It is likely to be a fairly loose material, half made up of masses of tiny particles, the other half of water and air. This curious assembly of inorganic constituents derives originally from rock which, being weathered by rainwater, atmospheric gases, ice, and roots, has slowly broken down into a form in which it can support multitudes of life-forms. These, in turn, enable it to support plants. Through a constantly self-reinforcing process, the soil becomes enriched with dead organic matter, some of which is called humus – the stuff that helps to make soil fertile.

The process of soil formation is slow. At best, even when sediments build up quickly, formation of 30 centimetres may take 50 years. More usually, when new soil is formed from parent rocks, one centimetre may need from 100 to 1,000 years. So to form soil to the depth of this page could take as long as 10,000 years. Unfortunately, reversing the process by human or natural disturbance is all too quick – soils can be degraded in a fraction of the time they take to form (see p. 37).

Of the Earth's total ice-free land surface, only 1.5 billion hectares (about 11 percent) can be readily cultivated. With human management, however, and

The fertile soil

Not all the soil which covers the Earth's ice-free land surface is suited for growing crops. In fact, of the total area of some 13 billion ha (about one-quarter of the globe), a mere 11% presents no serious limitations to agriculture. The rest is either too dry, too wet, too poor in nutrients (mineral-stressed), too shallow, or too cold.

As much as 28% of the world's land surface suffers from drought (not so surprising, when you consider that the Sahara occupies nearly a billion hectares). Mineral-stressed soils account for a further 23%. Soils that are too thin to be much use cover 22%, while waterlogged soils account for 10%. Permafrost soils (ground that is permanently frozen) cover 6% – and these do not include Antarctica or Greenland.

The illustration shows how the world's soil resources are shared out, each segment of the "globe" showing the percentage of total land area and proportions of soil types for the continent it represents. Notice how the fertile soil is far from evenly distributed, with Europe claiming the largest portion relative to its land area. The pie-charts on the continents show how the land is being used: for forest, grazing, cultivation, or "other" (wild, waste, or urban).

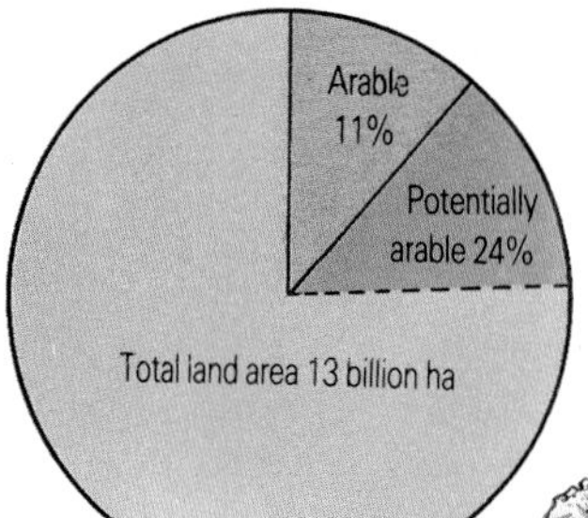

Arable 11%
Potentially arable 24%
Total land area 13 billion ha

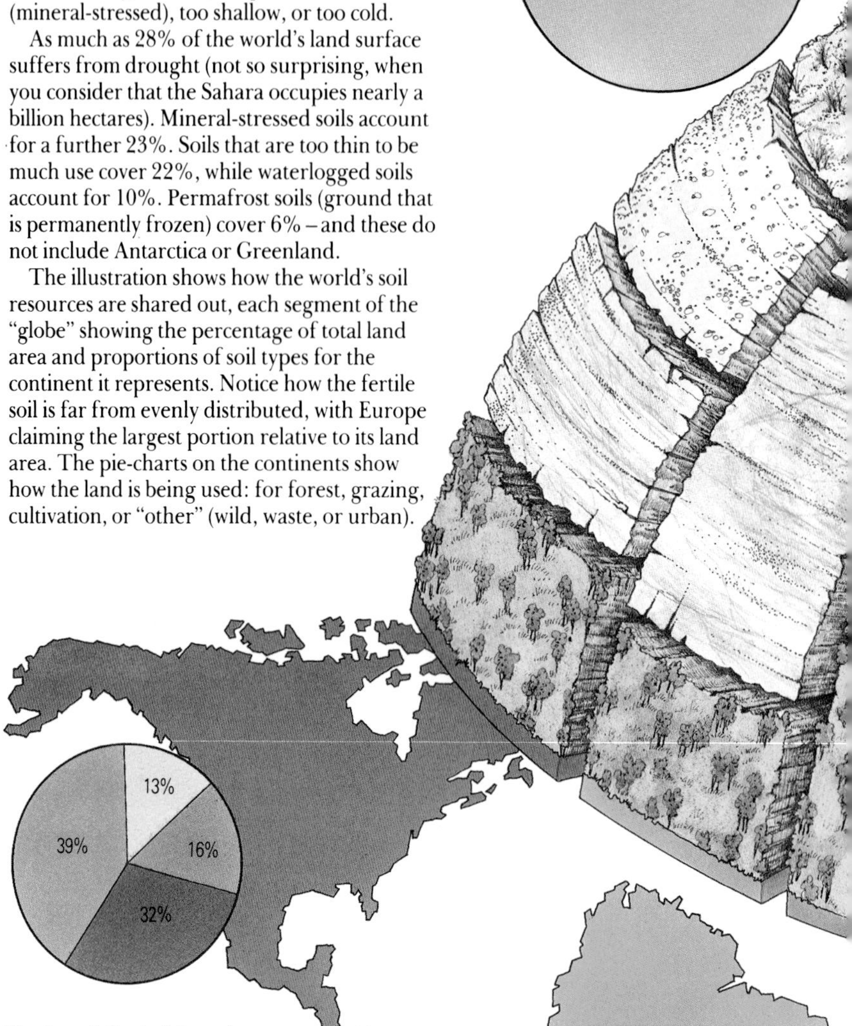

13%
16%
32%
39%

North and Central America
Soil suited to growing crops accounts for 22% of the land, and yet only 13% is actually used as cropland. Between 1945 and 1975, about 30 million ha of land in the US were lost under concrete and asphalt, half of this being arable land.

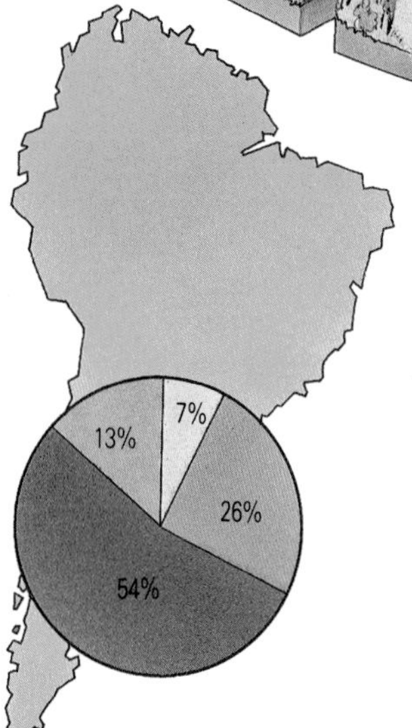

7%
26%
54%
13%

South America
The major limiting factor is mineral stress (47%), the areas affected generally being vast tracts of forest. (Tropical forest soils are typically low in nutrients.) Less than half of the fertile soil is used for crops.

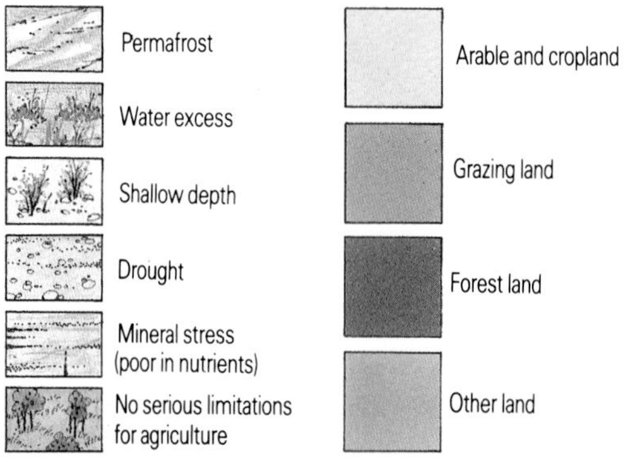

Permafrost

Water excess

Shallow depth

Drought

Mineral stress
(poor in nutrients)

No serious limitations
for agriculture

Arable and cropland

Grazing land

Forest land

Other land

How much land could we cultivate?
We cultivate only 11% of the Earth's surface, and yet some experts believe that as much as 24% could be used for crops.

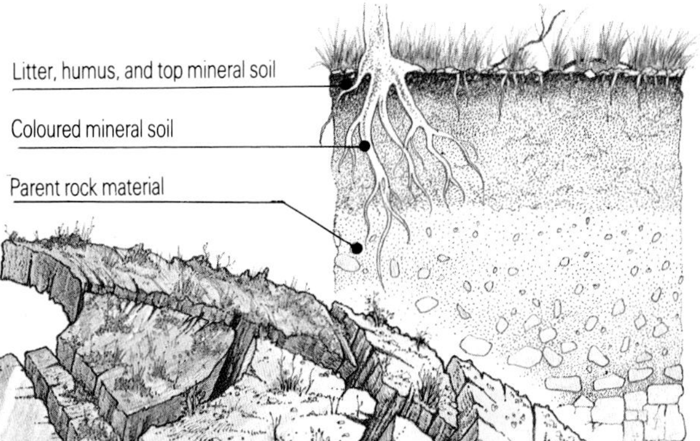

Litter, humus, and top mineral soil

Coloured mineral soil

Parent rock material

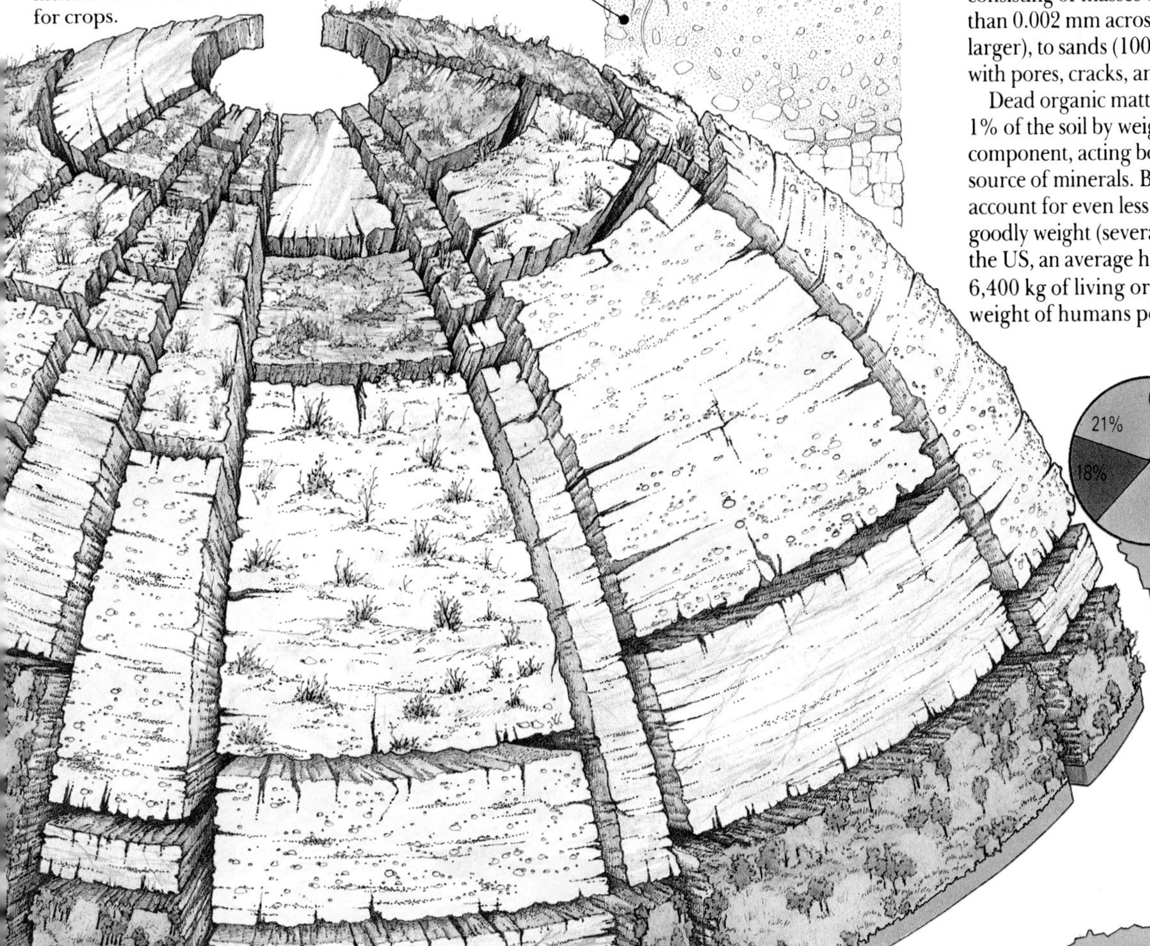

What is a fertile soil?
Structure and composition are key factors in determining soil fertility. Plant roots must be able to penetrate easily to obtain dissolved nutrients. A loam is a naturally fertile soil, consisting of masses of particles from clays (less than 0.002 mm across), through silts (10 times larger), to sands (100 times larger), interspersed with pores, cracks, and crevices.

Dead organic matter amounts to only about 1% of the soil by weight, but it is a vital component, acting both as a sponge and as a source of minerals. By contrast, living organisms account for even less – 0.1% – though still a goodly weight (several tonnes per hectare). In the US, an average hectare of soil supports 6,400 kg of living organisms, while the average weight of humans per hectare is a mere 18 kg.

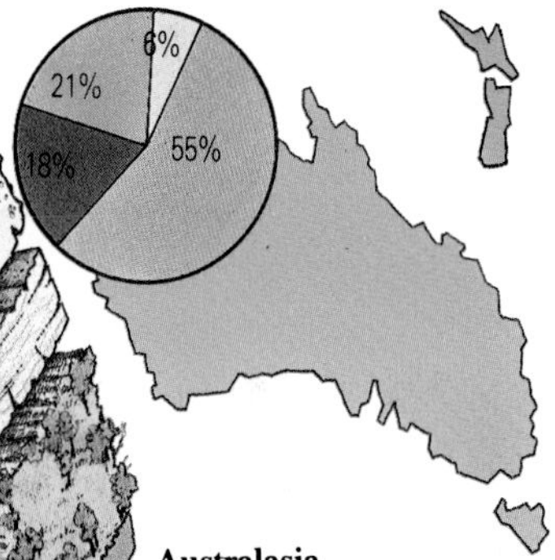

Australasia
As with Africa, drought is the main agricultural constraint, affecting some 55% of land area. Australians do not make full use of their good soil: 15% is fertile, yet only 6% is cultivated.

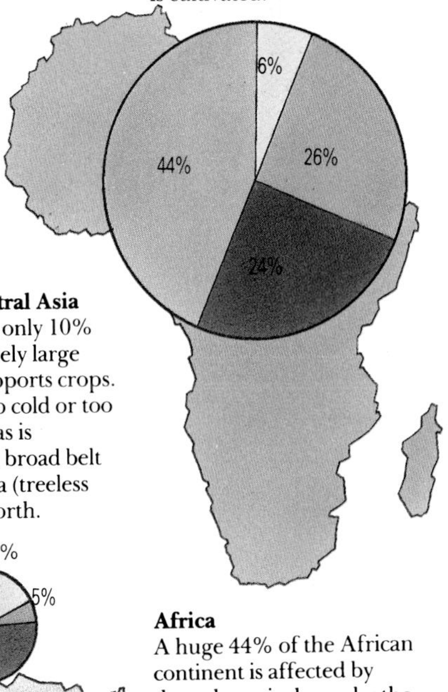

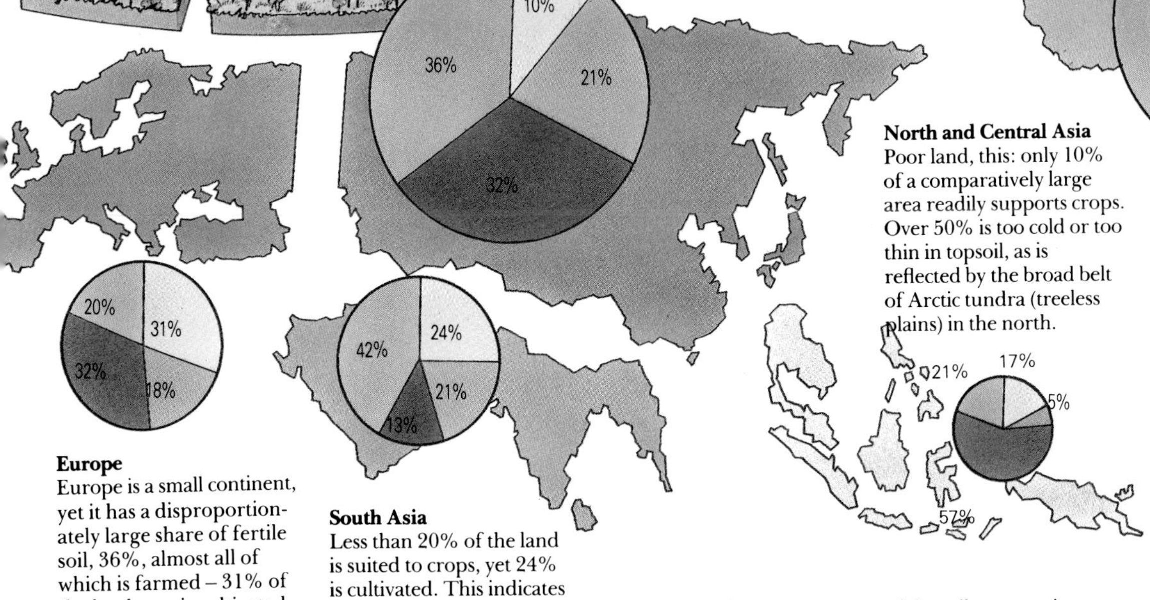

North and Central Asia
Poor land, this: only 10% of a comparatively large area readily supports crops. Over 50% is too cold or too thin in topsoil, as is reflected by the broad belt of Arctic tundra (treeless plains) in the north.

Africa
A huge 44% of the African continent is affected by drought, as is shown by the proportion of land designated "other land". However, 16% is considered suited to agriculture, yet only 6% is actually cultivated.

Europe
Europe is a small continent, yet it has a disproportionately large share of fertile soil, 36%, almost all of which is farmed – 31% of the land area is cultivated. One-third of all soils suffer from mineral stress.

South Asia
Less than 20% of the land is suited to crops, yet 24% is cultivated. This indicates working of unsuitable land (though irrigation can help) – a reflection of the needs of the huge population.

Southeast Asia
Fourteen percent of the land is fertile, yet 17% is given over to crops (cf. S Asia). Note that nearly 60% of the soils are nutrient-deficient, again correlating (as elsewhere in the tropics) with the proportion of forested land.

at very considerable expense, it is estimated that an extra 1.7 billion hectares could be brought under the plough.

Our green planet

In the end, or rather in the beginning, we are all plants. Without the green mantle for our planet supplied by over one-third of a million plant species, animal life as we know it (including *Homo sapiens*) would never have evolved. Millions of years ago, it was the rise of plant life that boosted the stock of oxygen in the atmosphere from a trace gas to the one-fifth proportion that fostered the outburst of animal life. Some biologists believe that the demise of one plant species may eventually lead to the extinction of up to 30 animal species, as the consequences reverberate up the food chains.

Plants convert sunlight into the stored chemical energy on which all animal life depends for food (and humans for fuel too). The enormous diversity of plants offers adaptations to every conceivable environment, from desert to tundra, the tropics having the richest speciation. We depend on this green wealth at every turn – from indirect benefits for soil and climate to direct supplies to our tables, factories, and hospitals.

How much plant life is there on the planet? And where does it grow most abundantly? Answers to these questions suggest where we might look to increase our growing of crops or our production of fibre. "Phytomass" is the scientific term used to measure an amount of dry plant material (undried plant matter is three or four times heavier), and it is expressed in tonnes per hectare. When we speak of dry animal matter, we use the term "zoomass"; phytomass and zoomass together comprise biomass. Of all biomass on the face of the planet, 99 percent is plant material.

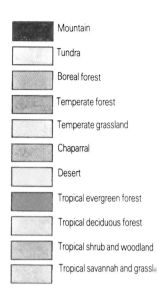

Mountain

Tundra

Boreal forest

Temperate forest

Temperate grassland

Chaparral

Desert

Tropical evergreen forest

Tropical deciduous forest

Tropical shrub and woodland

Tropical savannah and grassl...

The green potential

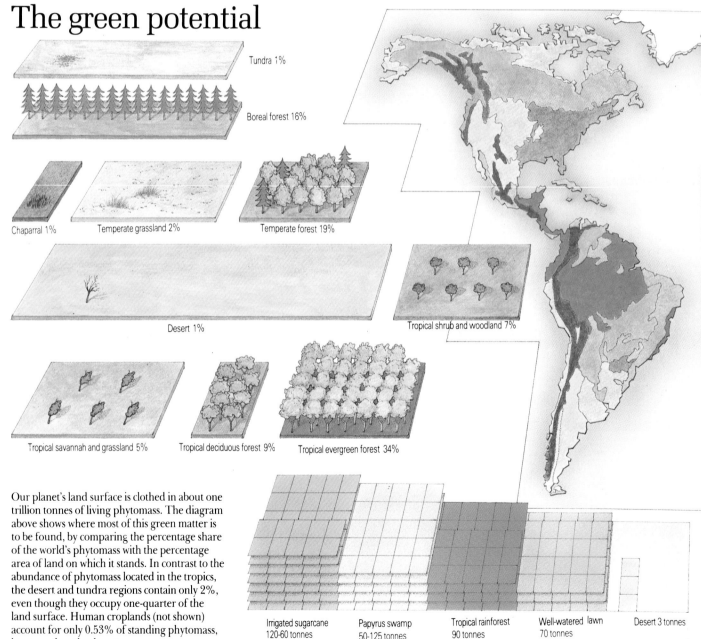

Tundra 1%

Boreal forest 16%

Chaparral 1%

Temperate grassland 2%

Temperate forest 19%

Desert 1%

Tropical shrub and woodland 7%

Tropical savannah and grassland 5%

Tropical deciduous forest 9%

Tropical evergreen forest 34%

Our planet's land surface is clothed in about one trillion tonnes of living phytomass. The diagram above shows where most of this green matter is to be found, by comparing the percentage share of the world's phytomass with the percentage area of land on which it stands. In contrast to the abundance of phytomass located in the tropics, the desert and tundra regions contain only 2%, even though they occupy one-quarter of the land surface. Human croplands (not shown) account for only 0.53% of standing phytomass, less even than the deserts.

Irrigated sugarcane 120-60 tonnes

Papyrus swamp 50-125 tonnes

Tropical rainforest 90 tonnes

Well-watered lawn 70 tonnes

Desert 3 tonnes

Not surprisingly, we find that forests contain more than three-quarters of all terrestrial phytomass – in fact some 950 billion tonnes. Of this amount, well over half (or about one-third of Earth's land biomass) is contained in tropical forests, even though they now cover only 6 percent of the land surface. Curiously enough, our cultivated crops amount to less than 7 billion tonnes of standing phytomass, a trifling 0.5 percent, even though they cover more of the land surface than tropical forests.

Comparing the amounts of total living phytomass is only one way of estimating the "green potential" of different ecosystems and regions. Another way is to consider the amount of new plant material generated each year. Again, forests are in the forefront, accounting for 37 percent of Earth's annual production of more than 133 billion tonnes. Tropical forests account for 23 percent. With year-round growth, tropical forests can produce as much as 90 tonnes of plant material per hectare per year, or almost twice as much phytomass as is generated by temperate forests (and a higher level of productivity than for any other vegetation type except for a few forestry plantations, water weeds such as water hyacinth, ultra-moist savannahs, and high-yielding crops such as sugarcane). Because organic matter is speedily decomposed, however, annual net increment in virgin tropical forest is usually nil. Thanks to intensive agriculture, our crops produce about 15 billion tonnes of phytomass each year (11 percent of the world's total). A good corn crop in the US can generate 15-20 tonnes of plant material per hectare per year, while white potatoes can yield almost 30 tonnes.

In general, the amount of new phytomass produced each year in moist parts of the world doubles as one moves from boreal to the temperate zones, and more than doubles as one moves from

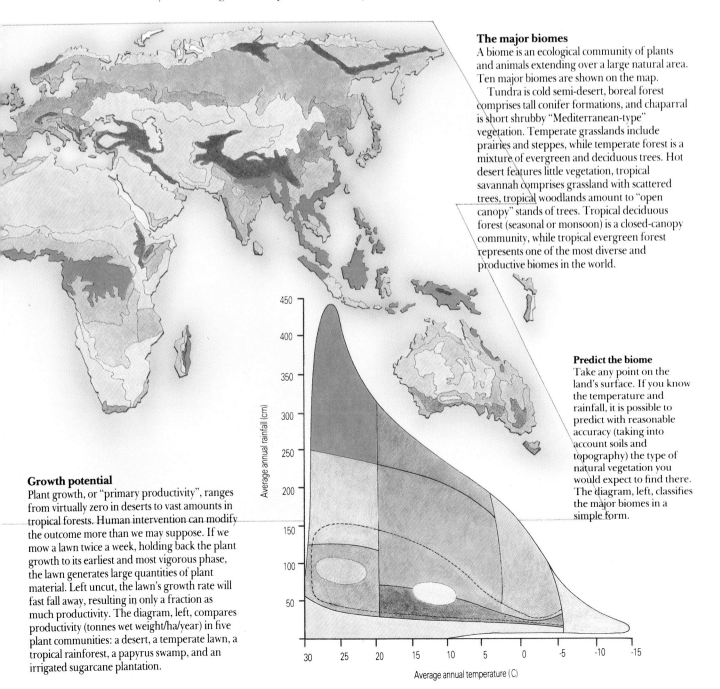

The major biomes
A biome is an ecological community of plants and animals extending over a large natural area. Ten major biomes are shown on the map.

Tundra is cold semi-desert, boreal forest comprises tall conifer formations, and chaparral is short shrubby "Mediterranean-type" vegetation. Temperate grasslands include prairies and steppes, while temperate forest is a mixture of evergreen and deciduous trees. Hot desert features little vegetation, tropical savannah comprises grassland with scattered trees, tropical woodlands amount to "open canopy" stands of trees. Tropical deciduous forest (seasonal or monsoon) is a closed-canopy community, while tropical evergreen forest represents one of the most diverse and productive biomes in the world.

Predict the biome
Take any point on the land's surface. If you know the temperature and rainfall, it is possible to predict with reasonable accuracy (taking into account soils and topography) the type of natural vegetation you would expect to find there. The diagram, left, classifies the major biomes in a simple form.

Growth potential
Plant growth, or "primary productivity", ranges from virtually zero in deserts to vast amounts in tropical forests. Human intervention can modify the outcome more than we may suppose. If we mow a lawn twice a week, holding back the plant growth to its earliest and most vigorous phase, the lawn generates large quantities of plant material. Left uncut, the lawn's growth rate will fast fall away, resulting in only a fraction as much productivity. The diagram, left, compares productivity (tonnes wet weight/ha/year) in five plant communities: a desert, a temperate lawn, a tropical rainforest, a papyrus swamp, and an irrigated sugarcane plantation.

The global forest

the temperate zones to the tropics. Similarly, ecological complexity increases towards the tropics, from "simple" communities with few species in the polar regions, to communities with great abundance and diversity at the Equator.

Forests and the biosphere

Majestic and diverse, the world's forests represent some of the most exuberant expressions of nature. Whether among the giant Douglas firs in Oregon, the aged oaks of Sherwood Forest in Britain, the vast fir forests of Central Europe, or in a rainforest in Amazonia or Borneo, we feel small and insignificant by comparison. Ranging across some 30 percent of the planet's land surface, forests are the climax ecosystems of a green and flowering world. They tend to support greater stocks of biomass, produce new biomass faster, and harbour greater abundance of species (both plant and animal) than any other ecological zone.

Not only are forests powerhouses of basic biospheric processes, notably photosynthesis and biological growth, creation of fertile humus, and transfer of energy, but their exceptional contribution to the biosphere goes much further. They play major roles in the planetary recycling of carbon, nitrogen, and oxygen. They help to determine temperature, rainfall and various other climatic conditions. They are often the fountain-heads of rivers. They constitute the major gene reservoirs of our planet, and they are the main sites of emergence of new species. In short, they contribute as much to evolution as all other biomes.

Among the many goods supplied by forests, the principal one is wood. Wood serves many purposes. It is one of the first raw materials that we use, and it is likely to be our last. It plays a part in more activities of a modern economy than any other commodity, and almost every major industry depends on forest products in at least one of its processes. A house wall built of wood requires about 20 percent less energy for heating and 30 percent less for cooling than a house made of other construction materials. Wood also serves a multitude of purposes as plywood, veneer, hardboard, particleboard, and chipboard. It is competitive too, since substitutes such as steel, aluminium, cement, and plastics need more energy in their production. Furthermore, we use much wood in the form of paper – a key medium of civilization. Having expanded by more than 50 percent since 1965, the world's industrial timber harvest now amounts to 1.7 billion cubic metres per year. An average American citizen uses more than twice as much wood as all metals together.

From a human standpoint, we can look upon forests as the great providers and protectors. They maintain ecological diversity for us, they safeguard watersheds, they protect soil from erosion, they supply fuel for about half the world's people, they provide wood for paperpulp and industrial timber,

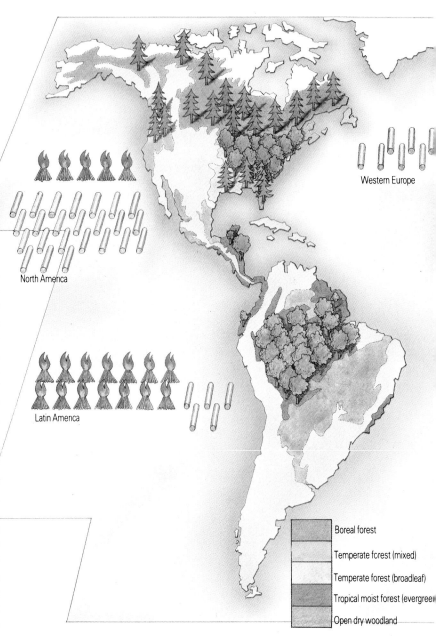

Western Europe

North America

Latin America

Boreal forest

Temperate forest (mixed)

Temperate forest (broadleaf)

Tropical moist forest (evergreen)

Open dry woodland

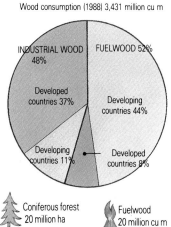

Wood consumption (1988) 3,431 million cu m

INDUSTRIAL WOOD 48%

FUELWOOD 52%

Developed countries 37%

Developing countries 44%

Developing countries 11%

Developed countries 8%

Coniferous forest
20 million ha

Fuelwood
20 million cu m

Broadleaved forest
20 million ha

Industrial wood
20 million cu m

Land where the crowns of trees cover more than 20% of the area is known as closed forest. More than half the world's closed forests comprise broadleaved trees, many of which grow in Latin America. Of coniferous forests three-fifths are located in the former USSR, and over one-quarter in North America.

The tree symbols on the map represent a fixed area of closed forest. The firewood and log symbols in the ocean areas indicate how much wood is used for fuel and industrial purposes; combined they give the annual woodcut for each region.

World wood consumption

We now consume more than 3.4 billion cubic metres of wood a year, enough to cover a fair sized city, such as Birmingham (UK), to the height of a 10-storey building. About 55% of this wood comes from broadleaved trees (generally speaking, hardwoods), and 45% from

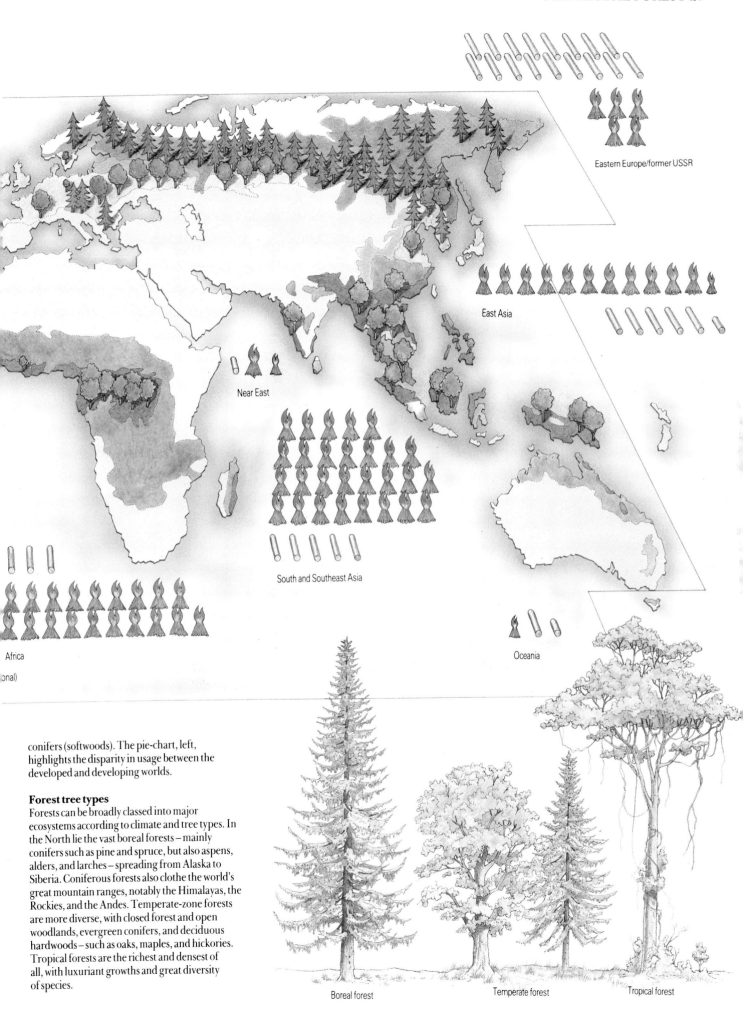

Eastern Europe/former USSR

East Asia

Near East

South and Southeast Asia

Africa
(onal)

Oceania

conifers (softwoods). The pie-chart, left,
highlights the disparity in usage between the
developed and developing worlds.

Forest tree types
Forests can be broadly classed into major
ecosystems according to climate and tree types. In
the North lie the vast boreal forests – mainly
conifers such as pine and spruce, but also aspens,
alders, and larches – spreading from Alaska to
Siberia. Coniferous forests also clothe the world's
great mountain ranges, notably the Himalayas, the
Rockies, and the Andes. Temperate-zone forests
are more diverse, with closed forest and open
woodlands, evergreen conifers, and deciduous
hardwoods – such as oaks, maples, and hickories.
Tropical forests are the richest and densest of
all, with luxuriant growths and great diversity
of species.

Boreal forest

Temperate forest

Tropical forest

and they are pleasing to the eye. Without them, our planetary home would be a lot poorer. Yet in certain parts of the tropics, we are losing this heritage at an alarming rate (see p. 38).

Powerhouses of the tropics

Tropical forests form a green band around the Equator, extending roughly 10 degrees north and south. This means that they account for only a small proportion, about 6 percent, of the Earth's land surface. Yet they comprise almost half of all growing wood on the planet, and harbour at least 70 percent, and perhaps even 90 percent, of all species – a genetic resource that increasingly serves our daily welfare via agriculture, medicine, and industry. They also comprise the most complex and diverse ecosystems on Earth.

A one-and-a-half hectare patch of forest may hold over 200 different trees alone. They grow in multi-layered profusion: tall "emergents" piercing the canopy; lianas, stranglers, and climbers with aerial roots festooning their buttressed trunks; lichens, mosses, and algae adorning every surface; and an array of fungi colonizing the forest floor. Almost every branch is hung with epiphytic ferns, orchids, or bromeliads, while smaller trees and shrubs compete for light and space below. This intricate plant life supports an even greater diversity of insect and animal life, much of it specialized, with life cycles linked to certain plants.

Yet for all their intrinsic interest, these forests remain almost unknown to us. Science has identified less than one in ten, perhaps one in one hundred, of their species. If you went into a tract of forest with a net, you would need only a few hours to catch an insect not yet known to science (and to be named after you). We now know more about certain sectors of the moon's surface than about the heartlands of Amazonia – and the moon will be around for a long while to come, whereas tropical forests are being disrupted and destroyed with every tick of the clock. Each time a small area of forest is cleared, several species, perhaps potentially valuable, are lost forever.

Most people (developers included) are surprised to learn that the soils that support the most luxuriant tropical forests are generally of low quality and unsuited to agriculture. Receiving very little of their nutrient supplies from the shallow, impoverished soil, tropical forests have built up stocks of key minerals above the ground, within the vegetation. When leaves fall, or a tree crashes to the ground, decomposer organisms recycle the nutrients within a few weeks, in contrast with the many months required in a temperate forest. Thus tropical forests have an almost leak-proof system to retain nutrients within their living structures. They flourish despite the soils, not because of them.

In so far as tropical forests constitute a kind of benchmark for life processes, we shall not understand life properly until we understand tropical

Tropical forests

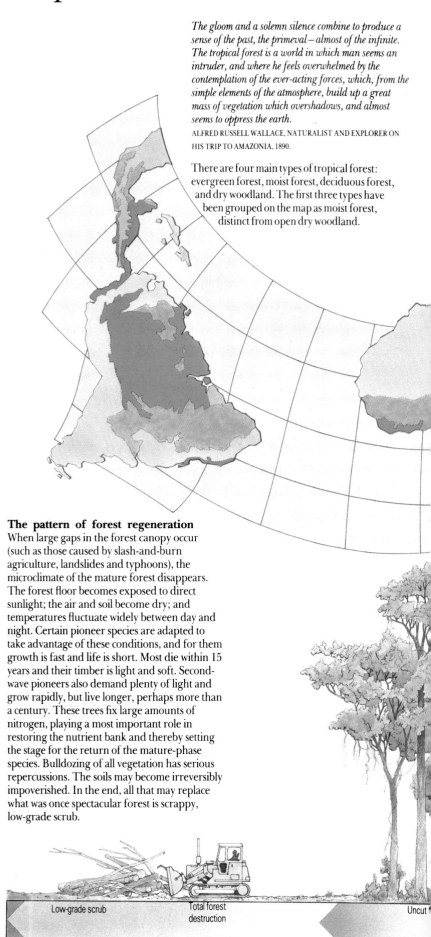

The gloom and a solemn silence combine to produce a sense of the past, the primeval – almost of the infinite. The tropical forest is a world in which man seems an intruder, and where he feels overwhelmed by the contemplation of the ever-acting forces, which, from the simple elements of the atmosphere, build up a great mass of vegetation which overshadows, and almost seems to oppress the earth.
ALFRED RUSSELL WALLACE, NATURALIST AND EXPLORER ON HIS TRIP TO AMAZONIA, 1890.

There are four main types of tropical forest: evergreen forest, moist forest, deciduous forest, and dry woodland. The first three types have been grouped on the map as moist forest, distinct from open dry woodland.

The pattern of forest regeneration
When large gaps in the forest canopy occur (such as those caused by slash-and-burn agriculture, landslides and typhoons), the microclimate of the mature forest disappears. The forest floor becomes exposed to direct sunlight; the air and soil become dry; and temperatures fluctuate widely between day and night. Certain pioneer species are adapted to take advantage of these conditions, and for them growth is fast and life is short. Most die within 15 years and their timber is light and soft. Second-wave pioneers also demand plenty of light and grow rapidly, but live longer, perhaps more than a century. These trees fix large amounts of nitrogen, playing a most important role in restoring the nutrient bank and thereby setting the stage for the return of the mature-phase species. Bulldozing of all vegetation has serious repercussions. The soils may become irreversibly impoverished. In the end, all that may replace what was once spectacular forest is scrappy, low-grade scrub.

Low-grade scrub   Total forest destruction   Uncut

Evergreen forest receives at least 4,000 mm of rainfall a year, with hardly any dry season. It is entirely closed at its canopy, and features an abundance of luxuriant vegetation. Occasionally standing 60 m high, the several distinct strata give an impression of "forest piled upon forest". In wetter areas, it is known as rainforest.

Moist forest receives at least 2,000 mm of rainfall a year (no more than 3 months with less than 100 mm). It is very similar to evergreen forest, except that it lacks the ecological complexity.

Deciduous forest receives 1,500 mm of rainfall a year, with 4-6 months virtually dry. Hence the forest loses its leaves for extended periods. Open dry woodland or wooded savannah receives less than 1,000 mm of rainfall, and often experiences prolonged drought.

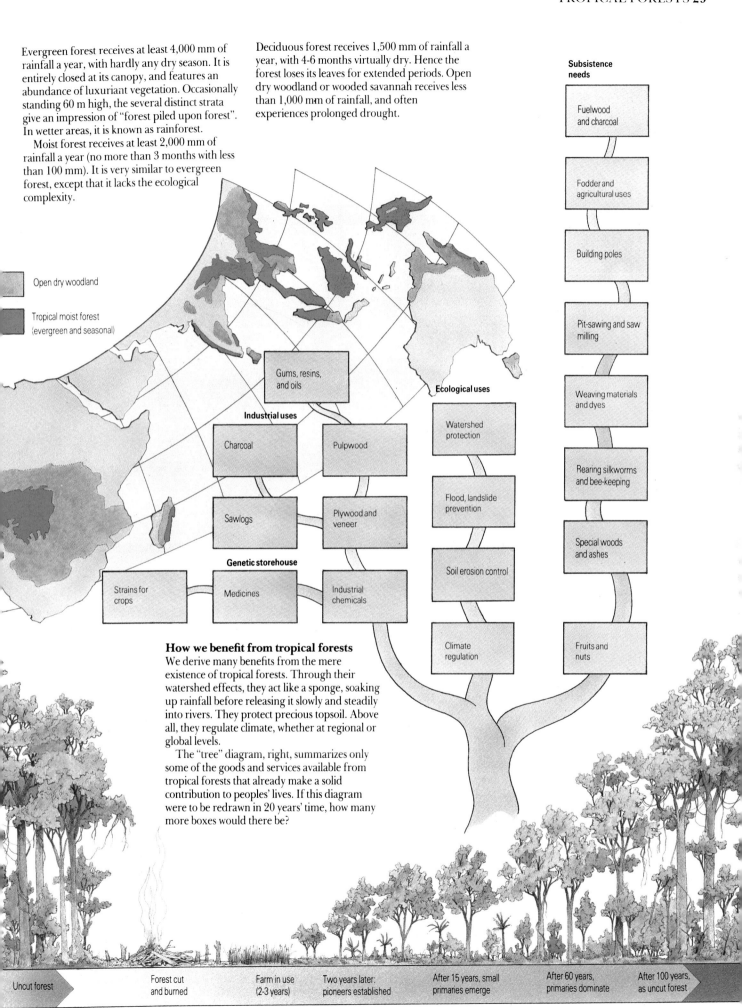

Open dry woodland

Tropical moist forest (evergreen and seasonal)

Subsistence needs

Fuelwood and charcoal

Fodder and agricultural uses

Building poles

Pit-sawing and saw milling

Weaving materials and dyes

Rearing silkworms and bee-keeping

Special woods and ashes

Fruits and nuts

Gums, resins, and oils

Industrial uses

Charcoal

Pulpwood

Sawlogs

Plywood and veneer

Genetic storehouse

Strains for crops

Medicines

Industrial chemicals

Ecological uses

Watershed protection

Flood, landslide prevention

Soil erosion control

Climate regulation

How we benefit from tropical forests
We derive many benefits from the mere existence of tropical forests. Through their watershed effects, they act like a sponge, soaking up rainfall before releasing it slowly and steadily into rivers. They protect precious topsoil. Above all, they regulate climate, whether at regional or global levels.

The "tree" diagram, right, summarizes only some of the goods and services available from tropical forests that already make a solid contribution to peoples' lives. If this diagram were to be redrawn in 20 years' time, how many more boxes would there be?

Uncut forest | Forest cut and burned | Farm in use (2-3 years) | Two years later: pioneers established | After 15 years, small primaries emerge | After 60 years, primaries dominate | After 100 years, as uncut forest

forests. They are revealing more about evolution than all other natural environments put together.

The growth of agriculture

The first great developments in agriculture took place some 10,000 years ago in a series of river basins, notably the Nile, the Euphrates/Tigris, the Ganges/Brahmaputra, and the Yangtse (see pp 146-7). In these tropical areas, with their year-round warmth and river-supplied water, people exploited the fertile floodplains to embark on an enterprise that ranks as a human advance to match mastery of fire and the art of writing – and much more important in terms of basic human survival.

Since this first exercise in agriculture, we have dug up a sizeable sector of our planet, one-and-a-half billion hectares in all (the US covers just under one billion hectares). The most productive areas to date have often been the temperate-zone lands, so-called because of their supposedly clement climates. The year-round warmth of the tropics is fine for crop plants, but it is also fine for weeds, pests, and diseases: the tropics lack that great herbicide and pesticide of temperate zones, known as winter. Moreover, many temperate-zone lands feature naturally fertile soils, whereas many tropical soils have lost their nutrients through millennia of tropical downpours that wash out crucial minerals. An area of just one hectare of naturally rich soil in East Anglia (UK) or Iowa can yield as much harvest in one year as ten hectares of naturally impoverished soil in Bolivia or Zambia.

At the same time, temperate-zone farmers, being members of the affluent world, can afford to maintain their soils' fertility by means of ever-growing inputs of synthetic fertilizer, plus capital-intensive machinery and other investments. In other words, temperate-land farmers can now engage in "industrialized agriculture", an option not generally available in the developing world.

On a global scale, our croplands have generated an adequate harvest, more or less, until about the middle of this century. Since that time, we have witnessed the growth of human numbers and of human aspirations, twin pressures that have caused us to concentrate on a handful of high-yielding crop varieties – especially wheat, rice, maize, and potato – to supply us with the bulk of our dietary needs. New territories to grow these crops were opened up from 1950-80: grain-growing areas, which now occupy 70 percent of all croplands, expanded by one-quarter. But then the growth in cropland slowed to a crawl. Overall, the world cropland base did not grow in the 1980s and the potential for profitable expansion in the near-future is limited as gains are offset by losses, due mostly to erosion. So much for the hopes of the Food and Agricultural Organization (FAO) to expand our global croplands by a further 15 percent by the turn of the century. Growth in food output must

The world croplands

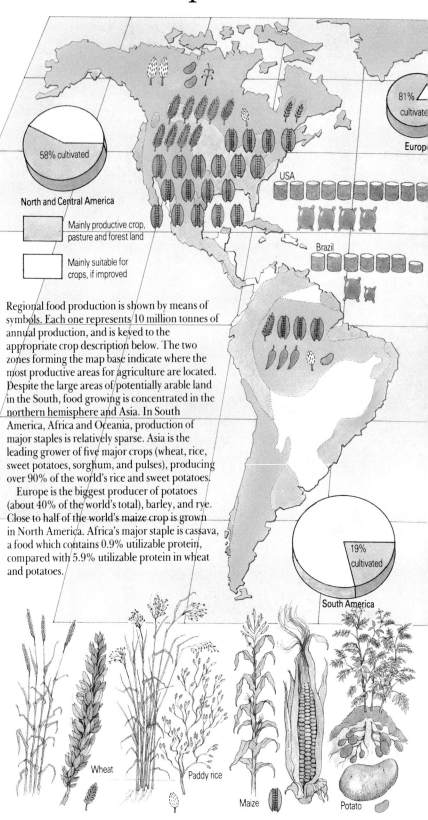

81% cultivated
Europe

58% cultivated

North and Central America

Mainly productive crop, pasture and forest land

Mainly suitable for crops, if improved

USA

Brazil

19% cultivated

South America

Regional food production is shown by means of symbols. Each one represents 10 million tonnes of annual production, and is keyed to the appropriate crop description below. The two zones forming the map base indicate where the most productive areas for agriculture are located. Despite the large areas of potentially arable land in the South, food growing is concentrated in the northern hemisphere and Asia. In South America, Africa and Oceania, production of major staples is relatively sparse. Asia is the leading grower of five major crops (wheat, rice, sweet potatoes, sorghum, and pulses), producing over 90% of the world's rice and sweet potatoes.

Europe is the biggest producer of potatoes (about 40% of the world's total), barley, and rye. Close to half of the world's maize crop is grown in North America. Africa's major staple is cassava, a food which contains 0.9% utilizable protein, compared with 5.9% utilizable protein in wheat and potatoes.

Wheat   Paddy rice   Maize   Potato

Wheat is the most important cereal in terms of world food production, providing a staple food for over a third of the world's population. It is grown principally in temperate climates and also in some sub-tropical regions. The protein content varies between 8 and 15%.
Rice is the leading tropical crop in Asia. Wet-rice cultivation allows for continuous cropping, so it supports high densities of population. Nutritionally, it is an excellent food, with a protein content of 8-9%.
Maize In the US, the largest producer of maize, the bulk of the crop is fed to livestock. As food for people, maize is a staple crop in S America and Africa. The average protein content is 10%.

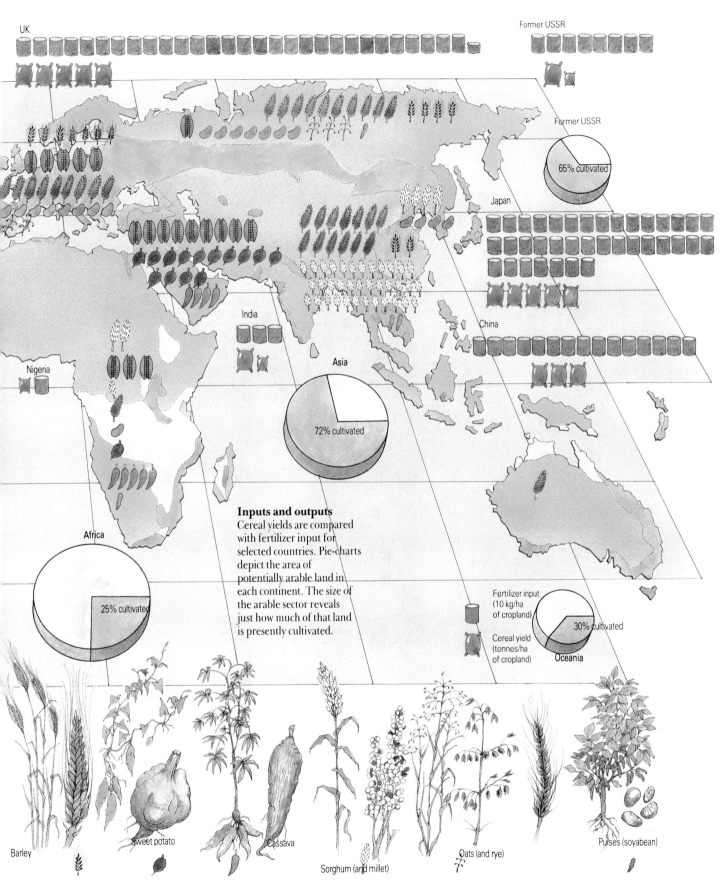

UK

Former USSR

Former USSR

65% cultivated

Japan

India

China

Asia

72% cultivated

Nigeria

Africa

25% cultivated

Inputs and outputs
Cereal yields are compared
with fertilizer input for
selected countries. Pie-charts
depict the area of
potentially arable land in
each continent. The size of
the arable sector reveals
just how much of that land
is presently cultivated.

Fertilizer input
(10 kg/ha
of cropland)

Cereal yield
(tonnes/ha
of cropland)

30% cultivated

Oceania

Barley

Sweet potato

Cassava

Sorghum (and millet)

Oats (and rye)

Pulses (soyabean)

Potato Potatoes grow
successfully in cool, moist,
temperate regions, and are
a staple carbohydrate in
many developed countries.
Barley The fourth most
important cereal crop, it is
used mainly for animal
feed and malting for beer

and whisky. In parts of Asia
and Ethiopia it is still an
important food crop.
Sweet potato Commonly
grown in wetter tropical
regions, sweet potatoes are
generally used as a
secondary rather than a
staple food. Their chief

food value is starch.
Cassava A very important
food crop in Africa, being
extremely resistant to
drought. It has a very low
protein content and should
be supplemented with
other high-protein foods.
Sorghum (and millet)

This tropical cereal is a
staple food in drier parts
of Africa and Asia. The
grain lacks gluten and
cannot be used for bread-
making.
Oats and rye Both these
crops prefer cool, damp
climates. Oats are grown

mainly for feeding
livestock, while rye's chief
use is for bread flour.
Pulses (soyabean) In
poorer regions, pulses may
form the principal source
of dietary protein.
Soyabeans may contain
between 30 to 50% protein.

therefore come from efforts to protect soil, improve irrigation, and restore the productivity of degraded land along with the high-yielding varieties of crops and fertilizers which have powered the increase in food output since mid-century – but all at a price.

Animals for food

People began to domesticate animals around the same time as they began to cultivate plants, or shortly thereafter. They probably started with dogs, using them for hunting in return for a dependable food supply. Then as humans learned the arts of crop husbandry, they found it more convenient to herd wild herbivores, and pen them close to settlements. From that early stage, growing crops and raising livestock advanced side by side.

Today we enjoy a range of domesticated animals that includes cattle, sheep, goats, pigs, water buffalo, chickens, ducks, geese, and turkeys. These creatures supply us with high-quality protein, whether in the form of milk or meat. They also supply us with hides, wool, and other material items. Even more important, they provide draught power. However widespread in the developed world the tractor may be, large numbers of people in developing countries still depend upon oxen and buffalo. In India alone, there are over 80 million draught animals – a power output equivalent to 30,000 megawatts. Among certain pastoralist peoples, such as the Masai of East Africa, livestock represents on-the-hoof wealth, and hence a source of social status.

Yet despite our appetite for meat and milk, the number of animal species we have domesticated is a good deal less than that of plants. The nine types of animals represented on the map account for almost all our animal protein from livestock. Production of this protein demands a grazing area of more than three billion hectares, a much greater area than crops (1.5 billion). Note, however, that not all animals graze all of the time: camels, and particularly goats, consume much foliage from shrubs.

In 1990, humankind consumed around 175 million tonnes of meat, or about 32 kilograms per person. Yet only one in four of the world's people eats a meat-centred diet – and it is the developed world that consumes the lion's share. In the world's premier meat-eating nation, the United States, per capita consumption is 112 kilograms a year, compared with 89 kilograms in Germany, 71 in the United Kingdom, 47 in Brazil, and 24 in China. In India, meat consumption per person per year is a miserable 2 kilograms, what the average US citizen consumes in a week. With 60 percent of the world's livestock, developing nations enjoy only around 20 percent of all meat and milk produced.

This maldistribution is all the more unfortunate in that domestic animals represent a sound way for developing-world people to broaden their food supply. Livestock eats not only grass, foliage, and other cellular plant material indigestible by humans; certain types, notably pigs, chicken, and

The world's grazing herd

North America

Central America

Cattle	Poultry
40 million	300 million
20 million	100 million
Pigs	**Mules/asses**
40 million	10 million
20 million	5 million
Sheep	**Goats**
40 million	40 million
20 million	20 million
Horses	**Camels**
10 million	10 million
5 million	

Buffaloes
40 million
20 million

South America

Dromedary camel

Bactrian camel

Anglo-Nubian goat

Aylesbury duck
Free-range chicken

Swaledale sheep

The world supports three times as many domestic animals as humans – more than 15 billion. (There are twice as many chickens on Earth as people.) While humans have doubled their numbers since mid-century, the number of domestic animals has tripled. There are more than 3 billion creatures known as ruminants, mainly cattle, sheep, goats, and buffalo, also camels and llamas. There are also more than 850 million pigs. China is home to two out of every five pigs in the world, while India boasts the most cattle – 15% of the global total.

Efficiency or wastefulness?

Most domesticated animals forage off plants that offer no sustenance to humans, thus they are not competing with humans. When we count in remote rangelands, forests, and other little-recognized stock-supporting territories, domestic animals make use of 6 billion ha of land, or almost half the planet's ice-free surface. They thereby mobilize much plant material to our benefit – and they do it with no adverse consequences for natural environments, except when their numbers rise to unsustainable levels. At the opposite end of the spectrum, however, one calorie of grain-fed steak costs at least 10 calories in its production – an absurdly inefficient way for us to nourish ourselves. Roughly 40% of the world's grain is fed to livestock – and in richer countries, such as the US, the figure is as high as 70%. The land on which animal feed is grown is known as "ghost hectarage" – it adds a further 40 million ha of land required to support livestock in the US alone.

Low-grade grazers
Marginal land grazers, browsers, and foragers efficiently convert otherwise unusable energy

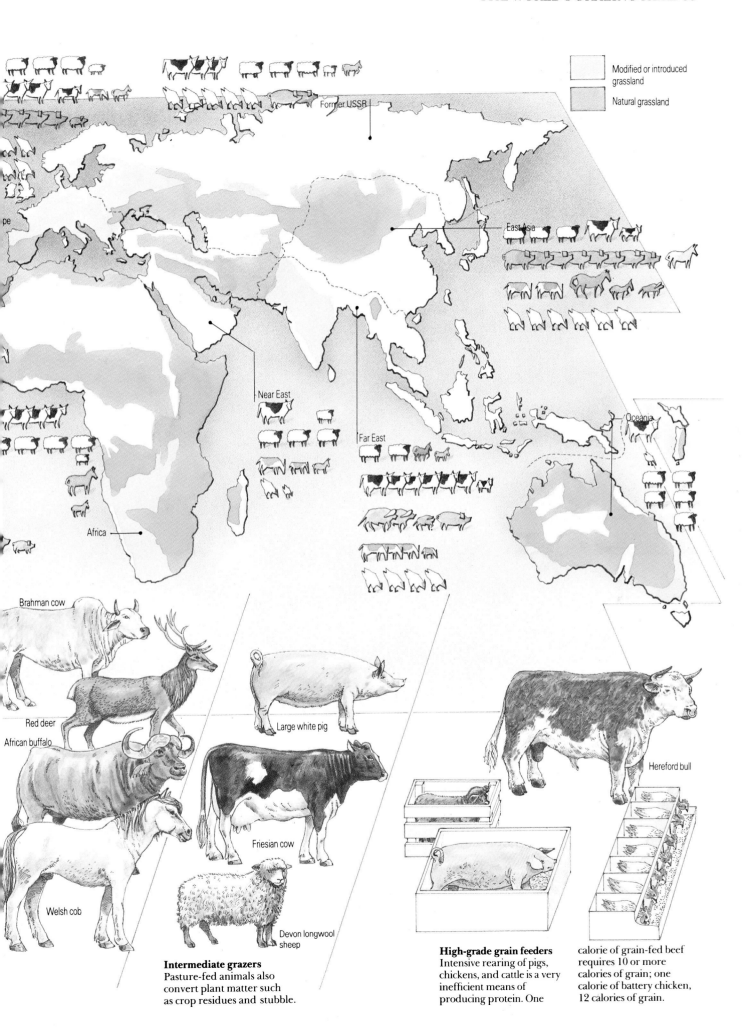

Modified or introduced grassland

Natural grassland

Former USSR

East Asia

Near East

Far East

Africa

Oceania

Brahman cow

Red deer

African buffalo

Large white pig

Welsh cob

Friesian cow

Devon longwool sheep

Hereford bull

Intermediate grazers
Pasture-fed animals also convert plant matter such as crop residues and stubble.

High-grade grain feeders
Intensive rearing of pigs, chickens, and cattle is a very inefficient means of producing protein. One calorie of grain-fed beef requires 10 or more calories of grain; one calorie of battery chicken, 12 calories of grain.

fish, consume all manner of material such as farmyard garbage and kitchen refuse.

The ability to feed ourselves

There is no doubt that we produce enough food to send everybody to bed with a full stomach. Yet tens of millions starve, and millions more are malnourished. The problem is that the Earth is less than "fair" in allocating its land resources. Some sectors are much better endowed with fertile soils than others; some sectors are much more vulnerable to natural injury; and some respond much better to constructive human manipulation. At the same time, we must recognize that we have been less than fair to our Earth. All too often, we have abused it: we have over-worked the soils, we have prodigally felled its forests, we have over-grazed its grasslands, and we have otherwise mistreated the many other gifts of nature. Our meagre sense of husbandry has brought us to a point where, when many more

The global larder

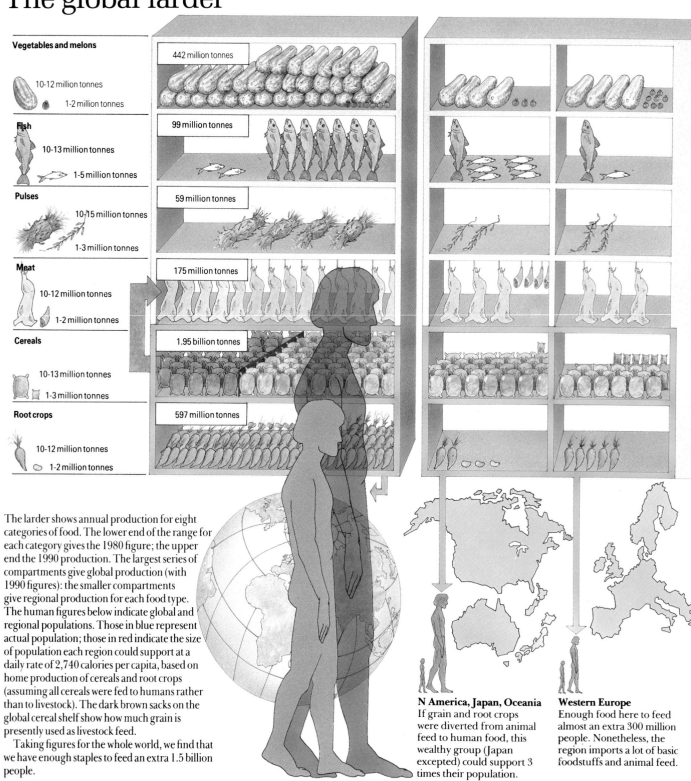

Vegetables and melons
10-12 million tonnes
1-2 million tonnes
442 million tonnes

Fish
10-13 million tonnes
1-5 million tonnes
99 million tonnes

Pulses
10-15 million tonnes
1-3 million tonnes
59 million tonnes

Meat
10-12 million tonnes
1-2 million tonnes
175 million tonnes

Cereals
10-13 million tonnes
1-3 million tonnes
1.95 billion tonnes

Root crops
10-12 million tonnes
1-2 million tonnes
597 million tonnes

The larder shows annual production for eight categories of food. The lower end of the range for each category gives the 1980 figure; the upper end the 1990 production. The largest series of compartments give global production (with 1990 figures): the smaller compartments give regional production for each food type. The human figures below indicate global and regional populations. Those in blue represent actual population; those in red indicate the size of population each region could support at a daily rate of 2,740 calories per capita, based on home production of cereals and root crops (assuming all cereals were fed to humans rather than to livestock). The dark brown sacks on the global cereal shelf show how much grain is presently used as livestock feed.

Taking figures for the whole world, we find that we have enough staples to feed an extra 1.5 billion people.

N America, Japan, Oceania
If grain and root crops were diverted from animal feed to human food, this wealthy group (Japan excepted) could support 3 times their population.

Western Europe
Enough food here to feed almost an extra 300 million people. Nonetheless, the region imports a lot of basic foodstuffs and animal feed.

mouths are clamouring to be fed, the resource base itself is falling into critical disrepair. Yet we certainly possess the technological skills and the economic capacity to supply a rightful place for the entire human family at Earth's feast table: all we appear to lack is the ultimate commitment – to say "Yes, we shall do it".

But "we shall do it" entails far more than simply ploughing up more land, or applying more sensitive agricultural skills. The basic problems are not so much technical and scientific, they are political and economic. Poor people are generally hungry because they lack the financial means to grow enough food, or to purchase it. While a transfer of relief food (and especially of grain used as livestock feed) from the developed world would help to relieve immediate hunger, it would tackle symptoms rather than problems. Poor people need to be able to feed themselves; and to do that, their entire lifestyles need to be upgraded.

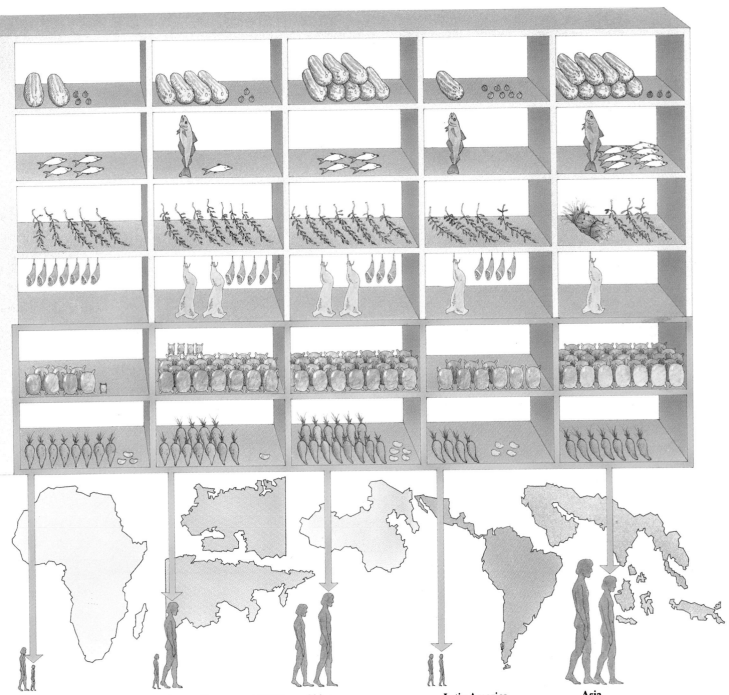

Africa
The only region to have grown steadily hungrier since 1970 – with 11% of the world's population trying to survive on 7% of its staple crop production.

Eastern Europe and USSR
A larder that would comfortably support its population, if all cereals were consumed by humans. Yet it depends on massive grain imports.

China
No mean achievement that this nation can grow enough food on 7% of the world's arable land to feed 22% of the world's population.

Latin America
This continent barely manages to feed itself, having a low production of cereals and root crops. Much of its arable land is under-utilized.

Asia
Despite a high rate of food production, the sheer size of the population of this region has stretched the land beyond its limits.

THE LAND CRISIS

With every day that goes by, as our numbers and expectations rise, we demand still more from our natural-resource base. In such a situation, it would be wise for us to redouble our sense of caring for the fragile planet that supports our material welfare. Quite the contrary, however. During the present decade, we may well degrade more forests, over-graze more grasslands, lose more land through urban development, and erode more topsoil than during the previous two decades. Unless we change our ways, we could inflict greater ravages on the Earth's ecosystem over the next few years than during the whole of this century up to the present. Far from using the Earth well, we are using it up – for all the world (so to speak) as if we had a spare planet parked out in space.

The threat of erosion

Few resource problems are so important, while so little publicized, as the disappearance of our soil. Each year, many billion of tonnes are being washed away into the sea or carried away with the wind.

There is no known way that we can replace our soil. If we wait upon natural processes, we shall wait for centuries, if not millennia. The disappearance of our soil threatens to undermine our agriculture, our very means of supporting ourselves. Yet because it is such a cryptic and "silent" problem, few of our leaders give it a fraction of the attention it deserves. It is hard to whip up public opinion about the issue. Europe, the continent least affected by erosion is estimated to be losing close to one billion tonnes a year, while Asia, the worst affected, could be losing around 25 billion tonnes. The US loses well over one billion tonnes a year (net of natural replacement) from its grainlands – equivalent to more than 300,000 hectares of crop-growing potential.

A similar sad tale can be told around the world, especially in the humid tropics with their thunderstorms. Violent downpours rip the topsoil from denuded hillsides, carving great gulleys in the landscape, while windstorms ravage semi-arid lands planted to crops. Ethiopia, though less than one-sixth the size of the US, is reputed to be losing at least as much topsoil each year. Half of all countries, and half of all arable lands, are suffering the problem at unacceptable levels. For ten years the cropland base has not expanded, and is unlikely to expand in the short-term as our gains in bringing new land under the plough are offset through soil erosion. As the population

The disappearing soil

If we are to sustain productivity, we must protect our soil. Soil degradation has many causes, but those which actively overtax the land are largely attributed to erosion and over-grazing. Soil is also being degraded as rising concentrations of salt in irrigated land change its chemistry.

Echoes of the Dust Bowl

In recent years, soil erosion exceeded soil formation on about one-third of US cropland, well over 60 million ha (see map). Despite memories of the devastating Dust Bowl of the 1930s, when 40 million hectares of arable land on the Great Plains were severely damaged, overproduction was encouraged at the expense of the environment. However, so great was public concern at the effects of topsoil loss and resentment of farm subsidy schemes that in 1986 Congress ushered in the Conservation Reserve Programme to give farmers an economic incentive to conserve the soil on their most erodible land. In return for an annual payment of about $120 per hectare, farmers convert cropland to grassland or woodland – before it becomes wasteland – for 10 years. At the end of the first stage of the scheme, in 1990, around 14 million hectares had entered the programme. Yet this represents only one-quarter of the highly erosion-prone land which requires protection.

The map, right, shows the cumulative effects of erosion in the US.

Global land loss

Each year about 11 million ha of arable lands are lost through erosion, desertification, toxification, and cropland conversion to non-agricultural uses. If this trend is allowed to continue unchecked, we shall have lost 275 million ha, or 18%, of our arable lands in the final quarter of the century. By 2025, the same amount again could disappear.

Severe – more than 75% topsoil lost

Moderate – 25-75% topsoil lost

Farmer and son in the face of a dust storm. Oklahoma, 1936

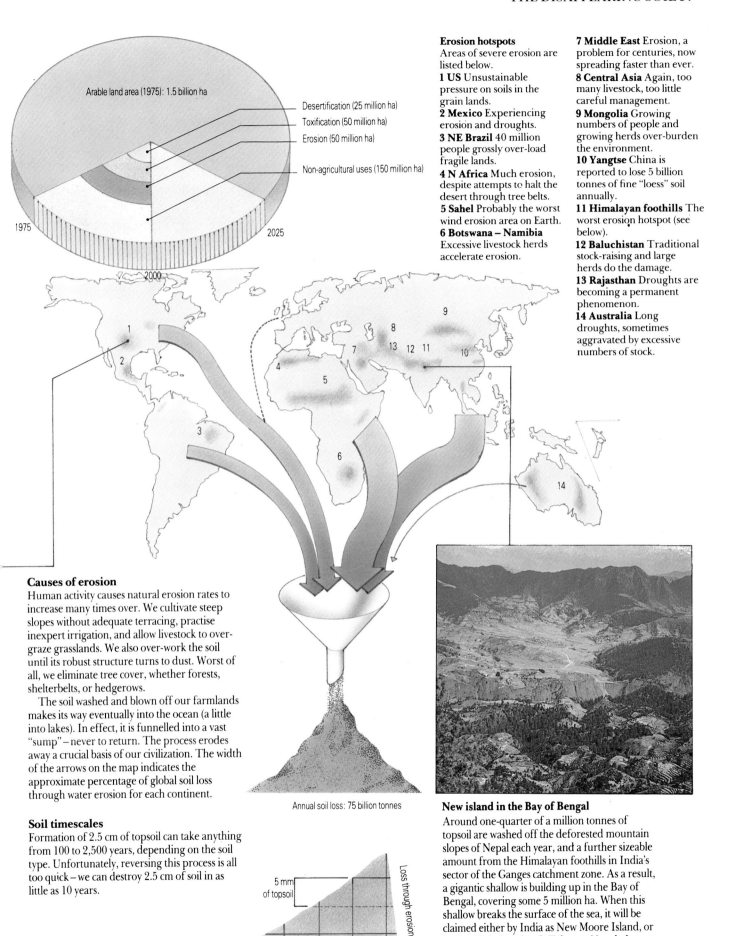

Arable land area (1975): 1.5 billion ha

Desertification (25 million ha)
Toxification (50 million ha)
Erosion (50 million ha)

Non-agricultural uses (150 million ha)

1975   2025

2000

Erosion hotspots
Areas of severe erosion are listed below.
1 US Unsustainable pressure on soils in the grain lands.
2 Mexico Experiencing erosion and droughts.
3 NE Brazil 40 million people grossly over-load fragile lands.
4 N Africa Much erosion, despite attempts to halt the desert through tree belts.
5 Sahel Probably the worst wind erosion area on Earth.
6 Botswana – Namibia Excessive livestock herds accelerate erosion.

7 Middle East Erosion, a problem for centuries, now spreading faster than ever.
8 Central Asia Again, too many livestock, too little careful management.
9 Mongolia Growing numbers of people and growing herds over-burden the environment.
10 Yangtse China is reported to lose 5 billion tonnes of fine "loess" soil annually.
11 Himalayan foothills The worst erosion hotspot (see below).
12 Baluchistan Traditional stock-raising and large herds do the damage.
13 Rajasthan Droughts are becoming a permanent phenomenon.
14 Australia Long droughts, sometimes aggravated by excessive numbers of stock.

Causes of erosion
Human activity causes natural erosion rates to increase many times over. We cultivate steep slopes without adequate terracing, practise inexpert irrigation, and allow livestock to over-graze grasslands. We also over-work the soil until its robust structure turns to dust. Worst of all, we eliminate tree cover, whether forests, shelterbelts, or hedgerows.

The soil washed and blown off our farmlands makes its way eventually into the ocean (a little into lakes). In effect, it is funnelled into a vast "sump" – never to return. The process erodes away a crucial basis of our civilization. The width of the arrows on the map indicates the approximate percentage of global soil loss through water erosion for each continent.

Annual soil loss: 75 billion tonnes

Soil timescales
Formation of 2.5 cm of topsoil can take anything from 100 to 2,500 years, depending on the soil type. Unfortunately, reversing this process is all too quick – we can destroy 2.5 cm of soil in as little as 10 years.

5 mm of topsoil

Loss through erosion

Years   10   20   30   40   50   60   70   80   90   100

New island in the Bay of Bengal
Around one-quarter of a million tonnes of topsoil are washed off the deforested mountain slopes of Nepal each year, and a further sizeable amount from the Himalayan foothills in India's sector of the Ganges catchment zone. As a result, a gigantic shallow is building up in the Bay of Bengal, covering some 5 million ha. When this shallow breaks the surface of the sea, it will be claimed either by India as New Moore Island, or by Bangladesh as South Talpatty. Nepal, the country which contributes most to the phenomenon, is not even being consulted.

continues to increase, so the global decline in grain area per person from 0.16 hectares in 1980 to 0.14 hectares in 1990 seems set to continue.

The decline of the tropical forest

Each year, around 17-18 million hectares of tropical forests and woodlands are eliminated from the face of the Earth. This is an area about the size of Washington State. We are also witnessing the degradation of further millions of hectares of forest a year. (A degraded forest is one that has been grossly disrupted, leaving behind an impoverished travesty of true forest.)

So pervasive and rapid is the depletion of tropical forests that these superb exemplars of nature may disappear within 20 years - except for isolated blocs in western Brazilian Amazonia, the Guyana hinterland, the Zaire basin, and parts of Indonesia and Papua New Guinea, plus a few relics in the form of parks and reserves.

Part of the problem is that people everywhere want more wood. An IIASA forecast of 1987 projects a 50 percent increase in industrial wood consumption over the next 40 years. In order to satisfy this growing demand, the commercial logger looks increasingly to tropical forests.

Logging in the tropics degrades some 4.5 million hectares of rainforest annually. Selective logging is the usual commercial practice, although this often heavily degrades the forest as seedlings and unharvested trees are damaged, and susceptibility to soil erosion is increased.

Eventually, after decades if not centuries, a heavily-logged forest ecosystem can regenerate. But it generally does not get the chance. The main damage done by loggers is unwitting. They lay down a network of timber-haulage tracks, allowing land-hungry settlers to penetrate deep into the forest heartlands. Slash-and-burn agriculture, which had previously represented a sustainable

Wood consumption

 Each symbol represents 200 million cu m of wood

 Deficit in fuelwood supply

 Fuelwood consumption

Industrial consumption

The shrinking forest

The demand for agricultural land, whether planned or unplanned, is the chief cause of the depletion of tropical forests. Unplanned agriculture, i.e. "spontaneous settlement" by slash-and-burn cultivators, is much more difficult to quantify than planned agriculture (for cash crop plantations, cattle ranches, and organized smallholder cultivation). The bottom graph demonstrates the rising demand for agricultural land in the tropics. By the mid-1950s more than 100 million ha of forest had been cleared; by the mid-1970s this figure had doubled to cope with the needs of expanding populations.

The rate at which the world's tropical forests are disappearing is shown in the central graph. During the period 1950-75, at least 120 million ha of closed moist forest were destroyed. By 1989 the rate of loss had increased by 90% and another 150 million ha had been eliminated. By the end of the century another 200 million ha could vanish. By 2015 will another 200 million ha have been eliminated?

Consumption: industrial vs fuelwood

The log and fuelwood symbols, right, show world consumption of wood for the period 1955-2015. In 1955 consumption of industrial wood exceeded that of fuelwood, but by 1975, the balance had tipped in favour of fuelwood, reflecting the demands of a growing developing world population. In fact, in 1975, there was a severe deficit in supply of fuelwood, shown here by dotted symbols. These shortfalls in supply have serious repercussions: burning dung and crop residues instead, rather than returning them to the fields, robs the soil of fertilizers and so reduces food output. In Africa and Asia at least 400 million tonnes of animal dung are burned each year. If this natural fertilizer had been used on croplands it could have produced an extra 20 million tonnes of grain.

Today, the share of wood consumption between fuel and commercial needs is split roughly 50-50. Soon, fuel needs could well decline way below half, due to inadequate supplies.

Loss of green cover
The three globes show the percentage of land covered by forest. In 1950, 30% of the land was covered by forest, half of which was tropical forest. By 1975, the area covered by tropical forest had declined to 12%. By 2000, we shall be lucky if tropical forests cover 5% of the land. This decline contrasts markedly with temperate forest, whose area remains constant around 20% (thanks to reforestation).

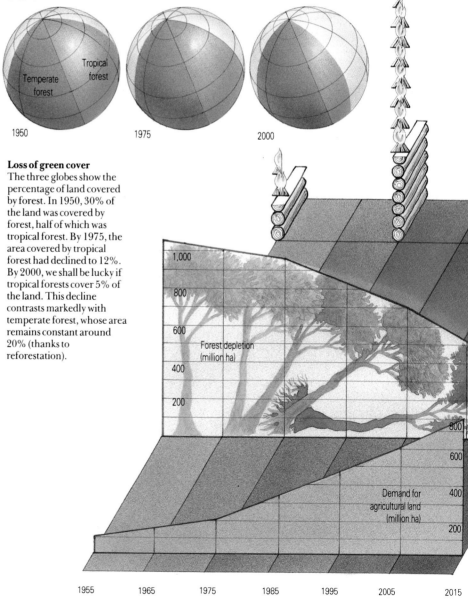

use of the forest (at low population densities), now poses the most serious threat. Small-scale cultivators, numbering up to 500 million, account for more than half of all deforestation and gross degradation. Having been squeezed out of their existing farmlands for various reasons – from maldistribution of land to the general inadequacy of rural development – these "shifted" cultivators see little alternative to their lifestyle, even though they realize it is harmful to their own prospects, and those of others. A similar sentiment applies to fuelwood gatherers, who, unless supplied with village woodlots and "tree farms" on suitable scale, see little option but to continue with their present course of destructive action. Fuelwood gathering depletes some two million hectares of tropical moist forest each year, and at least twice as much open woodland and scrub forest.

Not so hard placed is the cattle rancher, who sets light to at least 2.5 million hectares of forest in Latin America each year, mainly to raise beef for lucrative export markets in the developed world. Ranchers could double their output on existing pastures if they were to run their operations more efficiently. But government inducements encourage them to ranch extensively rather than intensively. When pastureland soils lose their residual fertility within half-a-dozen years and weeds over-run the holding, the rancher simply moves on to another patch of forest, and repeats the process.

In several senses, we all play a part in the decline of tropical forests. We seek specialist hardwoods like mahogany, teak, and ebony at unrealistically low prices, and we demand cheap beef from formerly forested pasturelands. True, some members of the global community are more culpable than others. But hardly anyone can claim that his or her hand is not, in some indirect way or another, on

Frontiers of attack
The red zones mark the areas of forest undergoing rapid depletion. Few areas (save inaccessible parts of western Amazonia and much of the Zaire basin) are immune.

the chain-saw and machete at work in tropical forests. Moreover, we shall all eventually suffer if tropical forests continue to disappear.

The effects of forest clearance

"Man has gone to the Moon but he does not know yet how to make a flame tree or a birdsong. Let us keep our dear countries free from irreversible mistakes which would lead us in the future to long for those same birds and trees."
PRESIDENT HOUPHOUET-BOIGNY OF COTE D'IVOIRE, A COUNTRY WHICH HAS LOST 90% OF ITS ORIGINAL FORESTS AND WOODLANDS.

When forest cover in a watershed is lost, the repercussions are far-reaching. The forests' sponge effect is lost, and the release of rainfall becomes erratic. Farmers in the valleylands of southern Asia are particularly vulnerable: rivers such as the Ganges, the Brahmaputra, the Irrawady, the Salween, and the Mekong no longer supply regular amounts of irrigation water – which causes the Green Revolution (see pp 56-7) to be less revolutionary than hoped.

City-dwellers suffer too. In the hinterland of Panama City and of Manila, capital of the Philippines, deforestation has caused so much injury to watershed functions that water supplies are threatened, bringing on a risk of contaminated water and pandemics. In Ecuador, Kenya, and Thailand, cities experience "brown-outs" caused by loss of hill forest: washed-off sediment leads to silting up of hydroelectric dams.

Clearance of tropical forests could also have severely adverse effects on the world's climate. In Amazonia, more than half of all moisture circulating through the region's ecosystem remains within the forest: rainwater is absorbed by plants, before being "breathed out" into the atmosphere. Were a large part of the forest to disappear, the remainder would become less able (however well protected) to retain so much moisture – and the effects could extend further, even drying out the climate for crops in southern Brazil.

Still more important, tropical forests help to stabilize the world's climate by absorbing much solar radiation: they simply "soak up" the sunshine. When forests are cleared, the "shininess" of the planet's land surface increases, radiating more of the sun's energy back into space (the "albedo" effect). An increase in albedo could lead to disruptions of convection patterns, wind currents, and rainfall in lands far beyond the tropics.

Although tropical forests do not significantly affect Earth's oxygen balance, they do play an important part in the carbon dioxide budget. When forests are burned, they release considerable quantities of carbon into the skies. The build-up of carbon dioxide in the atmosphere looks as if it is triggering a "greenhouse effect", bringing on drier climates for some, especially Americans (pp 112-3). What if the great grain belt of North America starts to become

Destroying the protector

As long as forest cover remains intact, rivers run clear and clean, and they run regularly throughout the year. When the forest disappears, the downstream effect is a regime of floods followed by droughts. Washed-off sediment not only causes river beds to silt up, but it also chokes hydropower dams and suffocates coastal fisheries.

Flooding in the Ganges Plain, as in Bangladesh above, provides a graphic example of the effects of deforestation. As the foothill forests are cleared for agriculture, the 500 million people in the valleys grow ever-more vulnerable to flooding. During the 1978 monsoon, India suffered losses of $2 billion, and hundreds of people were drowned.

Forest watershed removed, topsoil washed down

Flood plain

Silted-up river

Displaced forest peoples

There are several million people still following traditional lifestyles within tropical forests. Generally disregarded by the outside world, the first that we often hear about these peoples is when they shoot arrows at bulldozers. As recently as the early 1970s, the Tasaday tribe was discovered in a Philippines forest cut off by a 15-km strip from the outside world and pursuing a Neolithic lifestyle in isolation.

So fast are these groups being squeezed out of existence, that in Brazilian Amazonia, which featured 230 native groups with an estimated 2 million people only 500 years ago, there are now only half as many such groups, with a total below 50,000 persons. The Kayapos, for example, have been massacred in their thousands by illegal settlers, and have been forcibly transferred from their homelands. The photograph above shows the Megkronotis Indians (of the Kayapos) trudging through the ashes of a forest they refuse to abandon.

Scouring the land for fuel

According to UN estimates, more than 100 million people experience acute shortages of fuelwood. Many rural families now devote as much time to finding fuelwood as to any other activity (p. 108). As fallen wood becomes scarce, villagers lop branches, fell trees, even uproot stumps. Removing residues robs soil of nutrients, while clearing shrubs hastens erosion, leaving a barren landscape.

unbuckled, with less food not only for North Americans, but for dozens of countries that import grain from North America?

The advance of the desert

In several drier parts of the world, deserts are increasing at an alarming rate. In those few areas where the process is natural, we call it desertization. But where the desert encroaches through human hand, we term it desertification – an ugly word for an ugly process.

In fact, it is less than accurate to assert that the desert is advancing or encroaching. Rather a strip of additional desert is "tacked on" to the original. Lands far away from true deserts are also being degraded to a desert-like condition.

More than one-third of the Earth's land surface is semi-arid or arid. These drylands support some 850 million people, and produce substantial quantities of meat, cereals, and fibres. However, about three-quarters of these lands are already desertified to some degree; more than 135 million people live in severely-affected regions. Every year, desertification degrades 21 million hectares to a condition of near or complete uselessness. This is caused almost entirely by human misuse or overuse of the land. Overgrazing by livestock and deforestation for fuelwood are major causes, but the salinization of poorly-managed irrigated land is becoming a very important factor.

The cost of desertification, in terms of lost production alone, is around $26 billion a year. By contrast, the expense of rehabilitating the degraded lands, and halting the spread of deserts, is estimated at $4.5 billion a year. The ratio of loses to costs is therefore 5:1. Even though large sums are being spent, it is not enough and we are losing the battle against desertification.

Why don't governments invest the necessary funds on the grounds that it makes economic common sense? A main reason lies with the status of the people affected. They are "marginalized" people in two senses: they live in marginal lands; they are marginal to the politico-economic structures of their countries. National leaders know they represent no threat to the system if their needs are ignored.

Governments frequently funnel sizeable sums of money into drylands when trouble of a different sort erupts. For instance, in the war between Ethiopia and Somalia in the late 1960s, the superpowers poured vast funds into the zone in the form of military hardware, because the region was so close to tanker lanes from the Gulf. These funds would have been enough to rehabilitate degraded lands in the two countries – and halt desertification along much of the Saharan frontier.

As for the international community, it has not responded as urgently to the problem as the situation demanded. The Sahel disaster of the early 1970s prompted a 1977 UN conference which set the goal of the year 2000 to halt desertification.

The encroaching desert

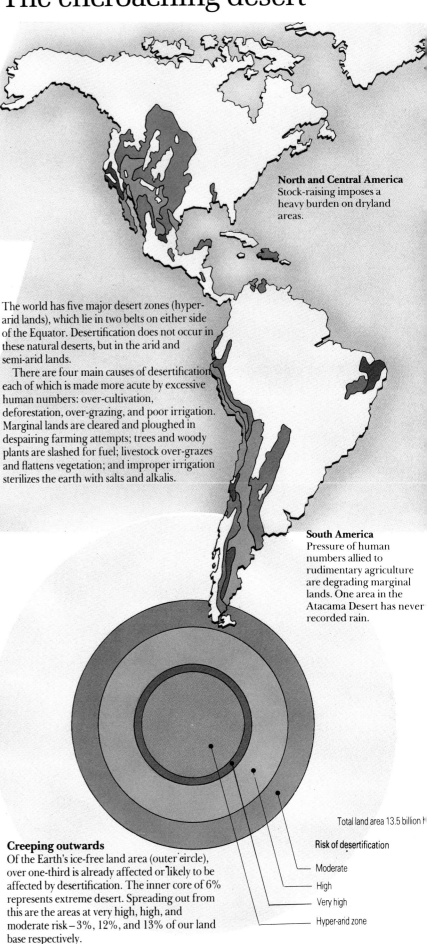

North and Central America
Stock-raising imposes a heavy burden on dryland areas.

The world has five major desert zones (hyper-arid lands), which lie in two belts on either side of the Equator. Desertification does not occur in these natural deserts, but in the arid and semi-arid lands.

There are four main causes of desertification, each of which is made more acute by excessive human numbers: over-cultivation, deforestation, over-grazing, and poor irrigation. Marginal lands are cleared and ploughed in despairing farming attempts; trees and woody plants are slashed for fuel; livestock over-grazes and flattens vegetation; and improper irrigation sterilizes the earth with salts and alkalis.

South America
Pressure of human numbers allied to rudimentary agriculture are degrading marginal lands. One area in the Atacama Desert has never recorded rain.

Total land area 13.5 billion h

Risk of desertification

— Moderate

— High

— Very high

— Hyper-arid zone

Creeping outwards
Of the Earth's ice-free land area (outer circle), over one-third is already affected or likely to be affected by desertification. The inner core of 6% represents extreme desert. Spreading out from this are the areas at very high, high, and moderate risk – 3%, 12%, and 13% of our land base respectively.

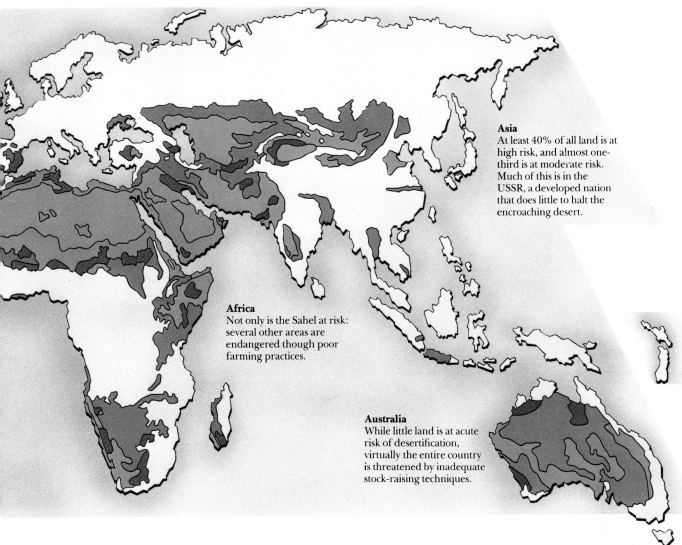

Asia
At least 40% of all land is at
high risk, and almost one-
third is at moderate risk.
Much of this is in the
USSR, a developed nation
that does little to halt the
encroaching desert.

Africa
Not only is the Sahel at risk:
several other areas are
endangered though poor
farming practices.

Australia
While little land is at acute
risk of desertification,
virtually the entire country
is threatened by inadequate
stock-raising techniques.

The Sahel disaster

Desertification was brought to the world's
attention by the Sahel disaster of the early 1970s.
The ostensible cause of the débâcle lay with the
drought that overtook the semi-arid zone dwellers
of the Sahel. But the main problem lay further in
the past. For the previous two decades, the region
enjoyed better-than-average rainfall.
Consequently, when cash crop cultivation in lands
further south (p. 47) started to expand along with
the indigenous populations, throngs of
pastoralists and cultivators moved north, toward
the fringe of the Sahara. Lands that had not been
dug for centuries were cultivated, and livestock
were crowded on to smaller areas of pasture. (The
photograph, right, shows the effects of over-
grazing in Niger.) Sadly, these migrant peoples
knew, from their tribal traditions, that the moister
phase would probably be transitory, which made
the drought all the more terrible when it finally
struck – more than 100,000 people and 3.5
million head of livestock perished in the early 70s
and disaster hit the region again in the early-to-
mid 1980s.

The crisis still persists right along the Sahel.
Human activities continue to reduce the
productivity of lands that speedily degrade
under stress.

Yet the situation will invariably worsen by then as the population of the drylands rises to well over one billion.

Our hungry world

There are two forms of malnutrition in our world. While tens of millions of people in the developing countries literally starve to death each year, health statistics from the developed countries reveal a growth in the incidence of illness caused by over-eating. An average developing-world person enjoys only about two-thirds as many calories as an average developed-world person, only half as much protein, and only one-fifth as much animal protein. When we count in all the grain fed to livestock, many a

developed-world citizen accounts for three times as much food as his or her counterpart in the developing world.

Thus the scandal of the waist-line: expanding excessively among the rich, and shrinking excessively among the poor. Even a domestic cat in Europe or North America receives more meat each day than many a person in the developing world.

More people are hungry now than ever before. Some 950 million people in lower-income countries (excluding China) do not eat enough for an active working life. This is 19 percent of world population, an increase since 1980, when 16 percent were not getting enough food. Most of these unfortunates live in sub-Saharan Africa and Asia.

"If we were to keep a minute of silence for every person who died in 1982 because of hunger, we would not be able to celebrate the coming of the 21st century because we would still have to remain silent."
FIDEL CASTRO, MARCH 1983

Hunger and glut

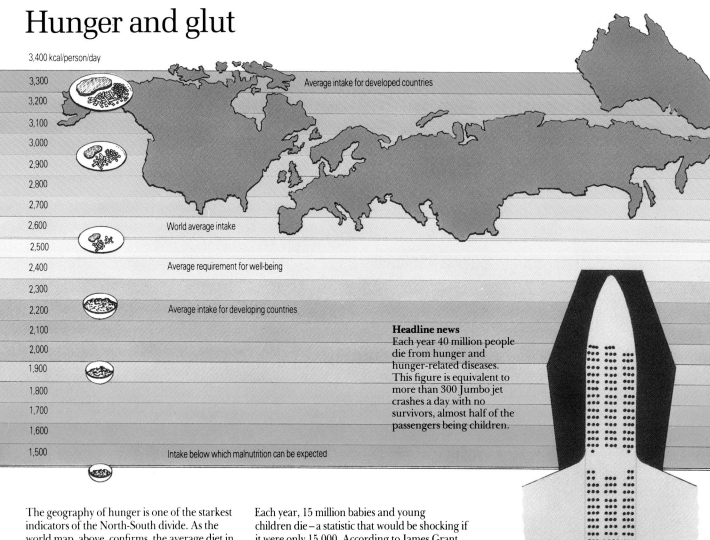

3,400 kcal/person/day

3,300 — Average intake for developed countries
3,200
3,100
3,000
2,900
2,800
2,700
2,600 — World average intake
2,500
2,400 — Average requirement for well-being
2,300
2,200 — Average intake for developing countries
2,100
2,000
1,900
1,800
1,700
1,600
1,500 — Intake below which malnutrition can be expected

Headline news
Each year 40 million people die from hunger and hunger-related diseases. This figure is equivalent to more than 300 Jumbo jet crashes a day with no survivors, almost half of the passengers being children.

The geography of hunger is one of the starkest indicators of the North-South divide. As the world map, above, confirms, the average diet in the North provides a good deal more than the essential average daily intake of calories. The average diet in the South, by contrast, not only falls below the world average calorific intake, but often also falls below the minimum intake required for survival, let alone health. Malnutrition is the concealed cause of many diseases, particularly among children. With their pot bellies and sunken eyes, they quickly become vulnerable to infection. Perhaps worst of all, diarrhoeal diseases drain from a child's stomach whatever nutriment it has been able to ingest – reinforcing basic malnutrition.

Each year, 15 million babies and young children die – a statistic that would be shocking if it were only 15,000. According to James Grant, head of UNICEF, we could save many of these children at an overall cost of $5 each, through programmes promoting immunization, breast-feeding, rehydration therapy (which counteracts diarrhoea) and improved child care generally.

In the North, malnutrition takes the form of over-consumption of sugars, fats, and animal products, resulting in obesity, heart disease, and diabetes. In the US alone, at least one-third of those aged over 40 can be classified as obese. In 1982, the UK spent´235 million on slimming aids – compared to just´50 million donated to private aid agencies like OXFAM.

1974-76   2000

Numbers
undernourished
Each figure
represents
5 million people.

More than 100% of
requirement

85-99% of requirement

Less than 85% of
requirement

The hunger gap
The average daily calorific
requirement for well-being
is estimated to be around
2,400 calories (for
developed and developing
countries). Developed
world citizens receive some
40% above this figure,
while the average Third
World person subsists on
10% less.

The map shows the
geography of world
hunger. It gives the
percentage of nutritional
requirements received in
selected countries.

In sub-Saharan Africa, the average amount of food available per head has actually declined since 1970. By fortunate contrast, the situation has apparently improved somewhat in Asia and Latin America.

But the innocent-seeming qualification "apparently" conceals a catch. The sharp disparity of hunger and glut occurs not only among countries, i.e. between Northerners and Southerners. It occurs *within* many countries of the developing world, where the top one-fifth of the population may be 10 to 20 times more affluent than the bottom one-fifth. In Kenya and the Ivory Coast, Brazil and Mexico, Iraq and the Phillipines, the privileged gorge themselves alongside the starving. In all these countries, the national statistics for average diets may suggest that things are not so bad; but in Kenya alone, 40 percent of the rural population suffer a deficit of 640 calories a day. Clearly, we must examine the overall statistics more carefully if we are to hear the protests of those who eat less in one day than a typical British person consumes by lunchtime (about 1,500 calories) – and tries to do many hours of hard labour off it.

Go to a global food conference, and you will see that many political leaders from the Third World (like their colleagues from the rich world) are, frankly, fat – and their lives will be shortened just as surely as will those of the hungry multitudes whom they earnestly discuss. Perhaps attendance at a global food conference should be made conditional upon a suitably trim figure?

By contrast, we find in a few countries, notably China and Sri Lanka, that while the average per-capita consumption of calories is low, hardly anybody starves. The food available is more equitably shared among the community.

The absurdity of hunger and glut within many developing countries is aggravated by excessive consumption in the North. So prodigal is an average American in the use of food, that one-quarter of it is never eaten; it rots in the supermarket or fridge, or is thrown away off the plate.

The consequences of hunger extend beyond physical suffering and misery. It reduces capacity to work, and increases susceptibility to disease. Among children, insufficient protein can retard development, physically and mentally.

Some may protest that the poor countries should grow more food for themselves. Well, so they should – and they would feel encouraged to do so if they did not allocate so much land to cash crops for export. As a first look at some solutions – how we can *manage* our affairs to the advantage of all – it is fitting that the cash-crop factor be examined.

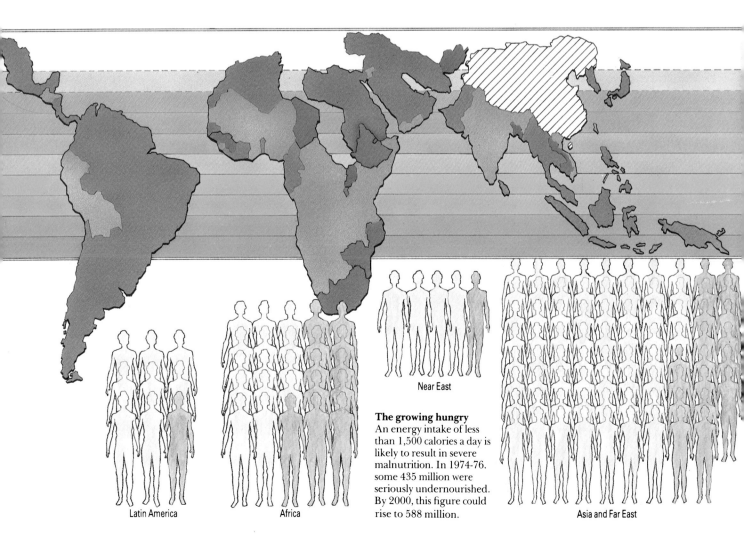

Near East

Latin America

Africa

The growing hungry
An energy intake of less
than 1,500 calories a day is
likely to result in severe
malnutrition. In 1974-76.
some 435 million were
seriously undernourished.
By 2000, this figure could
rise to 588 million.

Asia and Far East

MANAGING THE LAND

We live in a hungry world, faced with a deepening land crisis. The costly "escape" strategy of bringing ever more land under the plough will only worsen this crisis, and still barely feed our growing numbers if we carry on as now.

The solution lies not in how much land we have, but in how we use it. Let us look at some specific sectors, asking ourselves why we allow things to go wrong, and what we should do to set them right.

Cash crops trade: bonds or bondage?

Cash crops can be either boon or bane for developing countries. In some instances, they supply much impetus to the development process, by bringing in crucial foreign-exchange earnings which lubricate emergent economies. But all too often, because the land could be better employed in growing food for hungry local people rather than non-essential commodities for foreigners, they tend to slow down the development process.

It is difficult for developing economies to break free from the cash crop bind. For most developing nations, "going it alone" would be impossible: their domestic market is too small, their resource base too narrow. Unless they can trade in the world marketplace, they cannot purchase the tools and technology they need for development. And those countries that have no mineral resources of their own must find the means to earn foreign exchange as best they can. Even if they prefer not to follow a Western model of development, those developing countries that suffer increasing poverty and a deteriorating resource base need hard cash simply to purchase a more critical export from the developed world – grain.

Where does the answer lie? First, the developing nations must get their priorities right. In too many developing world countries, agriculture is subordinated to urban development and industrialization. A selective cash-crop strategy can avoid heavy dependence on a single export commodity, and place more emphasis on efficient food production, a better share-out of farmlands, and fairer distribution of income. Secondly, international agencies can change their approach to aid and development within the agricultural sector. Although some Third World countries have inherited colonial cash-crop links, and others have entered the vicious cash-crop circle by accident, all too many have been led there, even pushed there, by misconceived aid programmes. Not enough of these countries' leaders have the

The cash-crop factor

The major grain exporters
Grain produced in the North is distributed worldwide. More than 100 countries depend on these grain shipments.

North American imports

Western European exports

North American exports

Latin American exports

Latin American imports

Wheat
Coarse grains
Sugar
Soyabeans
Coffee
Cotton
Rice
Cocoa
Tobacco
Natural rubber

Cash crops generate funds for the producer countries through their export value in the world marketplace. Broadly speaking, they fall into two categories: essential food crops, such as cereals and legumes, which are mainly produced in the North and traded all over the world; and less essential crops, such as coffee, cotton, and tobacco, which are predominantly produced in the South for export to the North (reflecting colonial trading links). The map shows trade flows of ten major cash crops (based on 1980 import and export values). Some of the less essential crops command relatively high prices on world markets – for much of the time; and they generate more money than the land could produce through growing staple foods – for much of the time.

But the catch lies in the phrase "for much of the time". Commodity prices are fickle: when consumer demand soars, prices are good – whereupon many farmers are persuaded to go into the cash-crop business, switching from growing food to growing non-essential items. When the global economy turns down again, prices crash, and the farmer is left caught between a rock and a hard place.

What commodities can buy

In 1975, 1 tonne of coffee bought around 290 barrels of oil, yet in 1983 it could purchase only one-third as many; and the crop's capacity to service interest payments for international loans plunged by over 60%. Similarly cocoa and sugar, less valuable commodities, have declined in purchasing power by one-half and five-sixths respectively.

Commodity	Oil (barrels) bought by 1 tonne of commodity 1975	1983
Cocoa		
Coffee		
Sugar		

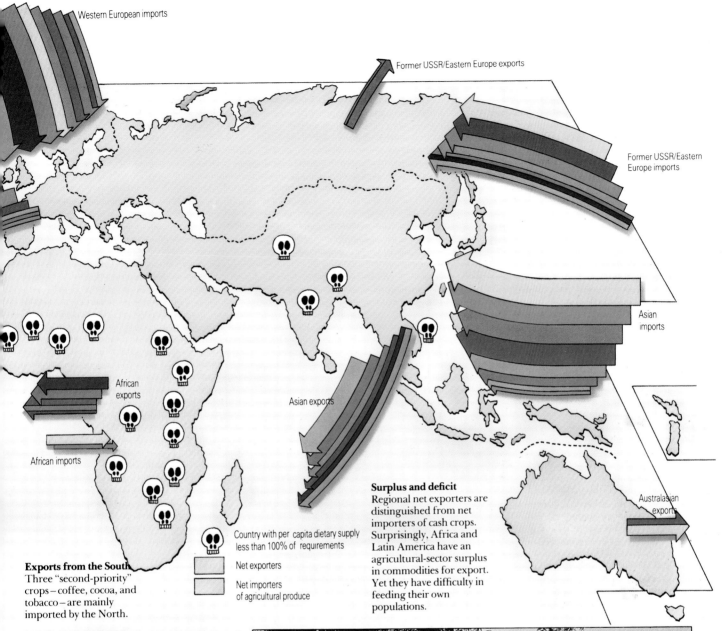

Western European imports

Former USSR/Eastern Europe exports

Former USSR/Eastern Europe imports

Asian imports

African exports

African imports

Asian exports

Australasian exports

Country with per capita dietary supply less than 100% of requirements

Net exporters

Net importers of agricultural produce

Surplus and deficit
Regional net exporters are distinguished from net importers of cash crops. Surprisingly, Africa and Latin America have an agricultural-sector surplus in commodities for export. Yet they have difficulty in feeding their own populations.

Exports from the South
Three "second-priority" crops – coffee, cocoa, and tobacco – are mainly imported by the North.

Peanuts in the Sahel
During their period of colonialism and two subsequent decades of independence, the Sahel countries have not given enough priority to growing food for local people. Instead, their emphasis has been on export crops such as cotton and peanuts. In Niger alone, the amount of land given over to peanut cultivation has tripled from 142,000 ha in 1954 to 432,000 ha in 1968, causing a large reduction in fallow land. Consequently nomadic pastoralists, who traditionally used this land for grazing in the dry season, moved north to previously unsettled land with poor carrying capacity. (Land carrying capacity is a measure of how many people the soil/climate can permanently support when the land is planted with staple crops.) Population pressures and inadequate farming practices have since combined to turn moderate-to-poor farmland into poor land, and poor land into useless land (see pp 42-3).

Thus at a time of famine in the Sahel, bumper cash-crop exports have been recorded. And as the real value of cash crops falls, the people who once ate what was grown on their own lands rely increasingly on imported food crops from the North. The photograph, right, shows food aid being distributed in the Sahel.

inclination, let alone the means, to reject aid that deflects their development strategies from their most urgent priority of feeding their peoples.

A streamlined food supply

As international trade expands and the world's economy grows more integrated, so the food of rich-world citizens tends to reach them from a "global farm" via a "global supermarket". At most major food stores in the West, consumers purchase products from all over the world. Similarly, millions of developing world people, together with large numbers of former Soviet citizens, consume grain from North America, Australia, and Argentina. True, only one-tenth of all food grown enters international trade. But we can note a similar integrative pattern within individual countries, too, as agriculture gives way to "agribusiness".

There is much to be said for a commercial system that enables us to enjoy abundant and diversified products from remote croplands and to benefit from those farmers who produce most food cheapest. We now produce sufficient food to feed the whole of Earth's human population, a feat that would have been impossible only 100 years ago.

That we manage to feed such multitudes is due in part to giant enterprises that dominate our food system. These enterprises handle not only the sale of food; they often control the growing, processing, and distribution. A sound achievement, as long as it means a streamlined operation, with more, better, and cheaper food at the end of the line.

There lies the snag. We face a situation in which the mega-corporations are coming to wield virtual monopoly power over key sectors of the food trade. And their desire for more control over

The global supermarket

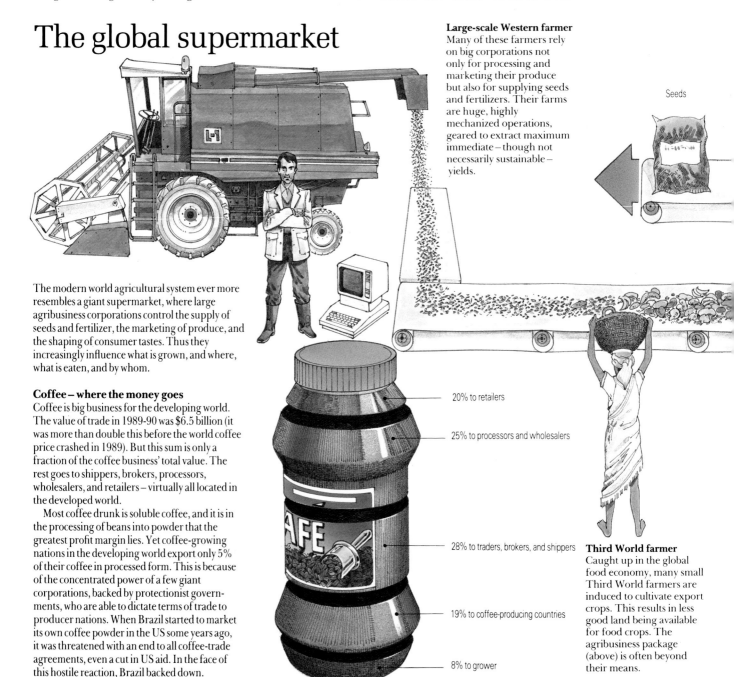

Large-scale Western farmer
Many of these farmers rely on big corporations not only for processing and marketing their produce but also for supplying seeds and fertilizers. Their farms are huge, highly mechanized operations, geared to extract maximum immediate – though not necessarily sustainable – yields.

Seeds

The modern world agricultural system ever more resembles a giant supermarket, where large agribusiness corporations control the supply of seeds and fertilizer, the marketing of produce, and the shaping of consumer tastes. Thus they increasingly influence what is grown, and where, what is eaten, and by whom.

Coffee – where the money goes

Coffee is big business for the developing world. The value of trade in 1989-90 was $6.5 billion (it was more than double this before the world coffee price crashed in 1989). But this sum is only a fraction of the coffee business' total value. The rest goes to shippers, brokers, processors, wholesalers, and retailers – virtually all located in the developed world.

Most coffee drunk is soluble coffee, and it is in the processing of beans into powder that the greatest profit margin lies. Yet coffee-growing nations in the developing world export only 5% of their coffee in processed form. This is because of the concentrated power of a few giant corporations, backed by protectionist governments, who are able to dictate terms of trade to producer nations. When Brazil started to market its own coffee powder in the US some years ago, it was threatened with an end to all coffee-trade agreements, even a cut in US aid. In the face of this hostile reaction, Brazil backed down.

20% to retailers

25% to processors and wholesalers

28% to traders, brokers, and shippers

19% to coffee-producing countries

8% to grower

Third World farmer
Caught up in the global food economy, many small Third World farmers are induced to cultivate export crops. This results in less good land being available for food crops. The agribusiness package (above) is often beyond their means.

our food-producing systems is expanding even further. Since 1970, a few giant petro-corporations have quietly taken over more than 400 small seed businesses; businesses that hitherto produced seeds with vast variety to suit diverse environments, tastes, and price ranges. By controlling the production of seeds, a petro-corporation can breed crops that need extra-large dollops of synthetic fertilizer, pesticides, and other petroleum-based additives, regardless of more desirable trends for future agriculture (viz. away from reliance on fossil-fuel inputs). What happens when the oil wells run dry? The corporation answers that it will tackle those problems as they arise, but meanwhile it is a private profit-making concern, not a public charity.

Probably even more insidious in its ultimately detrimental impact on agriculture is the support provided by giant corporations to the "farmers' lobby" in developed nations. The Common Agricultural Policy of the EC causes member governments to pay out $21 billion a year to encourage farmers to produce surpluses, such as the famous butter mountains and milk lakes. Similarly, US farmers are subsidized to the tune of $26 billion a year. These subsidies help depress the prices of produce from the developing world by undercutting competition. At the same time, the overproduction of food by the US and Western Europe leads to the dumping of subsidised, low-priced produce on poor countries, often putting local small farmers at a disadvantage, or out of business.

The global supermarket, dominated in part by giant corporations, is potentially a powerful medium that could serve us all. But much refinement on the part of governments, consumer organizations, and the like, is required if it is to serve the

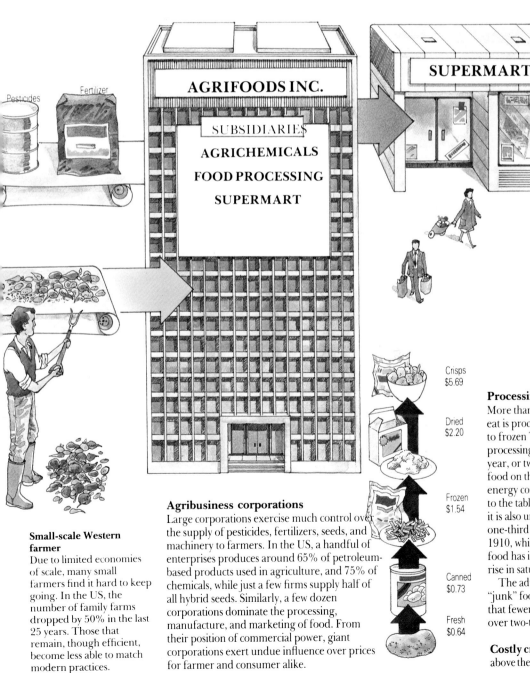

Small-scale Western farmer
Due to limited economies of scale, many small farmers find it hard to keep going. In the US, the number of family farms dropped by 50% in the last 25 years. Those that remain, though efficient, become less able to match modern practices.

Agribusiness corporations
Large corporations exercise much control over the supply of pesticides, fertilizers, seeds, and machinery to farmers. In the US, a handful of enterprises produces around 65% of petroleum-based products used in agriculture, and 75% of chemicals, while just a few firms supply half of all hybrid seeds. Similarly, a few dozen corporations dominate the processing, manufacture, and marketing of food. From their position of commercial power, giant corporations exert undue influence over prices for farmer and consumer alike.

Crisps $5.69

Dried $2.20

Frozen $1.54

Canned $0.73

Fresh $0.64

Consumers
The final stop in the system. Because of the influence of agribusiness, consumers are losing control over the quality and price of their food. Supermarkets may appear to offer a wide choice of products, but in reality their food sources are confined to a few crops.

Processing in the US food system
More than three-quarters of the food Americans eat is processed in some way, from waxed apples to frozen TV dinners. In energy terms, processing now costs more than $10 billion a year, or two-thirds as much again as growing the food on the farm, and almost one-quarter of all energy costs in getting food from the cropland to the table. Not only is this extremely wasteful, it is also unhealthy. Americans now consume one-third less fresh fruit and vegetables than in 1910, while their consumption of processed food has increased 3 times, together with a large rise in saturated fats and refined carbohydrates.

The advertising pressure in favour of more "junk" foods is better understood when we note that fewer than 50 US corporations account for over two-thirds of all food processing.

Costly crisps (left) See how the price per kg soars above the cost of the original potatoes.

best needs of customer and stockholder alike. A seven-year international boycott, led mainly by citizen activists, succeeded in persuading the giant Nestle corporation to agree to market its powdered milk more responsibly and promote it as an alternative rather than a substitute for mother's milk. This proves that progress can be made if enough citizens insist that their voices are heard. Problems with the Nestle agreement, however, led to the launch of a second boycott four years later.

The forest cash crop

More people everywhere want more wood. We presently consume 3.4 billion cubic metres per year, half of it as fuel. Within the next 40 years consumption could increase by 50 percent.

In the developed world, a large amount of wood is used for paper: around 200 million tonnes of pulp a year, compared with the developing nations' consumption of around 25 million. Developing nations have powerful incentives to develop their own sources of pulp, as they import virtually all their supplies at a cost that nearly wipes out their earnings from hardwood exports. As literacy increases, so the demand for paper increases: Brazil's consumption doubled during the 1960s, and almost doubled again during the 1970s.

As for industrial timber, Northerners increasingly seek specialist hardwoods, notably those from the tropics, for constructional and semi-luxury purposes (see right). By contrast, timber in the South is used mostly for essential purposes. At least one billion Southerners can be described as living in the Wood Age. Yet because prices for wood are determined internationally, and thus tend to be set by consumption in the North, Southerners find themselves squeezed out of the marketplace. When Northerners have to pay a little more for stylish veneer or a newspaper, there may be complaints about inflation, but the upshot does not generally affect living standards. For a person in the South, an increase in price often means doing without.

These imbalances are aggravated by the forestry policies of several developed nations. Japan, for example, fears for its economic security in world wood markets. Hence it adopts a "siege strategy" by building up its own forests stocks and growing more wood than it cuts, while depending heavily on foreign sources of timber. Within the context of Japan's own needs, this approach makes sense. But within a global setting, it illustrates the "tragedy of the commons" – and will ultimately paint Japan itself into an ever-tighter corner.

What can we do to improve the global wood situation? Plenty. But we need to act as a community of nations for the benefit of all humankind. Adequate funding should be provided for the establishment of commercial fuelwood plantations in the tropics, where the year-round steamy-warm climates are ideal for generating timber – thus increasing

Harvesting the forest

Forest products represent one of our most valuable categories of cash crops. As such, they are widely traded around the world, albeit with unequal patterns between the developed and developing worlds. This trade is likely to expand, especially in so far as two of the main consuming regions, Japan and Western Europe (outside Scandinavia), intend to continue to import huge volumes, even though they could grow more wood at home. (Britain, for example, which imports nine-tenths of its wood needs, could reduce its demand by planting more trees on under-utilized land.)

Demand for paper alone in the developed world is forecast to increase by a factor of 2.75 during the period 1975-2000, while developing nations, with their expanding populations, are likely to increase their consumption at least three-fold. The average Northerner consumes more than 150 kg of paper a year, compared with the Southerner's consumption of 5 kg.

The developed world also demands specialist hardwoods for numerous construction purposes, as well as for various luxury items such as parquet floors, fine furniture, decorative panelling, and weekend yachts. In Switzerland, whose forests are extensive and under-utilized, abachi wood is imported from the Ivory Coast to make superfine coffins.

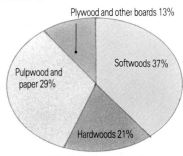

Trade in forest products
Unprocessed hard- and softwoods account for just over half of world trade. Pulp and paper make up about a third.

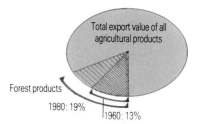

A valuable export
In 1960, forest products accounted for 13% of all agricultural exports. By 1980 the real-value share had increased to 19%.

Product demand in the North . .
Northerners use huge amounts of wood for constructing houses, office buildings, railway sleepers, pit props, and the like. They also consume a great deal of paper products (notably packaging materials), as well as luxury hardwood items.

. . and South
Southerners likewise consume much wood, but mostly for fuel. Almost all houses in the Third World use wood, but not as high-grade construction material, rather as simple building poles. As for paper products, the average Southerner consumes all too little – to the detriment of education and communications.

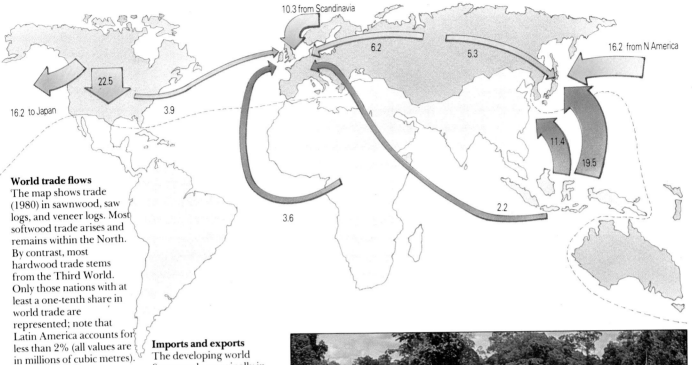

10.3 from Scandinavia

6.2

5.3

16.2 from N America

22.5

3.9

16.2 to Japan

11.4

19.5

2.2

3.6

World trade flows
The map shows trade (1980) in sawnwood, saw logs, and veneer logs. Most softwood trade arises and remains within the North. By contrast, most hardwood trade stems from the Third World. Only those nations with at least a one-tenth share in world trade are represented; note that Latin America accounts for less than 2% (all values are in millions of cubic metres).

Hardwoods
54.8 million cu m

Exports
86%

Imports
33%

Softwoods
94.1 million cu m

Exports
4%

Imports
12%

Pulpwood and paper
75.6 million cu m

Exports
4%

Imports
13%

Plywood and other boards
33.4 million cu m

Exports
15%

Imports
9%

Imports and exports
The developing world figures only marginally in global trade of forest products, except for its massive exports of hardwood. (It appears to play a fair part in imports of hardwood, but this is all from other Third World nations.) Despite its need for wood, the developing world's share of global trade is projected to rise only a little by 2000.

Hatched area shows the proportion of trade by developing countres

A source of revenue
When large numbers of huge logs are taken from a Southeast Asia forest (top right), at least half of the remaining trees are injured beyond recovery. Yet the logger does not care, as long as there is enough forest to last for the next few years. Governments could be much tougher in enforcing less harmful harvesting techniques, but the need for hard cash often means that they fail to see their forests as sources of future revenue. Fortunately a partial breakthrough is on the way, in the form of increased wood processing: the veneer sheet (bottom right) earns several times more foreign exchange than a raw log, thus relieving pressure to exploit every forest tract.

wood stocks and relieving pressure on virgin forest. In addition, developed nations should review their own forestry policies, with a view to producing more home-grown wood (including hardwood) and encouraging greater recycling of paper. There is no doubt about the Earth's capacity to supply us with sufficient wood; the question lies with our commitment to manage forests sustainably.

Planting tomorrow's trees

In the North, we claim that we know a good deal about how to manage forests. Virtually all of Europe's original forests have been replaced, largely by intensively-managed tree stands. In the South, ironically, much of the problem of deforestation lies with the streamlined logging techniques developed in the North. Mechanized operations in the tropics are often wasteful and destructive: 75 percent of the surrounding canopy can be damaged during the extraction of a few commercially valuable species. Unlike temperate forests, tropical forests cannot withstand such disruption because of their ecological complexity.

The major Northern development in recent years has been in genetic engineering. Trees grown from tissue culture can quickly reforest large areas of denuded land, and can be "engineered" with useful commercial qualities. However, industrial plantations, almost always consisting of genetically-uniform stands, are generally more vulnerable to pests and pathogens. Widespread disease and pest outbreaks are common throughout conifer plantations in the US, China, and central Europe.

In the South, because of the critical environmental role played by tropical forests, it is better to establish "tree farms" on lands already deforested than to harvest the natural forest. Tropical plantations which mix fast- and slow-growing species could be managed to provide fibre and timber. According to the Washington-based Worldwatch Institute, establishing such plantations on just 5 percent of the area of tropical forest already cleared could provide almost twice as much industrial wood as is currently harvested from all tropical forests.

Fuelwood plantations are also urgently required in the South to relieve pressure on natural forests. We need to increase five-fold the number of fuelwood trees planted per year, especially around farms and in village woodlots, at an annual cost estimated at $1 billion. The difficulties, however, are not all financial. Community forestry relies on the involvement of local people. If everybody's ideas are sought from the start, hopefully everybody will plant trees and tend them, and everybody will ensure that the harvesting system produces a regular supply of fuelwood.

Similar community efforts are needed to stabilize degraded watersheds. In China and parts of India there has been much success, due to the close coordination of planners and villagers.

Forests of the future

Tropical and temperate forests are so different in their biological makeup, that two fundamentally different approaches to their management are needed. Temperate forests are actually expanding slightly, due to reforestation in the North. Densely-settled zones such as southwest Germany are one-quarter covered with forests. Much land, however, is under-utilized.

Recycling paper

Developed countries could reduce their demand for paperpulp by at least one-quarter, simply through greater recycling. During World War II, most Northern countries recovered as much as half of their paper.

Forest management in the South

In most developing countries, forestry departments are understaffed and underfunded. Foresters see their main duty as keeping people out of forests, rather than helping them to establish woodlots, fuelwood plantations, and other village forestry projects. International agencies have only recently begun to take the preservation of tropical forests seriously, and are promoting forestry as an important aspect of rural development. The World Bank, for instance, has a central Environment Department as well as regional environmental divisions, and states that half of its loans have "environmental components". An environmental assessment process reviews the potential impact of proposed projects and the Bank can choose to withdraw aid if a project proves to be environmentally unsound.

Replanting watersheds
Nations are realizing the value of restoring tree cover on upland watersheds – a measure that benefits virtually everyone in the community. Although implemented on far too limited a scale as yet, this is an encouraging step forward.

Community forestry
Supported by international development agencies, many countries are encouraging their people to become actively involved in establishing fuelwood plantations. In Gujarat, India, schoolchildren may soon be raising as many tree seedlings as the government.

Forest management in the North

Destructive logging is by no means confined to the tropics. Old-growth forest in Canada and the western USA is being logged at rates surpassing the estimated sustained yields. The challenge is to halt the rapid mining of these forests and transfer our efforts to sustainably-managed secondary forests and plantations. In recent years, foresters in the former Soviet Union have planted about 1 million ha annually and aided natural regeneration on an equivalent area. In 1988 US government incentives resulted in farmers and woodland owners planting trees on 329,000 ha of land – a quarter of the total planted that year.

Of vital importance, also, is the more efficient use of wood. A variety of technologies from thinner saw blades to computerized scanners to detect defects in logs can be used to reduce manufacturing waste. In the US, for example, improving the efficiency of forest product manufacturing to Japanese levels would save enough wood to leave standing one out of four trees cut nationwide.

A sustainable land-use system

A strategy with much potential is agroforestry (shown above), which amounts to growing trees and food crops alongside each other. Forest land and marginal land, normally rated unsuitable for crops, can be utilized for the production of food. Certain tree species, notably the leguminous ones, fix atmospheric nitrogen in the soil, thereby helping to rehabilitate degraded forestlands.

The Chipko movement

In 1974, the women of Reni in northern India took simple but effective action to stop tree felling. They threatened to hug the trees if the lumberjacks attempted to fell them. The women's protest (known as the Chipko movement) saved 12,000 sq km of sensitive watershed. The photograph, right, exemplifies the fierce concerns felt by many local people for their forests.

Extractive reserves

Brazil's rubber tappers (above) have a vision of development centred around "extractive reserves": areas managed by local communities for uses compatible with the survival of the forest, such as tapping trees and gathering fruits and nuts. Since 1987, the Brazilian government has recognized reserves covering 3 million ha.

Slowly we are beginning to appreciate the manifold benefits tropical forests have to offer over and above temperate forests: benefits that will become available to us only when we manage the forests selectively and with understanding. We are also now recognizing their vital ecological role, especially in conserving soil, regulating the flow of water, generating rainfall, and moderating the local and global climate.

Preventing erosion and desertification

The loss of soil and the spread of deserts is by no means confined to the developing world. A large share of the 275 million hectares of arable land that we have lost in the final quarter of this century is in the developed world.

There are many ways to tackle the problem. We can establish shelterbelts of trees, practice better livestock husbandry, and plant resilient grass cover.

We can use a more caring agriculture, in good lands as well as poor ones. But the single best approach is to halt the drift of too many farmers into areas where they should not be. In essence, the problem lies with the people who are rendered landless, pushed to the fringes of their national societies – and with vulnerable lands which are simply not suited to agriculture. In short, we must keep marginal people out of marginal lands.

The key to this approach lies in the enhanced use of our good agricultural land. Put another way, we must try to produce three-quarters of the extra food we shall need in the future, not from virgin territories, but from *existing* farmlands.

There is no doubt that we can do it, as demonstrated by some notable instances. If more farmers in Java practised the intensive and sustainable agriculture that has been standard in neighbouring Bali (which features as many people per unit area

Terracing
Especially susceptible to erosion are hilly areas with fertile soils, notably volcanic zones with their intensively cultivated hill slopes. The answer is to construct terraces as exemplified by the 2,000-year-old rice paddies of the northern Philippines.

Managing the soil

The landscape above has been compressed in terms of scale to show the changes in land use as one moves from good agricultural land (left) to desert (right). The techniques suited to each category include soil-conserving practices and intensive use in high-potential lands; reduction of overload on arable lands to prevent topsoil loss; a sustainable mix of crops and livestock on intermediate farmlands; protection of the vulnerable dry lands and mountainous areas; and finally, attempts to halt desertification, and to rehabilitate other marginal lands.

Good land: intensified use

The best way to safeguard poor land is to make full use of the good land. The principal strategy is multiple use, growing crops in rapid rotation, and interplanting several crops at a time. The process is obviously enhanced by irrigation, despite its cost. Organic farming uses crop residues and other forms of mulch as "green manure". A few "garden farm" agro-ecosystems employ several dozen crops on just a couple of hectares, fostering renewable use of land.

Medium-value land

Moderately dry lands can be cultivated for crops that withstand long rainless spells, e.g. millet, sorghum, amaranth, certain beans, and fast-maturing maize. But to safeguard against erosion, the farmer should employ a series of additional crops that restore soil fertility, such as cowpea and groundnut, and allow for regular fallow periods.

as Java, often more), much of Java's devastated land could be rejuvenated into a garden isle like Bali. Other parts of Southeast Asia could be more productive if people were encouraged to follow similarly sustainable agricultural systems. In Amazonia, if newcomers adopted the traditional crop-growing strategies of long-established peasants in humid Latin America, the fertile floodplains of the Amazon, with their yearly enrichment of alluvial silt, could support millions of small-scale farmers in perpetuity. Only a very small fraction of the rainforests need to be utilized to produce a flourishing agricultural civilization rivalling those of historic times in the Ganges and the Mekong floodplains.

But if established croplands are to produce twice as much food, this will often mean a basic shift in agricultural strategies. The Green Revolution (pp 56-7) is being complemented by a Gene Revolution

(pp 60-61). Much better use must be made of scarce water through, for example, trickle-drip irrigation (p. 133). Priority should be given to growing food for local communities, with less emphasis on seductive diversions such as cash crops for export – a factor exemplified by the experience of the Sahel (pp 42 and 47). Of course, this will not appeal to people with a stake in the status quo, such as the local elites who profit from the export trade. These are the people who, during the Sahel disaster in the 1970s, were responsible for maldistribution of $6 billion of foreign aid. Only one percent went to forest programmes such as desert-halting shelter-belts, and only three or four percent went to traditional grain-growing agriculture.

In many other lands that are not yet so unfortunate as the Sahel but are heading in a similar direction, there is need for a basic switch in development planning if there is to be more

Hardy animals
Disease-resistant varieties of cattle and sheep that can survive drought and subsist off little forage are best suited to marginal lands. While they do not generate massive amounts of meat and milk, they cause less harm to the environment than do more exotic breeds.

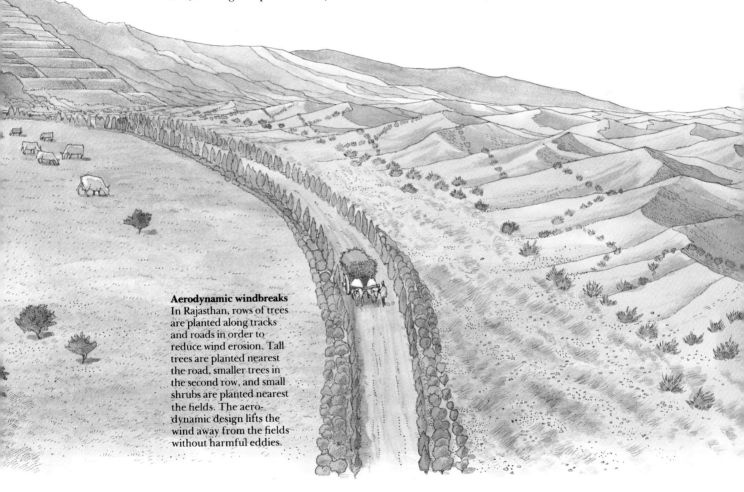

Aerodynamic windbreaks
In Rajasthan, rows of trees are planted along tracks and roads in order to reduce wind erosion. Tall trees are planted nearest the road, smaller trees in the second row, and small shrubs are planted nearest the fields. The aero-dynamic design lifts the wind away from the fields without harmful eddies.

Favoured trees
Many trees and shrubs are available in the wild to help us resist desertification. *Prosopis* species are not only highly resistant to drought: their pods are rich in protein and make good livestock fodder. The *Casuarina* tree grows fast in sand and, being a tall upright tree, it makes a first-rate windbreak. *Leucaena* trees grow very swiftly, are fine for fuelwood and, because they fix their own nitrogen, help to restore soil fertility.

Livestock lands
Dry savannahs and grasslands easily suffer through over-large herds of cattle, sheep, and goats. Plainly the aim must be to reduce excess numbers, though in many areas, persuading the herdkeeper to sell off part of the holding can be difficult. Livestock is often the chief means of storing wealth in Third World countries. A backup strategy lies in rotational grazing, which gives over-worked pastures time to recover. Devegetated lands can be improved by planting legume grasses, such as clover and alfalfa, which help to restore soil fertility.

Established desert
Natural desert is irreversible. It is in the borderlands of deserts that we need to apply caring agricultural techniques to stop lands which have been rendered agriculturally worthless from being "tacked on" to natural desert. Shrubs and bushes serve as barriers to hold back dunes; some woody plants, such as guayule and jojoba, produce rubber and liquid wax. Other anti-desert plants include various types of brushwood and hardy trees.

efficient use of existing croplands. There must be a redirection of emphasis from industry and manufacturing to agriculture, from favouring the top ten percent (the entrepreneurs who will be the "locomotives" of economic growth) to favouring the bottom 40 percent, the poorest of the poor. While this runs counter to much conventional wisdom concerning development, it runs parallel to the main thrust of World Bank policies since the late 1970s.

A costly revolution

The Green Revolution is one of the most remarkable advances ever seen in agriculture. In the developed world it started to work its wonders during the 1940s, leading to record crops in North America and Western Europe from the 1950s onwards. In the developing world there was a time lag, but from the early 1960s exceptional produc-

tivity became the norm in several parts of tropical Asia and Latin America. India, for example, produced only around 50 million tonnes of cereals and related food crops per year in the early 1950s; since then, the Green Revolution has taken off, pushing the total up, over three decades, to an extraordinary 167 million tonnes in 1987. Annual food production per person increased from 141 to 209 kilograms, while net imports of food were dramatically cut.

But there is a price attached to this success story. The Green Revolution consists principally of planting so-called "high-yield" varieties of rice, wheat, and maize, varieties which produce bumper harvests, and in some cases mature faster, so that the farmer can grow two or even three crops in a year. But they are not so much high-yield as high-response varieties. They do their job only when they receive stacks of fertilizer, pesticides, irrigation water, and

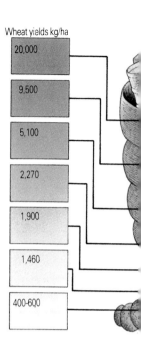

Wheat yields kg/ha

20,000
9,500
5,100
2,270
1,900
1,460
400-600

Green revolution?

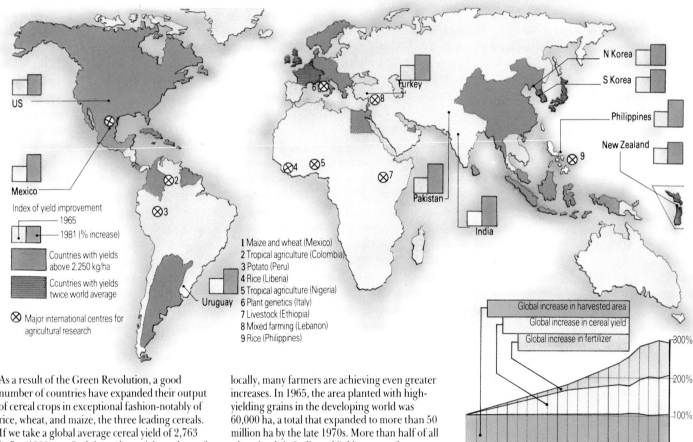

US

Mexico

Index of yield improvement
1965
1981 (% increase)

Countries with yields above 2,250 kg/ha

Countries with yields twice world average

⊗ Major international centres for agricultural research

Uruguay

1 Maize and wheat (Mexico)
2 Tropical agriculture (Colombia)
3 Potato (Peru)
4 Rice (Liberia)
5 Tropical agriculture (Nigeria)
6 Plant genetics (Italy)
7 Livestock (Ethiopia)
8 Mixed farming (Lebanon)
9 Rice (Philippines)

Turkey

N Korea
S Korea

Philippines

New Zealand

Pakistan

India

Global increase in harvested area
Global increase in cereal yield
Global increase in fertilizer

300%
200%
100%

Year 65   70   75   80

As a result of the Green Revolution, a good number of countries have expanded their output of cereal crops in exceptional fashion-notably of rice, wheat, and maize, the three leading cereals. If we take a global average cereal yield of 2,763 kg/ha (1990), we find that substantial numbers of nations now achieve harvests well over this, some of them double. These include such populous developing countries as Egypt, Indonesia, and China. In the developed world, which had a head start, we find that France, the UK, the Netherlands, and Japan are among those nations which have yields in excess of double the global average. But average cereal yields conceal the most dramatic success stories of the Green Revolution (see country indicators on map): India doubled its wheat yield in 15 years; in the Philippines, the rice yield has risen by 75%; and

locally, many farmers are achieving even greater increases. In 1965, the area planted with high-yielding grains in the developing world was 60,000 ha, a total that expanded to more than 50 million ha by the late 1970s. More than half of all wheat lands in India and Pakistan now feature Green Revolution varieties, and in the Philippines, half the area planted to rice. As a measure of the benefits in Asia alone, the increased output totalled at least 15 million tonnes in 1976, roughly double the harvest in 1971, with an economic value of at least $1.5 billion. Meanwhile, research into the new crops and agricultural techniques continues, though grossly underfunded. The map shows the location of some of the major international research centres for agriculture in the developing world.

Inputs and outputs

Using a base-line value of 100 for the early 1960s, we find that the global yield of cereals topped 200% in the late 1970s. The linkage with fertilizer is clear; a value of 285% for the late 1970s (the area harvested increased by 6% over the same period). Since then, yield has tended to level off. The Green Revolution has only given us a breathing space.

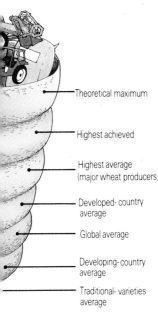

Theoretical maximum

Highest achieved

Highest average
(major wheat producers)

Developed-country
average

Global average

Developing-country
average

Traditional-varieties
average

other additives, together with good farming techniques. Fertilizer from fossil fuels, a key component of the package, was cheap during the Revolution's early years. But after OPEC imposed its first price jump in 1973, costs have soared. The situation has been aggravated by increasing demand. High-yield varieties need 70-90 kilograms of nitrogen per hectare, whereas the average amount available in most developing countries is only about 25 kilograms. If India's farmers were to apply fertilizer at the same rate as Dutch farmers, their needs would amount to one-third of global consumption.

Equally to the point, there has recently been a levelling off of increases in global grain yield. The problem lies with the "fertiliser response ratio", i.e. the return, in increased yields, produced by using additional fertilizer. Worldwide, this response ratio is decreasing, as the biological limits of the new hybrid strains are reached.

There have also been adverse economic and social repercussions. The high-yield varieties tend to be planted by those farmers who can afford the expensive additives. The bigger harvests generate greater incomes, which can then be invested in extra land. Slowly but steadily, the rich farmers broaden the gap between themselves and the poor farmers.

In short, the Green Revolution has achieved some remarkable breakthroughs, but it has many drawbacks and is plainly running out of steam. The challenges are enormous: after more than doubling our food supplies between 1950 and 1980, the following decade saw an increase of only one-fifth. During the 1980s food production per person peaked, then fell as agricultural output stagnated. Yet world food production must increase by 60 percent by 2025 just to maintain current nutritional levels for a projected population of 8.5 billion. It seems that we are largely

Cornucopia
The performance of high-yield crops depends on climate and technology. The best average yield for wheat is more than double the world average – but only slightly over half of the highest achieved yield, a quarter of the theoretical maximum. The Green Revolution has a good way to go yet.

Before and after in Muda
A multi-faceted landscape with a pluralist society (above) has been transformed into a homogenized landscape dominated by larger holdings (below). The landless resort to slash-and-burn farming in Malaysia's forests.

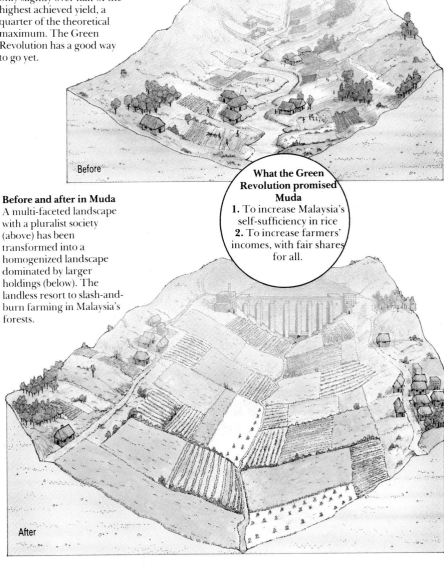

Before

What the Green Revolution promised Muda
1. To increase Malaysia's self-sufficiency in rice
2. To increase farmers' incomes, with fair shares for all.

After

Promise and reality: Muda River
However capable the Green Revolution has proven to be in agronomic terms, its economic and social success has been mixed at best. In northern Malaysia, the Muda River area reveals what can go wrong.

A $90-million dam allowed irrigated paddies to produce two crops of high-yield rice a year, instead of the traditional rice. By the early 1970s, output had almost tripled. Before the project, the area (which accounted for 30% of Malaysia's rice-growing land) supplied 30% of the country's rice; this rapidly rose to 50%. Previously Malaysia had been only half self-sufficient in rice; by 1974, it was 90%. But whereas average incomes for all categories of farmers increased, the wealthier categories enjoyed a boost of 150%, while poorer farmers experienced only a 50% advance. The gap between the rich farmer and the impoverished peasant increased from 900 Malaysian dollars per month to 2,350.

Worse still, the harvests failed to expand after 1974, due to the "plateauing" effect: the more fertilizer farmers applied to their crops, the more their returns diminished, until yields finally levelled off. In this second stage, real incomes fell for all farmers, but especially for the poor who, by 1979, dropped below pre-Revolution levels. Moreover, the richer farmers, finding themselves with disposable income, started to buy up land from the poorer farmers, causing the bottom sector of the community to take up tenant farming, or to be pressurized off the land altogether.

"To those that have shall be given ...": the gap between the well-to-do and the absolutely impoverished increases *within* the Third World, as well as between the Third World and the developed world. This regrettable pattern has been repeated in virtually all countries of the Green Revolution, with the signal exceptions of egalitarian societies such as China and Sri Lanka.

squandering the time the Green Revolution bought us in which to sort out larger problems, notably the runaway growth of population and the grossly inequitable distribution of land and food.

We need a new agriculture. We must adapt our present models, and devise another revolutionary advance in agriculture to complement the Green Revolution. Fortunately, we possess the germ of just such an advance (pp 60-61).

The Western agricultural model

American farmers are extraordinarily productive. Amounting to only about two in 100 of the US populace, they not only feed their fellow citizens, but provide more than half of all agricultural products on international markets. American farmers produce 15 percent of the world's wheat, 36 percent of sorghum, and 46 percent of maize on only 11 percent of the world's croplands. For decades, the world has benefitted from America's food surpluses. But the prospect of further unfaltering success is an illusion. While industrial agriculture is extraordinarily productive, it can also be extraordinarily destructive – in a slow, quiet manner, which means that the situation could become critical before it appears serious.

What are the costs of this advanced form of agriculture? The first and heaviest toll must surely be in soil loss. An astonishing 80 million hectares of US croplands have been rendered unproductive, if not ruined outright. The nation has lost at least one-third of its best topsoils, and erosion rates are now worse than ever, more than one billion tonnes per year.

Secondly, the prodigal application of synthetic fertilizers and pesticides is responsible for more than half of all US water pollution. Thirdly, the spread of irrigation, which accounts for more than 80 percent of US water use, is steadily depleting the country's groundwater stocks. Finally, transport costs incurred in the distribution of food amount to $5 billion a year.

Not surprisingly, the industrialization of agriculture in the US has turned many farms into sizeable commercial enterprises. Since 1950 there has been a steady trend toward large holdings, until now there are only half as many farmers. Due to the escalating costs of fertilizer, pesticides, fuel, and land itself, farmers have run themselves deeply into debt, with interest payments of $20 billion in the 1970s rising to $225 billion in the 1980s. Hence farmers feel obliged to try to reap ever-larger harvests each year, in apparent indifference to the environmental costs.

American agriculture of the past four decades epitomizes the successes, and the failures, of industrial farming worldwide. This agricultural model has been widely copied and adapted in Green Revolution areas. The success is self-evident, but the consequences of industrial agriculture are all too painfully apparent. More recently, US farmers,

Agriculture in the balance

Each year there are more mouths to feed – approaching 100 million more. That we have largely managed to feed them so far is an extraordinary tribute to the success of agriculture. Grain production rose from 624 million tonnes in 1950 to 1,628 million in 1984, and to 1,741 million in 1990. Much of this massive increase arises from Western-type "high-tech" solutions, using oil-based fertilizers and new hybrids planted in huge monocultures. But the 40-year success story conceals a major loss of momentum: the annual growth in global grain production since 1984 has been about 1%, while that of population has been nearly 2%. The result is that the global peak of 343 kg per person in 1984 declined to 329 kg in 1990. Admittedly, in 1988 droughts brought a reduction in the harvests of the world's three leading food producers, but the record harvest of 1990 did little to replenish world grain stocks. Are the scales tipping towards failure?

Each sack represents 50 million tonnes of grain

Eating oil
Since 1950, population growth and environmental degradation have reduced grain lands per head by more than a third, but per capita fertilizer use has increased five-fold. Grain production per person peaked in 1984; since then it has fallen.

Population

Grain per capita

5 kg fertilizer per capita

Area of harvested grain land per capita

	1950	1984	1990
Population	2.51 billion	4.8 billion	5.3 billion
Grain per capita	248 kg	343 kg	329 kg
Fertilizer	5 kg	26 kg	27 kg
Area of harvested grain land per capita	0.24 ha	0.15 ha	0.14 ha

Success   Failure

The costs of advanced agriculture

Many of today's farmers practise a form of "deficit financing". To meet urgent needs, from food to interest on loans, they are depleting our future land base, destroying our wild and cultivated genetic heritage, using up fossil energy and groundwater supplies, and applying toxic chemicals to the land. We lose nearly half our crops to resistant pests and bad storage, produce "milk lakes" we cannot dispose of, and use vital land to grow feed for livestock, to support unhealthy meat-rich diets. Meanwhile, poorer countries experience falling food sufficiency, inadequate research facilities, and growing social inequality.

30% of harvest lost to pests

Spreading monocultures
Vast areas are planted with identical crops strains, all vulnerable to the same pathogens – and all helping to erode genetic diversity (p. 156).

Race against nature
Intensive spraying often fails to eradicate the pests which thrive on monocultures, taking over a third of our harvest. Resistant and mutant forms are always one jump ahead.

The oil farmers
If every nation expended as much oil per head in agriculture as the US, current world oil reserves would be emptied in a dozen years.

Lost genes
Traditional crop varieties are swept away by the bulldozer and replaced by the new hybrids. Yet these old forms hold vital genetic material.

70% of global water use

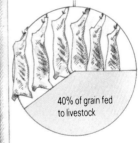

40% of grain fed to livestock

SURPLUS

The new dust bowls
Deforestation, over-cropping, over-ploughing, or over-grazing of dry lands destroy the soil – the farmer's capital.

Thirsty crops
Irrigation is vital to the miracle grains. Agriculture now takes 73% of global freshwater demand.

Wasteful diets
Nearly 40% of the world's grain is fed to livestock for the unhealthy meat-rich diet of the North.

Surplus and waste
Huge EEC and US surpluses, "mountains" of butter and sugar, "lakes" of milk and wine, cost $53 billion in 1983.

Food production per capita
(1969-70 = 100)

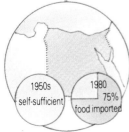

108% 110% 95%

Middle-income India and China Other low-income

'70 '80   '70 '80   '70 '80

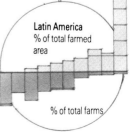

1950s self-sufficient   1980 75% food imported

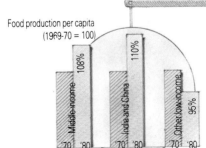

Latin America
% of total farmed area

% of total farms

Leaving the poor behind
Even before world grain production per capita peaked, many low-income countries had been experiencing a decline in food production per person.

Relying on grain imports
Egypt, with its fertile floodplain in the Nile, used to be self-sufficient in grain. Sadly, it now imports over 75% of its food requirements.

Land concentration
Wealth breeds wealth: large holdings increasingly crowd out small-holders (who are generally more productive), as in Latin America, above.

Biased research and development
Green Revolution research has tended to ignore crops grown by small farmers in Latin America and Africa – such as cassava, sorghum, and millet.

and many in the rest of the world, have begun to retrench, pulling back from the unsustainable marginal lands they have been farming, and slowing the massive growth in fertilizer use. Both these trends have contributed to the general levelling off of grain output, but it is clear that a wholly new strategy is necessary.

The new agricultural revolution

We need a new revolution in agriculture, both scientific and political. Fortunately the germ of the scientific one already exists. But the political will and means to apply it are yet half-formed. Hitherto our approach has been to "bend" the environment to suit our crops. Now, thanks to plant breeders

and genetic engineers, we can bend the plants instead – manipulating crops to flourish in harmony with their environments, rather than in spite of them. Instead of pouring in fertilizer, water, weedkillers, and pesticides, we can grow plants that fend for themselves – desert-dwellers like jojoba, new arid-land staples like the morama bean or buffalo gourd, crops tolerant of extreme temperatures; even strains of wheat, barley, and tomatoes that permit seawater irrigation.

We already make use of legumes which produce their own fertilizer via the symbiotic nitrogen-fixing bacteria in their roots. By the turn of the century, we may manage to transfer this capacity to other plants, whereupon farmers need not fear the rising

More research
Increased funding and new research centres are badly needed, especially in low-income areas, to develop locally adapted strains and cultivation techniques relevant to small and large-scale farmers.

Towards a new agriculture

Ecology has shaped integrated pest management, water-conserving irrigation methods, new organic fertilizers, and new uses of crop residues and green mulches to protect land and improve energy efficiency. Conservation tillage prevents soil loss; now employed on 40 million ha in the US. Future crops will be more self-sufficient and high-yielding, tailored to their setting. But research support is slim – the principal international crop research centres funded by governments devoted only 5% of its 1989 budget to biotechnology research. The farms of the future will need more research and greater protection of the genetic resources which provide the raw material for the Gene Revolution.

Managing pests
Integrated pest control has been enormously successful in trials. It aims not to wipe out pests, but to keep them at tolerable levels, by applying a range of "natural" restraints – mixed planting; clearing pest reservoirs such as stagnant pools; introducing natural predators; laying bait; releasing sterile pest males; even using hormones to interfere with maturation.

Yeheb

Conservation tillage
Minimum-till farming is energy-efficient and protects the soil. Crop residues and stubble left on the land retain nutrients and prevent erosion. Next year's crop is sown in shallow, restricted furrows, or drilled in without turning the soil.

Hairy potato

Plants with a future
Food plants adapted to harsh lands are offering us new solutions.
The Somalian Yeheb bush
This arid-land native has nutritious peanut-sized seeds, and could be a staple crop of the deserts.
The hairy wild potato To keep aphids at bay, this wild plant mimics the alarm scent of aphids. These pest-deterrent devices could one day be bred into crops.

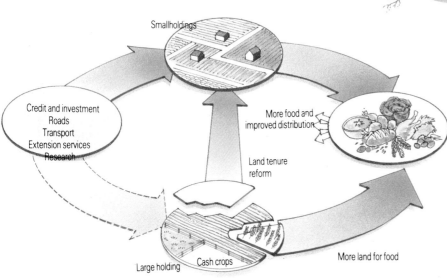

Smallholdings

Credit and investment
Roads
Transport
Extension services
Research

More food and improved distribution

Land tenure reform

Large holding   Cash crops   More land for food

Food first in the South
If the new agriculture is to lift the South on its wings, it must be truly revolutionary. A sharp turnround is called for, from over-emphasis on industry to farming, from export cash crops to domestic supply, and from commercial landholder to small farmer. To achieve food sufficiency, small farmers must be backed by regional co-operation, for however willing and productive they may be, they need a package of facilities to succeed. These include: credit and better prices; improved marketing, transport and advisory services; appropriate research; and security of tenure and access to good land. The best example of "food first" is China, now feeding 22% of the world's population on 7% of the world's arable land.

cost of synthetic nitrogen. Crop breeders are also exploiting disease resistance in wild species to develop crops with built-in defences against viruses and pests.

These breakthroughs are the centre-pieces of the forthcoming Gene Revolution, which promises to dramatically improve the production of our major crop species. To succeed, it will require greater emphasis on research than ever before, and a new set of economic strategies to take agriculture out of the "Cinderella" status. However, unlike Green Revolution technologies, Gene Revolution research and marketing are being undertaken mainly by the private sector. This may have implications for the speed at which the Gene Revolution spreads.

Most of all, the Gene Revolution requires political changes; a revolution so profound that every individual will be affected. For as long as rich-world citizens, by their high meat diets, continue to support the inequitable meat/grain connection, land will be over-worked, and people will go hungry. And as long as certain developing nations emphasize industry, urbanization, and cash crops at the expense of food sufficiency and the small farmer, the same applies.

Given proper back-up, the small farmers (who today make up over half of the world's poor) can feed themselves. Given the right advice, rich citizens can improve their diets. Given our new scientific understanding, the land can supply a sustainable harvest to feed us. It is merely a matter of getting our priorities right.

Mixed cropping
Interplanting and crop rotation maintain soil balance, reduce pest invasions, and by providing green cover below row plants prevent evaporation and erosion. Legumes, which fix nitrogen, restore fertility if planted between maize or permanent crops.

Gene banks
Preserving crop diversity in cold storage (pp 162-3) and running local seed libraries allow farmers a choice of solutions to cultivation problems.

Greenhouses
Cultivation under glass or in huge plastic tunnels can provide valuable cash crops from lands otherwise unsuited to farming, but is energy intensive. Israel farms the desert in this way to earn export cash, and so import staples.

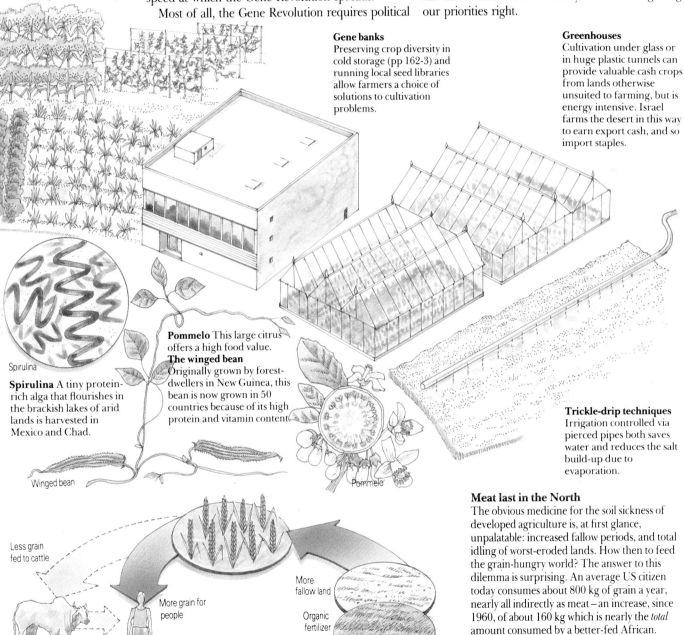

Spirulina

Spirulina A tiny protein-rich alga that flourishes in the brackish lakes of arid lands is harvested in Mexico and Chad.

Pommelo This large citrus offers a high food value.
The winged bean
Originally grown by forest-dwellers in New Guinea, this bean is now grown in 50 countries because of its high protein and vitamin content.

Winged bean

Pommelo

Trickle-drip techniques
Irrigation controlled via pierced pipes both saves water and reduces the salt build-up due to evaporation.

Less grain fed to cattle

More grain for people

Low-meat diet

More fallow land

Organic fertilizer

Meat last in the North
The obvious medicine for the soil sickness of developed agriculture is, at first glance, unpalatable: increased fallow periods, and total idling of worst-eroded lands. How then to feed the grain-hungry world? The answer to this dilemma is surprising. An average US citizen today consumes about 800 kg of grain a year, nearly all indirectly as meat – an increase, since 1960, of about 160 kg which is nearly the *total* amount consumed by a better-fed African. Merely by reverting to diets of 20 years ago, rich-world citizens could improve their health, release large amounts of grain, and free some land. The Norwegian government pioneered this approach, with farm incentives and public education. Even individually, many rich-world people are turning to a more meat-free diet, and herein lies much hope for the world.

OCEAN

Introduced by the crew of the
Rainbow Warrior II, and Steve Sawyer
Executive Director of Greenpeace International

Our first major vessel, used in many confrontations with whalers, and ocean polluters and dumpers, was named Rainbow Warrior, after an ancient North American Indian legend. This predicted that "when the Earth is sick and the animals disappear, the Warriors of the Rainbow will come to protect the wildlife and to heal the Earth". They would come from all races, creeds, and colours, putting their faith in deeds, not words.

The whole world knows about the tragedy that occurred in Auckland harbour, New Zealand, in July 1985. The Rainbow Warrior, while on route to Moruroa in French Polynesia to lead a protest against continued French nuclear testing and pollution of the marine environment, was bombed and sunk by French Secret Service agents, and one crew member drowned.

Four years to the day of the tragedy, a second Rainbow Warrior was launched, and today she sails with another crew, but with the same spirit as always and with the same dedication.

Greenpeace's first campaign was against the testing of nuclear weapons, and this continues to be an important priority near to the heart of the organisation. With the end of the Cold War, we almost dared to hope that the nuclear powers would take the opportunity to stop ravaging the environment with nuclear military technology, and, as a first step, end nuclear testing.

Therefore, Greenpeace resolved to send our most potent symbol, the new Rainbow Warrior, back to the French nuclear test site at Moruroa and to mount an active compaign in the United States to protest at the resumption of nuclear weapons tests. As before, we based our campaign both on protection of the environment and on a commitment to peace. The precise nature and extent of the environmental damage at Moruroa as a result of the tests remain unclear, and we called once again for an independent scientific assessment. But equally, if not more importantly in today's political climate, we call for an end to the tests as a gesture towards peace, understanding, and the realization that the massive problems the world faces in the 1990s are not helped by the continuation of the anachronistic posturing and threats of the Cold War. Although the ship was arrested by French Navy forces, we managed to contribute to the debate in France which led to the suspension of nuclear testing in 1992.

Following this campaign, the Rainbow Warrior sailed to Rio to be present during the "Earth Summit", and to bear witness to the fact that the international community would not stand for an "Earth Summit" which did not seriously address the fundamental threats to the environment of planet Earth.

Actions, unfortunately, often speak louder than words. But the words and images which follow in this section explain why our protest is so urgent. We can no longer use the world ocean as a dustbin. We must view it as a living ecosystem, a vital and integral part of our planet's workings.

The original Rainbow Warrior.

THE OCEAN POTENTIAL

Our planet should not be called Earth but Ocean – at least seven-tenths is covered with seas.

We know little about this water planet. Although the oceans are just as diverse as the land, and interwoven with human history, we tend to see them as barriers, as alien spaces. In reality, however, ocean ecosystems are continuous – or rather, a single ecosystem, a world ocean with land masses as the true barriers, though gradients of temperature and salinity separate the oceans into a multiple series of discrete regions.

It takes a "leap of the imagination" to perceive the integral role of the ocean in our planet's workings. The interdependent circulatory systems of ocean and atmosphere determine climatic flows right around the globe. At the same time, the great accumulations of seawater, almost entirely placid beneath the surface, exert a major stabilizing influence on climate: their very bulk produces a "flywheel" effect, a powerful buffer for what would otherwise be drastic fluctuations in our weather. The oceans also serve as a great reservoir of

Warm currents
1 Irminger
2 Norway
3 Gulf Stream
4 North Equatorial
5 Equatorial Counter Current
6 South Equatorial
7 Brazil
8 Kuroshio
9 Alaska
10 Agulhas
11 East Australian

The world ocean

The ocean is a single, dynamic medium, its waters constantly on the move under the influence of the sun's heat (which provides the initial thrust), the Earth's rotation, and solar and lunar tides. The major currents, great "travelators" of the ocean, deliver huge masses of water over long distances, providing a continuous interchange between warm equatorial water and cold water from the poles. The consequences for climate, marine ecosystems, and human fisheries are vital. The warm Gulf Stream, for instance, surging along faster than any ship, transports 55 million cu m per second – 50 times more water than all the world's rivers. Without it, the "temperate" lands of northwestern Europe would be more like the sub-Arctic. The Peru current and the Benguela of southwest Africa produce marine bonanzas as they bring in nutrient-rich waters dragged to the surface by offshore winds – a superabundance of plankton, fish, and seabirds. The map shows the major warm and cold surface currents, and the densities of human populations – a great proportion of them living in coastal zones.

The sea-floor's wealth

The dramatic ridges and trenches of the ocean abyss express its history, for the sea floor is in motion. Just as the continental "plates" drift on a hot, partially molten underlayer, so the ocean bed, too, is made up of moving plates. The great mid-oceanic ridges mark where plates are separating: the molten underlayer pushes up through the rift to form ridge material, while the sea floor spreads steadily away from the ridge axis. At the ocean margins, deep trenches occur as the oceanic and continental plates collide and one is drawn down under the other. Understanding sea-floor plate interactions helps scientists predict the locations of valuable minerals for future exploitation.

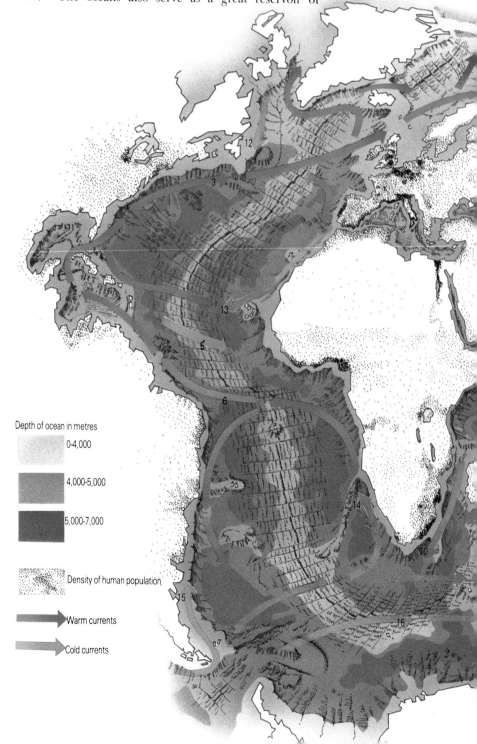

Depth of ocean in metres

0-4,000

4,000-5,000

5,000-7,000

Density of human population

Warm currents

Cold currents

dissolved gases which helps to regulate the composition of the air we breathe.

The oceans give no more of a liquid covering to the globe than a film of moisture on a football, yet their depth and grandeur are remarkable – Mount Everest could be readily lost in the Mariana Trench. Underwater landscapes are extraordinarily varied, with spectacular geologic features. Along the edges of land masses are barely sloping continental shelves, accounting for about eight percent of the ocean's expanse. Fed by sediments washed off the land, these shelves are often very fertile, and support abundant fisheries. Twice as extensive are the continental slopes, with gradients 4-10 times steeper, reaching down as far as 500 metres. Eventually these drop away to 3,000 metres or

more, until they meet the "foothills" of the abyssal floor. This vast plain, sometimes with scarcely a pimple in thousands of kilometres, elsewhere scarred and rugged like terrestrial "badlands", is broken at the centre by mighty ridges, stretching the length and breadth of the four great oceans. Matching them in scale are giant canyons and long, narrow trenches, located along continental margins and associated with island arcs, the deepest plunging 11,000 metres below the surface.

Little known, and apparently remote, the ocean is a remarkably rich "resource realm" of the planet. With a better understanding, we might learn to draw sustainable benefit from its fisheries, minerals, and energy, and from its services as a gigantic weather machine. The ocean ecosphere constitutes a

Cold currents
12 Labrador
13 Canary
14 Benguela
15 Falkland
16 West Wind Drift
17 West Australian
18 Oyashio
19 Californian
20 Peru (Humboldt)

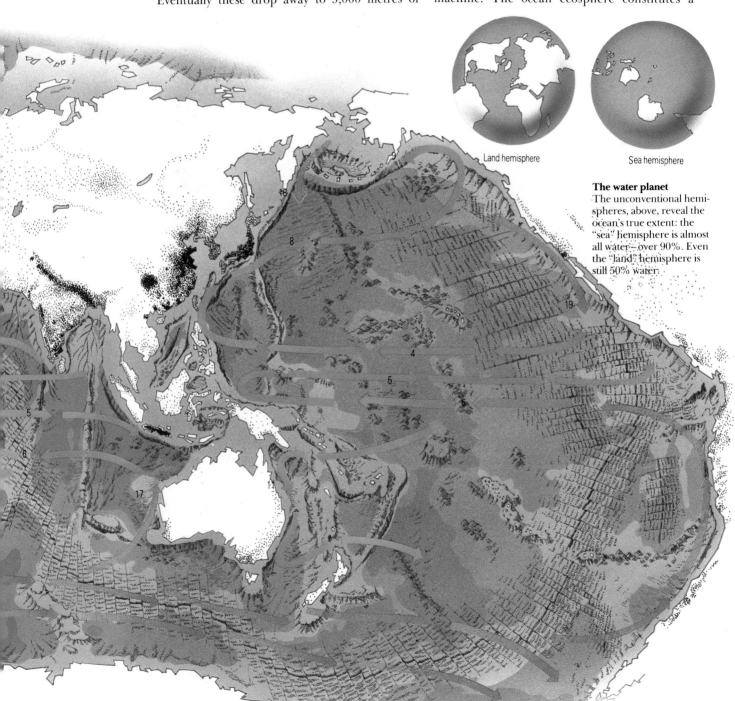

Land hemisphere

Sea hemisphere

The water planet
The unconventional hemispheres, above, reveal the ocean's true extent: the "sea" hemisphere is almost all water – over 90%. Even the "land" hemisphere is still 50% water.

magnificent frontier of scientific research – as exciting to many scientists as our exploration of the Sea of Storms on the waterless surface of the Moon.

The living ocean

The ocean is where life began. More than three-and-a-half billion years ago there evolved simple single-celled algae and bacteria very similar to those that form the basis of life in today's oceans. Collectively known as phytoplankton, from the Greek words meaning "drifting plants", this microflora exploits. the energy of the sun and nutrients in the water to manufacture complex molecules of living tissue. Being dependent on sunlight, phytoplankton flourish in a thin layer at the surface of the ocean, a biosphere that extends downwards for 100 metres at most.

Diverse communities of animal "grazers" feed off these rich blooms, especially the diminutive zooplankton – many of them, like the phytoplankton, single-celled creatures. Especially numerous are the radiolarians, whose exquisite silica skeletons are important constituents of deep-sea oozes. Other larger zooplankton include flatworms, small jellyfish, swimming crabs, and various shrimps.

These two categories of plankton are so abundant that they are estimated to generate, respectively, 16 billion and 1.6 billion tonnes of carbon (the basic material of living tissue) each year. Feeding off the plankton is an array of larger creatures, with an annual productivity of around 160 million tonnes. These species, especially the prolific herring family, supply us with two-fifths of our fish. By contrast, fish that eat other fish, such as members of the cod family, generate only about 16 million tonnes of carbon a year, and supply us with about one-eighth of the fish we eat.

Just as on land, life is unevenly distributed in the oceans – the marine world has its equivalents of deserts and rainforests. In certain sectors the sea floor is covered with extensive sand stretches, which, while no more lifeless than the Sahara, are distinctly impoverished compared with the rest of the ocean. At the other extreme, "rainforests" flourish – particularly in coastal wetlands, estuaries, and reefs in upwelling zones. The Great Barrier Reef of northeastern Australia, for instance, harbours more than 3,000 animal species.

Finally, there is the deep ocean – pitch-dark and near freezing, yet far from lifeless. Our preliminary probings of this remote realm reveal more than 2,000 species of fish, and at least as many invertebrates, many of them grotesque and primitive in form – the result of adaptations to an unpromising environment. But plainly they are altogether unprimitive, in that they subsist, indeed flourish, in such conditions.

The ocean is three-dimensional, and its plant wealth is not "fixed" – phytoplankton drift far and wide. Here, then, is an ecosphere fundamentally different from that on land, yet one so fertile that it

The living ocean

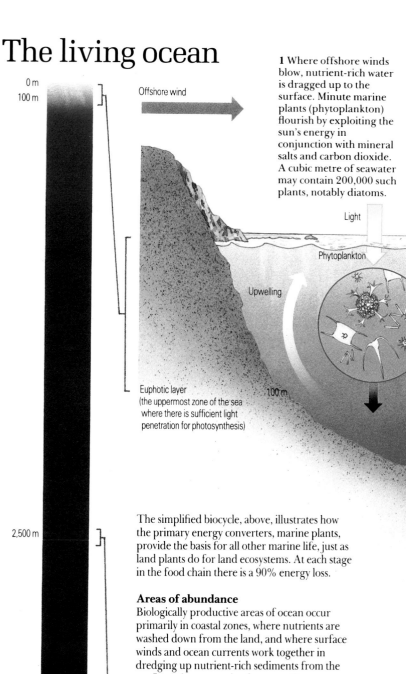

1 Where offshore winds blow, nutrient-rich water is dragged up to the surface. Minute marine plants (phytoplankton) flourish by exploiting the sun's energy in conjunction with mineral salts and carbon dioxide. A cubic metre of seawater may contain 200,000 such plants, notably diatoms.

Offshore wind

Light

Phytoplankton

Upwelling

100 m

Euphotic layer
(the uppermost zone of the sea where there is sufficient light penetration for photosynthesis)

0 m
100 m

2,500 m

5,000 m

The simplified biocycle, above, illustrates how the primary energy converters, marine plants, provide the basis for all other marine life, just as land plants do for land ecosystems. At each stage in the food chain there is a 90% energy loss.

Areas of abundance

Biologically productive areas of ocean occur primarily in coastal zones, where nutrients are washed down from the land, and where surface winds and ocean currents work together in dredging up nutrient-rich sediments from the sea floor. One example of an upwelling occurs at the Great Newfoundland Banks, where a meeting of warm and cold currents creates a maelstrom of seawater over a 150,000-sq km submerged plateau, generating phytoplankton in exceptional abundance. Flourishing off the plants are huge shoals of sardine-like fish, capelin, which in turn support many millions of cod, plus gannets, kittiwakes, and razorbills, also seals and humpbacked whales. Regrettably, this fishery has been grossly over-harvested by the ultimate predator, humankind (see pp 82-3).

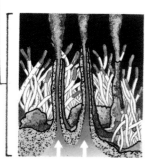

Sulphur-based life-forms

In 1977 a new deep-sea ecosystem was discovered. Like land animals around a waterhole, strange-looking worms, clams, and blind white crabs cluster around hot vents in the ocean floor, dependent on bacteria with the unique capacity to metabolize hydrogen sulphide.

2 Zooplankton includes a great diversity of marine life-forms, some feeding off phytoplankton, some preying on each other. Especially numerous are arrow worms, sea gooseberries, and copepods. Added to these are free- swimming larvae of shore- and bottom-dwellers such as worms, crabs, and echinoderms.

3 The "small fry" zooplankton in turn are consumed by squid, jellyfish, and shoaling fish such as herring and anchovies. Baleen whales too exist on a zooplankton diet, mainly copepods and krill. The basking shark, one of the world's largest fishes, also sustains itself on a largely plankton diet.

4 Next in the chain, feeding on the smaller shoaling fish, are medium-sized predators such as tuna. These bulky, fast-swimming fish are in turn eaten by marlin and sharks. Similarly, most seal species feed on small fish while themselves serving as prey to larger creatures – leopard seals and sharks.

Antarctic Convergence
The band of the Southern Ocean between 50° and 60°S is known as the Antarctic Convergence. Counter-rotating currents of cold and sub-Antarctic water travelling northwards (the East and West Wind Drifts) here pass under and interact with subtropical water moving south. The resulting turbulence drives nutrient-rich water to the surface, making the area highly productive, with great swarms of a crustacean known as krill forming the staple food for penguins, seals, squid, and whales.

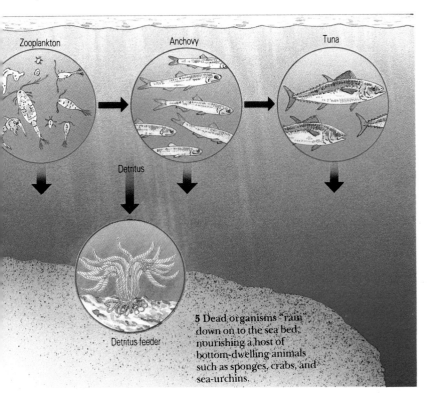

Zooplankton   Anchovy   Tuna

Detritus

Detritus feeder

5 Dead organisms rain down on to the sea bed, nourishing a host of bottom-dwelling animals such as sponges, crabs, and sea-urchins.

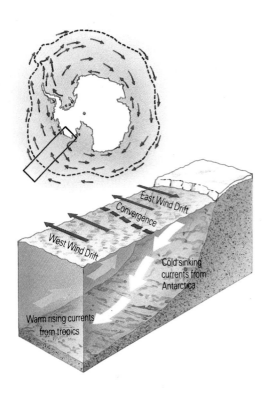

East Wind Drift

Convergence

West Wind Drift

Cold sinking currents from Antarctica

Warm rising currents from tropics

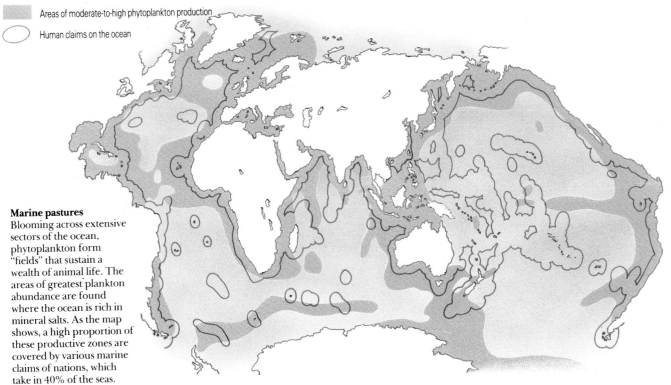

Areas of moderate-to-high phytoplankton production

Human claims on the ocean

Marine pastures
Blooming across extensive sectors of the ocean, phytoplankton form "fields" that sustain a wealth of animal life. The areas of greatest plankton abundance are found where the ocean is rich in mineral salts. As the map shows, a high proportion of these productive zones are covered by various marine claims of nations, which take in 40% of the seas.

supports around one-third of the planet's biomass. At present, we harvest about 0.2 percent of marine production. Plainly, we could make better use of the abundance of the living ocean.

Shores, reefs, and islands

The narrow coastal fringes of the world's ocean are at once its most productive and most vulnerable zones. Their shallow waters, saturated with sunlight and richly supplied with nutrients, provide the basis of most of our fisheries. Coastal and island ecosystems also serve as a great meeting ground between land and sea; large numbers of people live here, whether in traditional fishing communities or in cities. Many human activities exploit the wealth of these ocean borderlands.

The vital margins

Coastal ecosystems represent an extremely valuable resource – one that is increasingly threatened by human interests (pp 80-81). Their rich life depends on high levels of primary production – generation of the vegetable matter which provides the "base energy" of all food chains. Phytoplankton and sea plants are the primary producers of the oceans, their abundance and growth rates varying among ecological zones. The pie-chart (below) shows the proportional areas of open ocean, continental shelf, and coastal sectors, while the columns compare their primary productivity per unit area, illustrating dramatically the concentration of marine wealth in the ocean margins. Algal beds, reefs, and estuaries are more than 16 to 18 times as productive as the open ocean, mangroves well over 20 times.

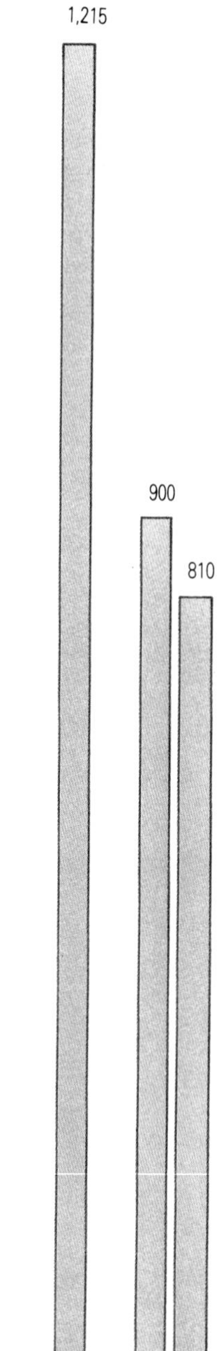

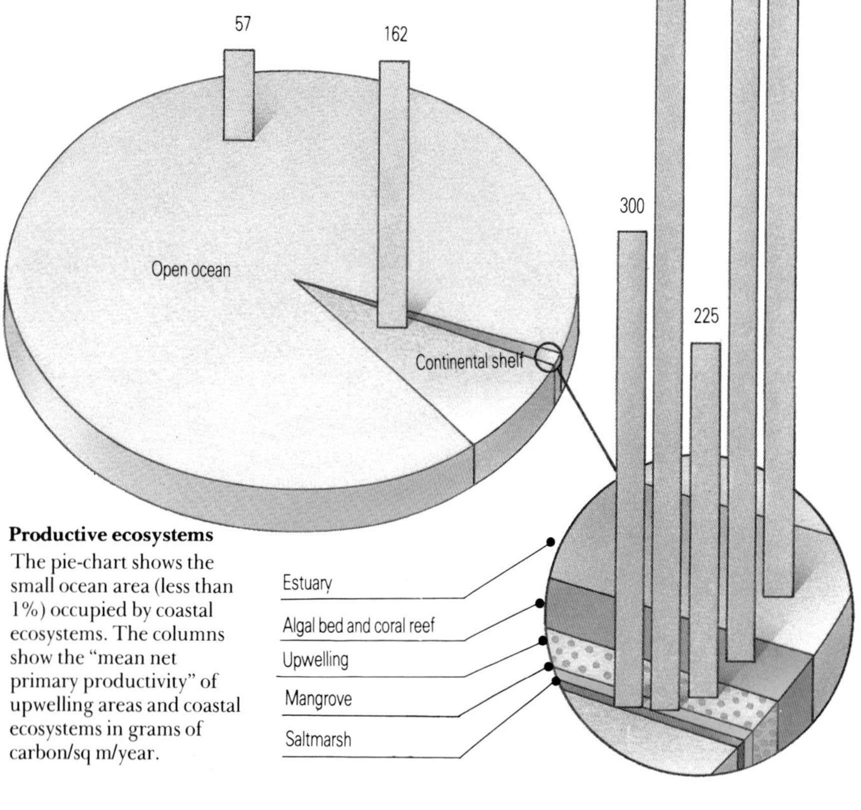

57

162

Open ocean

Continental shelf

Productive ecosystems
The pie-chart shows the small ocean area (less than 1%) occupied by coastal ecosystems. The columns show the "mean net primary productivity" of upwelling areas and coastal ecosystems in grams of carbon/sq m/year.

1,215

900

810

300

225

Estuary

Algal bed and coral reef

Upwelling

Mangrove

Saltmarsh

The four vital ecosystems for humankind and for marine life-forms are saltmarshes, mangroves, estuaries, and coral reefs. Saltmarshes are tidal wetlands of temperate zones, mangroves the tropical equivalent; their predominant offshore plants are seagrasses, true flowering plants that bloom beneath the sea. Seagrass meadows of the tropics are grazed by sea turtles, dugongs, and manatees; in the temperate zones they form rich winter food reserves for ducks and geese. In both zones the vegetation acts as a nutrient trap for shellfish such as shrimps, and for many finfish. In addition, the plants serve to filter out pollution, to mitigate the beating of storm waves and powerful currents, and to prevent erosion of the coastline. Estuaries, with their plentiful supplies of fertile silt washed down from rivers,

Mangrove wealth
Mangroves fringe more than half of all tropical shores, and harbour huge quantities of fish and shellfish, notably prawns and oysters. In Indonesia, mangrove fish have been farmed since the 15th century. Today the most important species is milkfish, raised in over 35,000 ha of culture ponds. Also reared are mullet, groupers, snappers, and sea perch. The simplest form of prawn culture involves netting mature prawns (right) as they leave mangroves for offshore spawning grounds. Between mangroves and dry land we sometimes find useful wetlands. Along the Malay Peninsula, for instance, grows the Nypa swamp palm, which supplies local people with fruit, sugar, vinegar, alcohol, and fibre.

cover twice the area of saltmarshes and mangroves. These river embayments or "tidal ponds", where seawater and freshwater communities intermingle, are very productive and feature vast numbers of annelid worms, crustaceans, and molluscs. Whenever we sit down to a dish of crabs, oysters, mussels, or prawns, we can thank the extreme productivity of estuaries. Further, they serve as nurseries for ocean fish such as silverside, anchovy, menhaden, and catfish. In one of the finest fishing areas in the world, the continental shelf of the eastern US, at least three-quarters of fished species spend a portion of their life cycles in estuaries.

Most diverse of all ecosystems are tropical coral reefs. They may also be the world's oldest ecosystems, in so far as they appear to be the only ones that have survived intact since the emergence of life. They feature more plant and animal phyla (major categories of life) than any other ecosystem; in their limited expanse coral reefs support one-third of all fish species. They also reveal more mutual-benefit relationships, or symbioses, between organisms than any other community. The more we learn about these successful and enduring symbioses, the better we shall know how to regulate relationships between plants and animals on land.

Other coastal ecosystems supply us with a host of products for our material welfare. Rocky shores feature algae of many shapes and sizes; seaweeds alone supply alginate compounds that contribute to literally hundreds of end-products, such as plastics, waxes, polishes, deodorants, soaps, detergents,

Saltmarsh

Mangrove

Coral reef

Continental shelf

Upwelling zone

Open ocean

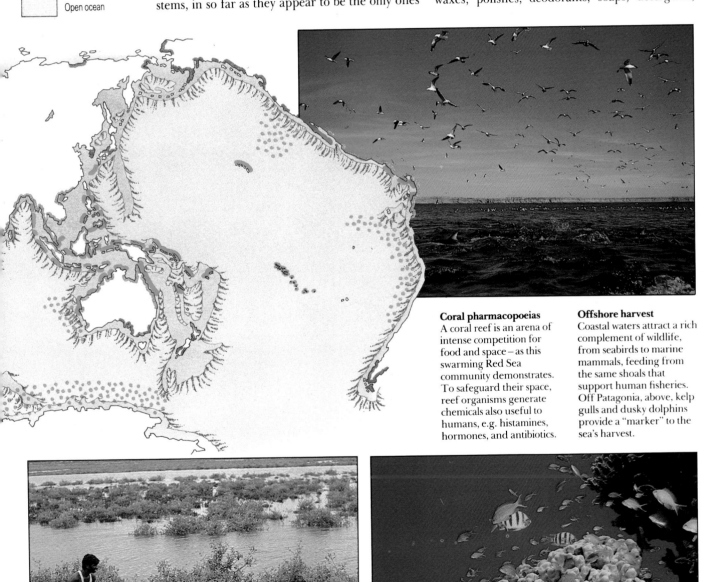

Coral pharmacopoeias
A coral reef is an arena of intense competition for food and space – as this swarming Red Sea community demonstrates. To safeguard their space, reef organisms generate chemicals also useful to humans, e.g. histamines, hormones, and antibiotics.

Offshore harvest
Coastal waters attract a rich complement of wildlife, from seabirds to marine mammals, feeding from the same shoals that support human fisheries. Off Patagonia, above, kelp gulls and dusky dolphins provide a "marker" to the sea's harvest.

shampoos, cosmetics, paints, dyes, lubricants, food stabilizers, and emulsifiers.

Filling our nets

We have fished the rich offshore waters since earliest times, allowing upwelling currents and seasonal migrations of the shoals to bring the wealth of the sea to our nets. This marine catch represents more than 85 percent of the world's total fishery. Even today, despite territorial limits and deep-sea fleets, the global shoal remains a renewable resource which nations must share, because neither the fish, nor the ocean ecosystems on which our catch depends, recognize human boundaries.

In 1950 the global recorded fish harvest was 20 million tonnes. During the following two decades we expanded our catch at the extraordinary rate of 6-7 percent a year. This dramatic gain was due to the increased exploitation of existing stocks, discovery of new stocks, and development of improved fishing technology. But it was not to last. After 1970, as more and more stocks were depleted, the average annual growth fell to 1-2 percent. Since 1979, however, the overall trend has been strongly upward, much of it due to increased landings of just a handful of species, such as Peruvian anchovy and South American sardine, whose populations are highly variable.

In four decades we have quintupled the global harvest to a recorded annual catch of 99.5 million tonnes, reaching the limit of sustainable yields, estimated by the UN Food and Agricultural Organization as 100 million tonnes a year. "Sustainable yield" refers to the *optimum* annual catch that can be derived indefinitely from harvested species, without causing a stock failure. Fishery biologists tell us that when we start to exploit a virgin stock, the fish respond by reproducing more abundantly due to an apparent food surplus. Yet the same biologists tell us that often they do not know how many fish were there in the first place. Our lack of knowledge is made worse by our highly-efficient fishing methods and the demands of the money markets, which call upon the exploiters to recoup their investments within just a few years, regardless of how long the fishery resource may take to recover. Result: our view of what might be a sustainable yield can be 10 times different from Nature's.

We harvest five main groups of marine species. Northerner's menus are largely made up of demersal fish (primarily bottom-dwellers) including cod, haddock, sole, and plaice. Pelagic fish (surface-dwellers) include herring, mackerel, anchovy, tuna, and salmon. Together, these two groups account for more than 72 million tonnes a year. Crustaceans, such as lobsters, shrimps, and other shellfish, provide more than 4 million tonnes and cephalopods, including the octopus and squid families, yield about 2.5 million tonnes. As for marine mammals, notably whales, their potential

The global shoal

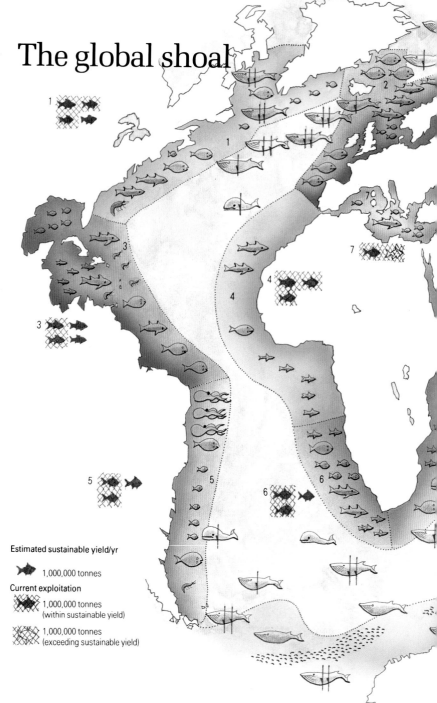

Estimated sustainable yield/yr

🐟 1,000,000 tonnes

Current exploitation

1,000,000 tonnes
(within sustainable yield)

1,000,000 tonnes
(exceeding sustainable yield)

The map shows the actual (netted) and potentially sustainable catches of fish and shellfish (1985-7) by fishing region, as well as the distribution and exploitation levels of whale stocks. As the netted symbols show, several areas are already fished to the limit of their sustainable potential, or even beyond it. The map also shows, for each fishing region, a breakdown of the catch into fish (demersal and pelagic), crustaceans, and cephalopods. The Southern Ocean also harbours vast swarms of tiny shrimp-like krill with a biomass estimated at 250-600 million tonnes. At present we catch less than 0.5 million tonnes a year. With superior management, and a better understanding of its vital role in the ocean ecosystem, Antarctic krill could become an important world source of protein. While seafood amounts to only 2% of our diets, it is much more important in terms of animal protein, a crucial element in our food. The seas provide 14% of our animal protein direct to the table and, in the case of developed-world citizens, a good deal more as fishmeal fed to livestock and as fertilizer.

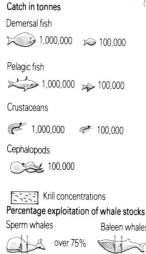

Catch in tonnes

Demersal fish

🐟 1,000,000   🐟 100,000

Pelagic fish

🐟 1,000,000   🐟 100,000

Crustaceans

🦐 1,000,000   🦐 100,000

Cephalopods

🦑 100,000

Krill concentrations

Percentage exploitation of whale stocks

Sperm whales   Baleen whales

over 75%

25-75%

less than 25%

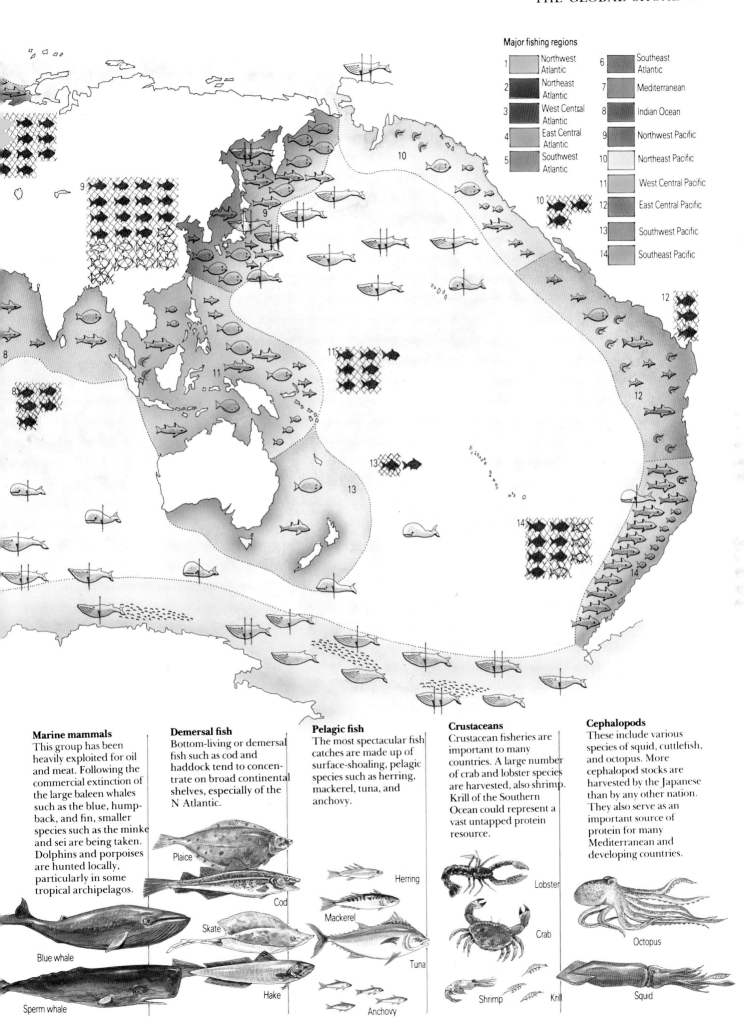

Major fishing regions

1 Northwest Atlantic
2 Northeast Atlantic
3 West Central Atlantic
4 East Central Atlantic
5 Southwest Atlantic
6 Southeast Atlantic
7 Mediterranean
8 Indian Ocean
9 Northwest Pacific
10 Northeast Pacific
11 West Central Pacific
12 East Central Pacific
13 Southwest Pacific
14 Southeast Pacific

Marine mammals
This group has been heavily exploited for oil and meat. Following the commercial extinction of the large baleen whales such as the blue, humpback, and fin, smaller species such as the minke and sei are being taken. Dolphins and porpoises are hunted locally, particularly in some tropical archipelagos.

Blue whale
Sperm whale

Demersal fish
Bottom-living or demersal fish such as cod and haddock tend to concentrate on broad continental shelves, especially of the N Atlantic.

Plaice
Cod
Skate
Hake

Pelagic fish
The most spectacular fish catches are made up of surface-shoaling, pelagic species such as herring, mackerel, tuna, and anchovy.

Herring
Mackerel
Tuna
Anchovy

Crustaceans
Crustacean fisheries are important to many countries. A large number of crab and lobster species are harvested, also shrimp. Krill of the Southern Ocean could represent a vast untapped protein resource.

Lobster
Crab
Shrimp
Krill

Cephalopods
These include various species of squid, cuttlefish, and octopus. More cephalopod stocks are harvested by the Japanese than by any other nation. They also serve as an important source of protein for many Mediterranean and developing countries.

Octopus
Squid

has been squandered, and is now tailing away to the merest fraction of what it might have been with rational management.

The new technological revolution

Ocean technology is an infant phenomenon as yet. But by the end of the century it could be supplying us with sizeable flows of critical minerals and energy. The "us", however, probably comprises very few nations – those with the funds and expertise to gain a head start in the nascent industry.

The wealth of the oceans is beyond doubt. Continental shelves harbour nearly half of the Earth's oil and gas resources, while seawater itself contains over 70 elements, some very important. One cubic kilometre of seawater holds about 230 million tonnes of salt; it also contains around a million tonnes of magnesium and 65,000 tonnes of bromine. We have extracted salt by evaporation for more than 4,000 years; today we also extract magnesium and bromine by chemical processes.

Not yet harvested but much more important are manganese nodules on the sea floor. These potato-shaped objects have a high metal content, primarily manganese (25-30 percent), but also nickel (1.3 percent), copper (1.1 percent), cobalt (0.25 percent), plus molybdenum and vanadium – all valuable in steel alloys. Although most manganese deposits are spread thinly over large areas, often at great depths, certain concentrations could in the future support commercial operations.

Most accessible are deposits of silver, copper, and zinc in Red Sea muds, worth $7 billion. Still more readily available is uranium in seawater – over 4 billion tonnes of it, much more than on land. Japan, with a strong commitment to nuclear energy and no domestic stocks of uranium, eventually aims to extract 1,000 tonnes a year.

As for ocean energy, tidal, wave, and thermal projects all have potential, and are beginning to be exploited.

Much of the ocean's mineral wealth lies beyond national jurisdictions of any sort, and should count as part of the common heritage of humankind (see pp 90-91). The global community already shares certain technologies both between nations and through international consortia (the oil industry is an obvious example). Yet there will still be winners

Ocean technology

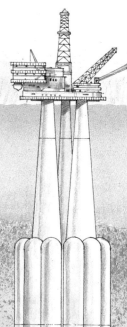

Dredging
Used to mine many materials, from sand and gravel to diamonds, dredgers also help to keep shipping lanes open. Technology is increasing the depth of operation from the present average of 30 m.

Wave energy
The Lanchester Clam is one of a variety of UK designs to extract energy from waves. Early commercial application is unlikely unless energy prices increase hugely.

Spiral wave power
The Dam Atoll is a conceptual design from the US, with potential applications from cleaning up oil spills and reducing wave erosion to desalination and power generation (1-2 megawatts).

CONDEEP® concrete gravity platform
These massive floating structures store thousands of tonnes of crude oil in enclosed submarine concrete tanks. At work in the North Sea is the Stratford B platform, the heaviest object ever moved by humans, with a displacement weight of over 900,000 tonnes. It can operate at the edge of the continental shelf at depths of over 140 m.

The frontier of the marine technology revolution is steadily being pushed deeper into the oceans in the search for critical resources, despite the obstacles of safety, cost-effectiveness, and territorial rights. The illustration demonstrates a range of present ocean technology from inshore to deep water, the coastal and continental shelf operations being the most established. Dredging, for instance, is a big-business operation, mining sand, gravel, and shell deposits for cement for the construction industry. An estimated 500 billion tonnes of gravel alone lie on the Atlantic shelf of North America. The oil and gas industries were pioneers in exploiting coastal mineral deposits, beginning off California in 1891. They moved into deep-water oil technology in the 1960s, boosting development in the 1970s as the cost of land-based crude oil and natural gas soared. The guyed (deep-ocean) tower shown here could operate down to 300 m; another rig system, known as the "tension leg platform", could operate down to 500 m. Beyond this depth, platforms become unstable, so technologists may opt for moving the whole production plant to the sea bed instead. The scheme illustrated aims to house an entire production facility in 5 interlocking concrete chambers on the sea bed, down to 1,000 m. Deepest of all are deposits of manganese nodules: while these are not mined so far, potential recovery areas at depths of 4,000-5,000 m are being explored with underwater TV systems and other techniques.

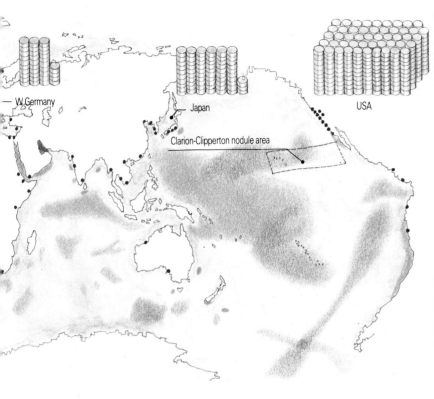

W.Germany

Japan

Clarion-Clipperton nodule area

USA

Ocean wealth and investment

A comparison of annual dollar investment in ocean technology clearly reveals the handful of nations leading the field (left). The map also indicates major reserves of mineral-rich nodules and sediments, active seawater extraction plants, and areas with OTEC potential. In many countries investment has slowed due to soaring costs and uncertain profits. But Japan has doubled its expenditure since 1980, concentrating on uranium extraction, nodule mining, and OTEC. Current sea-bed investigations by other countries include a US survey of polymetallic sulphides along the rift zone off Oregon (the only known large deposit lying within the proposed US exclusive economic zone), and the French exploration of manganese nodule deposits lying in their economic zone off Tahiti.

Main areas of manganese nodules

Metal-rich sediments

Mineral extraction plants from seawater

Offshore phosphorite deposits

OTEC potential (areas with temperature gradients of 22° C or more)

Annual investment 1980 (each coin $20 million)

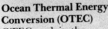

Ocean Thermal Energy Conversion (OTEC)
OTEC exploits the temperature difference between the warm surface and cool bottom layers in tropical seas to provide power. The process works like a giant refrigerator in reverse, evaporating and condensing a working fluid to drive a turbine that produces electricity.

Manganese nodule mining
The recovery of mineral-rich nodules from depths as great as 5,000 m has proved feasible in pilot tests, but is extremely costly. Commercial operation will depend on metal prices rising substantially; it also awaits a satisfactory resolution of present sea-bed ownership disputes by the United Nations Conference on the Law of the Sea (UNCLOS).

Guyed tower
This anchored structure is one of a new generation of deep-water oil platforms. Block 480 platform, for example, in the Gulf of Mexico will be placed at a depth of 300 m and pinned to the sea floor by a spoke-like array of cables, each more than 900 m long.

Deep-sea oil production
Designed to extract oil at depths where platform recovery is uneconomic, these structures will be gas- or nuclear-powered.

and losers in the race for ocean wealth: most nations simply lack the technology.

The ends of the Earth

Remote and ice-bound, the polar environments used to be among the least disturbed on Earth. But things are changing: now that their natural riches are recognized, the familiar tug-of-war between protection of unique ecosystems and exploitation of critical assets dominates their future.

We tend to think of the two polar zones as being similar. Yet they are quite different. The Arctic is essentially an enclosed sea, the smallest of the oceans, as well as the only one to be almost entirely landlocked. The Antarctic, by contrast, comprises a vast open ocean surrounding an ice-covered land mass twice the size of Western Europe.

One-third of the Arctic is underlaid with continental shelves from surrounding territories; it supports some of our richest fishing grounds, now accounting for roughly one-tenth of the global catch. A single haul of the net can bring in 100 tonnes of fish. In Antarctica, by contrast, much of the sea is overlain with ice shelves. The region contains nine-tenths of all ice on the planet, the ice cap averaging 2,300 metres deep, compared with a few dozen metres at the North Pole. Some nations, notably the water-poor and oil-rich countries of the Middle East, have even considered towing outsize bergs from Antarctica.

The polar zones are not the lifeless deserts we might expect. The Arctic fringes break out into spectacular blooms each summer, supporting large numbers of herbivores, from lemmings to reindeer and caribou. The territories north of the Arctic Circle also support two million humans, including over 800,000 aboriginal peoples.

The Antarctic continent has only two flowering plant species (plus a number of mosses and lichens), no native vertebrates, and (alone among continents) no long-standing human settlements. Yet the surrounding ocean, one of the most productive zones on Earth, generates massive summer outbursts of marine phytoplankton, fostered by exceptionally rich nutrient-bearing upwellings. These support vast populations of krill, which provide the food base for eight species of whales and the main support of 40 bird species, from penguins (90 percent of the bird biomass) to albatrosses.

The mineral resources of the Arctic are known to be vast and already reserves are being exploited. As for Antarctica, there may be great oil deposits lying under the narrow continental shelf, but there is no conclusive evidence that these reserves exist.

In October 1991, nations with an interest in Antarctica signed the Environment Protocol to ban minerals exploitation for 50 years. It remains to be seen if the accord has any "teeth", but the continent's mineral resources may be considered as "frozen stakes" for the next half century.

Polar zones

Main areas of distribution of:

▢	Cod
▢	Alaska pollack
▢	Capelin
🦭	Fur seal
🐋	Harp and hooded seals
▨	Coal field
⛏	Mining
⛽	Oil and gas production
⬅	Major ice drifts

Arctic resources

The Arctic, with its extensive continental shelves over which shoaling fish congregate, contains some of the richest fisheries.

Fish Over half the catch consists of cod, haddock, and Pacific pollack. Capelin is the main pelagic fish taken on the Atlantic side.

Seals Every spring thousands of harp and hooded seals gather in the freezing waters of the North Atlantic and Arctic Oceans, where they haul out of the ice to give birth. For centuries the pups have been clubbed for their valuable fur coats: 70 million killed in the last 250 years. In 1983 the EC banned sealskin imports, knocking the bottom out of the market which eventually led to the end of the large-scale hunt. Older seals are still hunted, 70,000 in 1991, allegedly to protect fisheries.

Minerals Two-thirds of the former USSR's gas reserves lie north of the Arctic Circle and Alaska supplies 25% of United States' oil. Recoverable Alaskan oil reserves are estimated at 3-9 billion barrels and there is great pressure for its development, following fears about the security of Middle East oil supplies in the wake of the 1991 Gulf War. The Arctic holds some of the world's largest deposits of coal, iron, copper, lead, and uranium.

Main areas of distribution of:

- Antarctic cod
- Toothfish
- Krill concentration
- Krill distribution
- Coal-bearing area
- Potential oil and gas areas
- ⓒ Mineral occurrences

Antarctic resources

Commercial fishing in the Southern Ocean has yet to prove profitable: while there is a wealth of plankton, the waters are deep with no broad continental shelf (unlike the Arctic).

Fish Of the 100 or so species of the region, only a few, such as the Antarctic cod toothfish and Patagonian hake, have been extensively trawled.

Whales Japan, Norway, and Iceland have made regular attempts to lift the moratorium on commercial whaling which came into force in 1986. Japan still continues its annual hunt for about 300 minke whales in Antarctic waters for "scientific research".

Minerals Geologists believe there may be over 900 major mineral deposits – but few in ice-free areas.

The world's largest coalfield may lie in the Transantarctic Mountains. US researchers believe there is oil in the continental shelf. However, under the terms of the Environment Protocol (1991) "any activity relating to mineral resources, other than scientific research, shall be prohibited" for at least 50 years.

Environment Antarctica's pure environment holds vital interest for scientists studying the Earth's evolution, atmosphere, and climate. Cores drilled out of the ice yield data on temperature and carbon dioxide changes over centuries, even millennia, as well as radioactive and toxic pollution increases since 1945.

Krill

Harvesting of the massive swarms of krill in the Southern Ocean began in 1976 and reached a peak of 528,200 tonnes in 1981-82. Current catch levels average about 400,000 tonnes per year, taken mostly by the former USSR and Japan, although Chile, South Korea, and Poland are experimenting. Many other fleets are likely to follow.

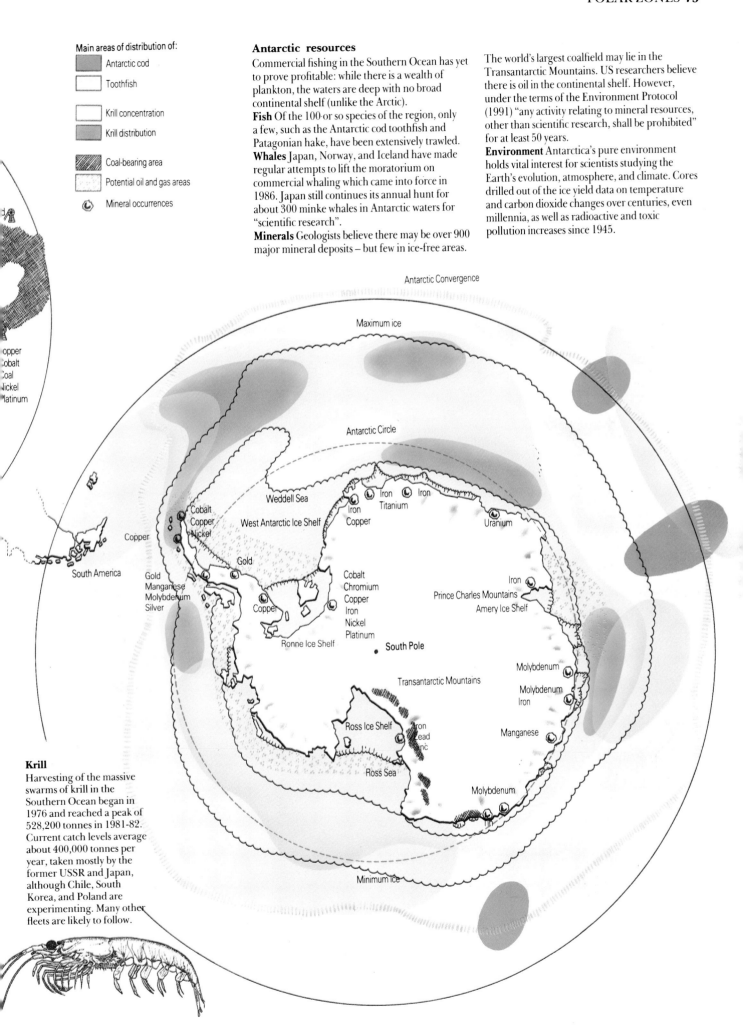

Antarctic Convergence

Maximum ice

Antarctic Circle

Weddell Sea

West Antarctic Ice Shelf

Iron · Iron
Iron · Titanium
Iron
Copper

Uranium

Cobalt
Copper
Nickel

Gold

Copper

South America

Gold
Manganese
Molybdenum
Silver

Copper

Cobalt
Chromium
Copper
Iron
Nickel
Platinum

Iron

Prince Charles Mountains

Amery Ice Shelf

Ronne Ice Shelf

South Pole

Transantarctic Mountains

Molybdenum

Molybdenum
Iron

Ross Ice Shelf

Iron
Lead
Zinc

Manganese

Ross Sea

Minimum ice

Molybdenum

THE OCEAN CRISIS

The empty nets

The ocean offers abundant resources. But through ignorance and misunderstanding we are placing this wealth in jeopardy – causing gross impoverishment of many fisheries, near extinction for most large whales, widespread pollution of fish-rich waters, and degradation and disruption of many coastal habitats and fish nursery grounds.

Declining fish stocks

Although the marine fish catch has quintupled since 1950, increases in levels of production have severely stretched many fisheries, especially those in coastal regions. Over the last quarter century many traditional fisheries have collapsed, some spectacularly, with a general reduction in stocks. Among the casualties are Atlantic cod, haddock, Atlantic herring, capelin, and Southern African pilchard. To compensate for these declines, landings of many other species have increased, such as the Japanese and South American sardines, accounting for most of the rise in the marine catch during the 1980s. But populations of sardines and other shoaling pelagics are notoriously unstable and have shown large fluctuations because of environmental factors.

In many fishing regions the harvest reaches or exceeds the estimated sustainable yield. Part of the problem lies with the technology employed by large, long-distance fleets, such as purse-seining – fine nets to "vacuum" the sea. Part lies with our ignorance of marine ecosystems. The main cause, however, has been the expanding demand for fish and fish products from Northerners.

As Western Europeans and North Americans eat more meat, they create a booming demand for animal-feed supplements, including fishmeal. Almost a third of the global fish catch now goes into meal and oil, mostly to feed the North. Other Northerners simply want more fish. Japan, for instance, relies on the oceans for 60 percent of its animal-protein supply (compared with a global average of 14 percent).

In addition, after over-using their local fisheries, the northern nations have been venturing further afield to abuse stocks off western Africa and elsewhere in the tropical developing world.

The global harvest has reached the limit of sustainable yields estimated by FAO as 100 million tonnes annually. Environmental pressures such as chemical pollution and the destruction of nursery grounds then have an increased effect on the

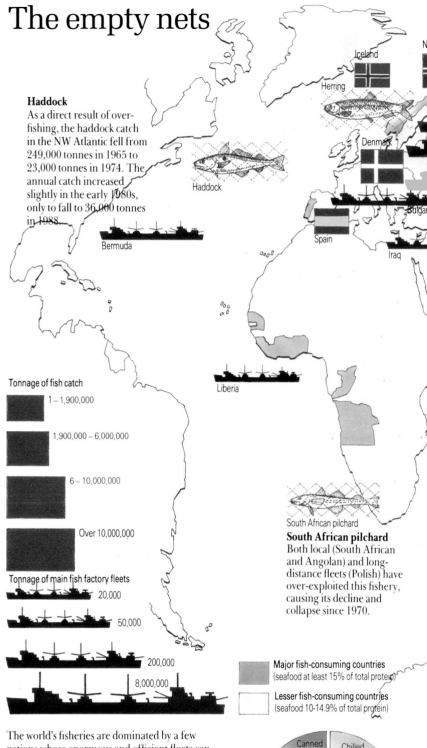

Haddock
As a direct result of over-fishing, the haddock catch in the NW Atlantic fell from 249,000 tonnes in 1965 to 23,000 tonnes in 1974. The annual catch increased slightly in the early 1980s, only to fall to 36,000 tonnes in 1988.

Tonnage of fish catch

1 – 1,900,000

1,900,000 – 6,000,000

6 – 10,000,000

Over 10,000,000

Tonnage of main fish factory fleets

20,000

50,000

200,000

8,000,000

South African pilchard
Both local (South African and Angolan) and long-distance fleets (Polish) have over-exploited this fishery, causing its decline and collapse since 1970.

Major fish-consuming countries (seafood at least 15% of total protein)

Lesser fish-consuming countries (seafood 10-14.9% of total protein)

The world's fisheries are dominated by a few nations whose enormous and efficient fleets can take the lion's share of the harvest. As the case studies show, fishery collapses have become more frequent and severe. Different-sized flags denote the catch sizes of the top fishing countries for 1989: 11 nations, each reaping over 2 million tonnes annually, account for 67% of the global total catch, with the former USSR heading the list at 11.3 million tonnes, followed closely by China and Japan. A further 7 countries each take more than 1 million tonnes. Ship symbols denote national tonnage of industrial-scale fishing fleets (the issue is complicated by the use of flags of convenience, notably by Japan). The major fish-consuming nations are indicated according to their level of dependence on fish protein in their diets.

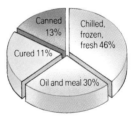

Canned 13%

Chilled, frozen, fresh 46%

Cured 11%

Oil and meal 30%

How we use the catch
Of the total world catch, nearly a third is used for feed and fertilizer. Roughly 11% is cured (salted, smoked), 13% is canned, and 46% is either chilled, frozen, or consumed fresh. The pie-chart shows the catch for 1989.

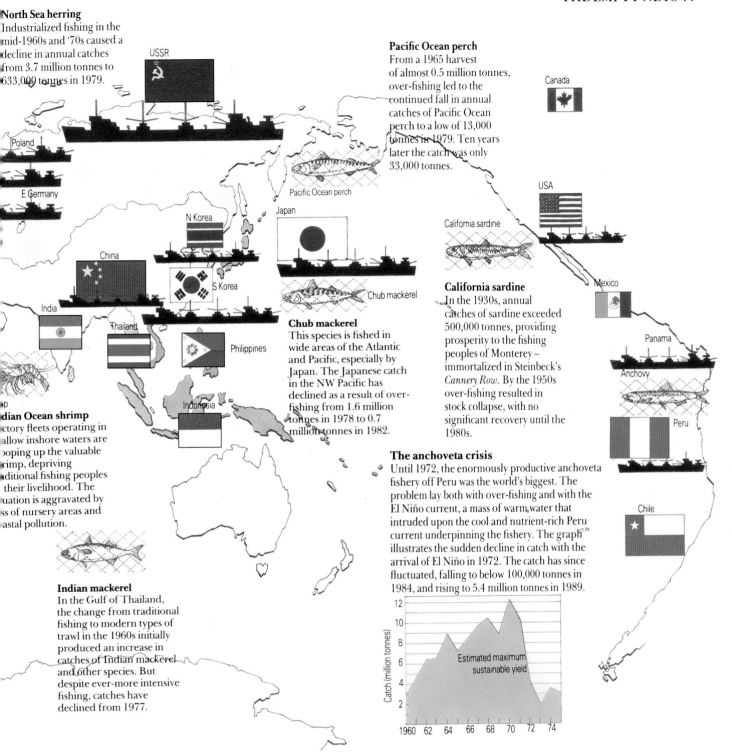

North Sea herring
Industrialized fishing in the mid-1960s and '70s caused a decline in annual catches from 3.7 million tonnes to 633,000 tonnes in 1979.

Pacific Ocean perch
From a 1965 harvest of almost 0.5 million tonnes, over-fishing led to the continued fall in annual catches of Pacific Ocean perch to a low of 13,000 tonnes in 1979. Ten years later the catch was only 33,000 tonnes.

Pacific Ocean perch

Canada

USA

California sardine

California sardine
In the 1930s, annual catches of sardine exceeded 500,000 tonnes, providing prosperity to the fishing peoples of Monterey – immortalized in Steinbeck's *Cannery Row*. By the 1950s over-fishing resulted in stock collapse, with no significant recovery until the 1980s.

Chub mackerel

Chub mackerel
This species is fished in wide areas of the Atlantic and Pacific, especially by Japan. The Japanese catch in the NW Pacific has declined as a result of over-fishing from 1.6 million tonnes in 1978 to 0.7 million tonnes in 1982.

Mexico

Panama

Anchovy

Peru

The anchoveta crisis
Until 1972, the enormously productive anchoveta fishery off Peru was the world's biggest. The problem lay both with over-fishing and with the El Niño current, a mass of warm water that intruded upon the cool and nutrient-rich Peru current underpinning the fishery. The graph illustrates the sudden decline in catch with the arrival of El Niño in 1972. The catch has since fluctuated, falling to below 100,000 tonnes in 1984, and rising to 5.4 million tonnes in 1989.

Chile

Indian Ocean shrimp
Factory fleets operating in shallow inshore waters are scooping up the valuable shrimp, depriving traditional fishing peoples of their livelihood. The situation is aggravated by loss of nursery areas and coastal pollution.

Indian mackerel
In the Gulf of Thailand, the change from traditional fishing to modern types of trawl in the 1960s initially produced an increase in catches of Indian mackerel and other species. But despite ever-more intensive fishing, catches have declined from 1977.

Catch (million tonnes)
Estimated maximum sustainable yield
1960 62 64 66 68 70 72 74

Mechanized fishing vs local fisheries
Technological advance in fishing has brought its problems – most evident in the waters off developing countries. Indiscriminate fishing by long-distance foreign fleets is now on the decline. But misuse of aid offered to establish national fisheries, intended to help local communities, has resulted in the same over-exploitation – depleting the fish stocks and depriving the poorer coastal peoples of their livelihood. The confrontation between capital-intensive fisheries and coastal fishing communities affects many millions of people.

Small-scale, community based fisheries account for almost half the world fish catch for human consumption, and employ 95% of the people in fisheries. Moreover, they use only 10% of the energy of large-scale corporate fisheries. But governments often favour the large-scale over the small because the former have more effective lobbies and are sources of foreign exchange.

Factory fishing off West Africa
Africa's coasts feature some prolific fisheries. Each year they provide 3 million tonnes of fish, of which 80% comes from a relatively short upwelling sector along the western seaboard from the Strait of Gibraltar to the Zaire river. Industrial fishing fleets have exploited this zone with sophisticated catching gear and cold-storage facilities – long-distance equipment characteristic of advanced nations rather than of local African enterprises. One such giant vessel, serviced by as many as 15 catcher ships, can in one day take about 1,000 tonnes and process the whole lot into fishmeal.

In the mid-1960s, West Africans watched half their catch being taken away by developed nations. By the mid-1970s, the 22 African nations concerned had increased their catch substantially – but witnessed their share decline to one-third of all fish taken, the bulk being accounted for by the USSR, Spain, France, Poland, Japan, and 14 developed nations.

One tonne of fishmeal fed to livestock in Europe produces less than half a tonne of pork or poultry, far less again if used as fertilizer to grow animal-feed grains. Were the West African fish resources to be used for direct human consumption by the people of West Africa, they would represent an additional 12 kg of animal protein per person per year – a 50% increase for many people.

resource productivity. Global warming could also seriously affect fisheries. As surface waters warm, plankton distribution and ocean currents could be affected.

Sources of marine pollution

The seas are a sump. They continuously absorb vast quantities of silt and minerals washed down from the land. Now, however, we are asking them to accept growing amounts of human-generated materials as well, from sewage sludge, industrial effluent, and agricultural run-off, all with their chemical contaminants, to radioactive wastes.

The oceans can do a good job for us as a gigantic "waste treatment works". The question is, how much waste can they safely handle? That is to say, what sorts of waste are they fitted to absorb, where can they best accommodate it, how long will they take to degrade it through natural processes – and what level of adverse consequences are we prepared to accept?

These critical factors are not receiving nearly enough attention. Each year we dump hundreds of new chemicals into the seas, to go with the thousands already there, and with next to no idea of their potential impact. Human-made toxic substances are being detected in deep ocean trenches, even as far as Antarctica. This phenomenon is the result of global circulatory systems, processes of which we have hardly any understanding.

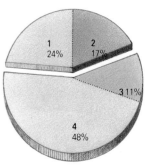

Waste discharge and run-off

1 municipal 2 industrial

3 Atmospheric

4 Transportation including tanker accidents)

Polluting the oceans

The oceans receive the brunt of human waste, be it by deliberate dumping or by natural run-off from the land. At least 83% of all marine pollution derives from land-based activities. The illustration below shows a major conveyor of these pollutants, a river flanked by both agricultural land and industry, a scene that reflects much of the industrialized North. As the river becomes polluted, a high percentage of waste is flushed downstream and deposited in the biologically productive estuarine waters and coastal zones. Here, the poisons enter marine food chains, building up their concentrations in higher species. This process of bioamplification was sharply illustrated in Japan in the early 1950s by Minamata disease – methyl mercury poisoning due to eating tuna with heavy concentrations of mercury in their tissues (the wastes originated from a coastal factory). By 1975, there were 3,500 known victims.

The New York Bight
The severe pollution of the New York Bight has been caused by years of waste discharges and dumping. The pie-charts give the proportion of waste elements that ended up in the waterways either from the land or barging of wastes during the 1980s. Thankfully, the dumping of waste has now been banned.

☐ To estuary

■ Direct to bight

▨ To bight via dumping

Discharge (tonnes per day)

Carbon 2,600

Oil and grease 870

Nitrogen 520

Iron 230

Copper 13.8

Lead 12.7

Mercury 0.3

PCBs 0.014

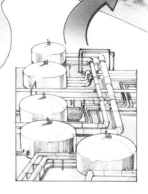

Outflow from estuary

Barging

Agricultural run-off
Pesticides and herbicides, not readily bio-degradable, are persistent pollutants. As they pass through marine food chains, their effect is concentrated. Nitrates from fertilizers over-enrich water, causing algal growth and eventual deoxygenation.

Urban centres
Municipal drainage systems pour out domestic and industrial sewage, contaminated with toxic chemicals, heavy metals, oil, and organic nutrients. Construction sites release enormous amounts of sediment into rivers.

Industry
Much of the complex mix that goes into industrial waste ends up in the sea. Included in this mélange are partially bio-degradable food wastes, heavy metals, and persistent pesticides. It often takes a human casualty to alert us to the source of the pollution.

Nuclear facilities
Radioactive effluent is discharged into coastal waters from nuclear-fuel reprocessing plants such as those at Sellafield (UK) and Cap de la Hague (France). Both facilities have been implicated in sickness and deaths of local people.

Oil refineries
Oil terminals tend to be sited along coasts, often built on valuable saltmarsh or near productive estuaries. Accidental oil loss and seepage from refineries contribute some 100,000 tonnes of oil annually to ocean pollution.

Oil in the ocean ecosystem
Some 3.2 million tonnes of oil enter the ocean each year via many routes: from the atmosphere, from waste discharges, by natural seepage from the sea bed, and from oil production and transport at sea. The pie-chart shows the proportional input of oil into the ocean from various human sources.

Chemical run-off caused by humans into the oceans is much greater than Nature's contribution – mercury two-and-a-half times the natural rate, manganese four times, zinc, copper, and lead about 12 times, antimony 30 times, and phosphorus 80 times. As for oil, human-caused pollution – often by wanton carelessness, or even deliberate discharge – accounts for four-fifths or more of the total volume entering the sea, some 3.2 million tonnes a year. We hear much about oil-killed birds and other marine life. Fortunately, they generally recover their numbers within a few years. The worst damage is more insidious: certain components of oil are toxic, others are carcinogenic, and they tend to persist for extended periods of time.

Heavy metals such as mercury, lead, cadmium, and arsenic, and chemicals such as DDT and PCBs must rank high on the list of harmful pollutants. We have learned to our cost of the effects of mercury through the Minamata episode in Japan (see below, left) and more recent deaths in Indonesia; and we have discovered too late the impact of DDT and PCBs, through reproductive failures among birds of prey and other wildlife.

The most significant factor of all is that at least 85 percent of ocean pollution arises from human activities on land, rather than at sea, and that 90 percent of these pollutants remain in coastal waters – by far the most biologically productive sector of the oceans. The wanton destruction that is

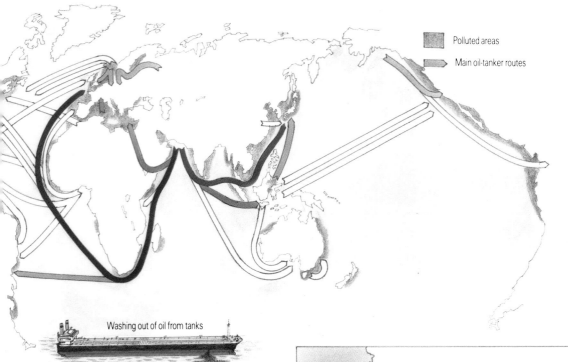

Polluted areas

Main oil-tanker routes

Washing out of oil from tanks

Sewage dumping

Nuclear dumping

Global marine pollution
The global circulation of ocean currents and the continuity of marine life means that no part of the ocean is exempt from pollution: the classic example is the discovery of DDT in the fat of Antarctic penguins, thousands of kilometres from source. The map (left) illustrates pollution hotspots – coastal areas close to industrial conurbations such as the North Sea, and regions with high populations such as the sea off Rio de Janeiro, Brazil and the Java Sea, Indonesia. The map also shows oil-tanker routes, the heaviest traffic being from the Middle East to Europe. Most tanker accidents occur along congested routes, close to the coast.

Sea-borne pollution
Oil spills, such as the *Exxon Valdez* disaster in 1989, receive much media coverage but represent only a fraction of ocean pollution. More extensive is the deliberate dumping of wastes. The dumping of sewage sludge – organic, industrial, and human waste – is potentially dangerous due to the mass of bacteria and viruses contained in it. However, the ocean disposal of hazardous and toxic waste is being phased out in a number of regions. The dumping of radioactive waste began in 1946 and continued until its banning in 1982. The drums and storage tanks used have subsequently proved leaky.

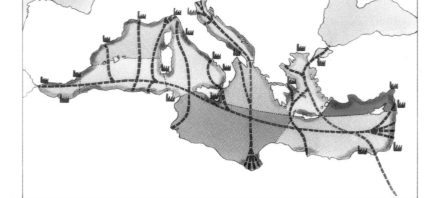

The polluted Mediterranean
The Mediterranean is a notorious pollution black-spot. Of the 100 million people along the coast, nearly 50% live in towns and cities, contributing a heavy sewage load into a sea that is not easily flushed clean. The worst areas are near Barcelona, Marseilles, Genoa, Piraeus, and Naples. Oil tankers regularly cross the Mediterranean, the heaviest pollution occurring at loading bays and refineries along the coast of Libya and Tunisia.

Oil-tanker routes

High sewage pollution

Industrial centres

Areas of heavy oil pollution from land-based sources

occurring in these vital zones will have serious consequences, not only for human welfare, but for the entire marine realm.

Our vulnerable coastline

"The lasting benefit that society derives from wetlands often far exceeds the immediate advantage their owners might get from draining or filling them. Their destruction shifts economic and environmental costs to other citizens . . . who have no voice in the decision to alter them."
PRESIDENT CARTER'S ENVIRONMENTAL MESSAGE OF 1977

The destruction of marine habitats is most serious in coastal zones. Saltmarshes, estuaries, mangroves, and coral reefs, all areas of great beauty and vital to our welfare, are especially vulnerable to human disruption and degradation.

Coastal cities often look to nearby wetlands as a cheap way to dispose of rubbish, from industrial refuse to household garbage. Equally damaging is the dredging of offshore sand and gravel that destroys fish-spawning grounds. Along US coasts, dredging, together with filling operations, has eliminated more than 20,000 square kilometres of valuable wetlands. More than a quarter of US important coastal wetlands has been lost.

Just as important and especially susceptible are estuaries, "crossroads" between land and ocean ecosystems, and hence locations of much human activity. So productive are estuaries in terms of fish life that they are estimated to generate almost 80 million tonnes of fish a year. At least three-quarters of all fished species in the eastern United States spend part of their life in estuaries. Yet many of these prime habitats are being degraded at a rate that eliminates entire communities of fish. Of 80,000 square kilometres of US estuarine waters, one-third is now closed to shellfishing because of habitat disruption. Overall the destruction of offshore habitats costs the country more than $80 million a year through loss of commercial fisheries alone.

Another cause of habitat loss lies with eutrophication brought on by sewage and fertilizer run-off. The added phosphates and nitrates cause a "bloom" of algae; as they decompose they use up much of the oxygen, choking out other life (except bacteria that produce hydrogen sulphide by breaking down sulphates in their search for oxygen).

In the tropics, similar pressures are contributing to the destruction of both mangroves and coral reefs. In the Philippines, for example, 44,000 square kilometres of coral reefs supply one-tenth of the commercial fish catch, but is in decline due to severe habitat degradation. Philippine mangroves have also declined from about 5,000 square kilometres in the 1920s to only 1,400 square kilometres today. More than half of this loss is due to the cutting of mangroves to create culture ponds to rear fish and prawns. Despite the designation of reserves for mangroves and reefs, the pressure

Destruction of habitat

Almost two-thirds of the human population live in one-third of the world's land adjacent to coasts. Of the ten largest metropolitan areas, nine (accounting for some 125 million people) border estuarine regions – Tokyo, São Paulo, New York, Shanghai, Calcutta, Buenos Aires, Seoul, Bombay, Rio de Janeiro. In several cases these conurbations have all but obliterated coastal wetlands – biologically productive areas which are also important as pollution filters and as natural buffers between land and sea. The US, for example, is rich in estuaries. Those along the Atlantic and Gulf coasts are especially important as a fishery and shellfish resource. As many as 95-98% of commercial fishery species of these regions spend their early life feeding in the rich, warm, and sheltered estuarine waters. But for how long can they sustain their productivity, in the face of a proliferation of industrial complexes which are not only sited on wetlands, but spill their wastes into the fertile waters?

Just as worrying, if often less dramatic than pollution, are the effects of dredging and other forms of land management. The sludge dredged from the waterways of south Louisiana, as shown in the photograph below, forms a "levee", or continuous raised spoil bank, which effectively impounds many marshes. Natural drainage channels are either interrupted or cut – leading to widespread habitat destruction.

The Mississippi Delta
Louisiana's 1.2 million ha of saltmarshes account for some 40% of US coastal wetlands – a fraction of their original extent. Apart from natural erosion, the discovery of oil in 1901 led to construction of a vast network of waterways (below). This activity, together with land drainage, has adversely affected the wetlands, notably the shrimp nursery areas and menhaden fisheries of the Gulf of Mexico.

Chesapeake Bay
Chesapeake Bay on the US Atlantic coast is one of the world's most productive ecosystems and has long provided oysters, crabs, and fish in abundance. But today the Bay is becoming depleted, due to the effects of industrial pollution and agricultural fertilizers.

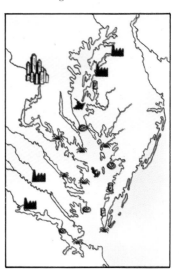

Causes of mangrove destruction
Clear felling
Agriculture
Coastal development
Salt ponds
Mining
Waste

Coral reef destruction

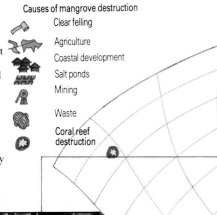

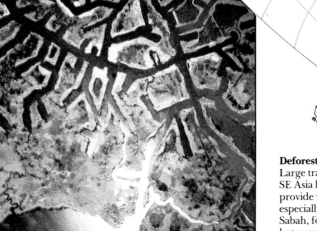

Deforestation
Large tracts of mangrove in SE Asia have been cut to provide woodchips, especially for the Japanese. Sabah, for example, has lost over 122,000 ha of mangroves (40% of the total mangrove area).

Conversion to agriculture and aquaculture

In Asia and Africa, pressure on arable land has led to clearance of mangroves for agricultural land (e.g. rice cultivation in W Africa). Impoundment of mangroves for aquaculture is a more traditional activity; over 1 million ha of mangroves have been cleared for fish farming in the Indo-Pacific alone. The natural values of mangroves are seldom taken into account before converting these coastal ecosystems into agricultural or aquacultural "factories".

Salt pond construction

A classic example of conflicting resource needs in coastal zones lies with the clearance of mangroves for salt/brine evaporation ponds, as in NW India, W Africa, and Malaysia.

Waste disposal

Garbage and solid waste are often dumped in mangroves, while sewage effluents and hazardous chemicals contribute to their poisoning.

Mining

Mine waste can cause smothering of mangrove roots. In northern Puerto Rico, for example, mining for sand and an airport development have led to the destruction of a large tract of mangrove.

Coastal development

The clearance of mangroves for domestic and industrial development is a major problem, particularly in high-income countries. In southern Queensland, Australia, large tracts of mangrove have been destroyed for housing and canals. Similarly, much of the mangroves on the estuary of the Singapore river have been lost to housing and industry.

Mangrove forest destruction

Worldwide, approximately 165,000 square kilometres of mangrove remain – 58,000 in Asia, 39,000 in Africa, and 68,000 in the Americas. Some 58% of the mangroves in Indomalaya have been lost, often by their conversion into brackish saltwater ponds for raising prawns and fish. In Africa, 55% has been lost. The map pinpoints areas of widespread mangrove loss, keyed to the most damaging factors involved. The photograph, above, shows mangroves killed by oil spills.

The assault on coral reefs and lagoons

The widescale destruction of coral reefs is a tragedy for humans and marine realm alike – coral reefs shelter island waters, harbour enormous numbers of species, and could one day provide us with a pharmacopoeia of new drugs (pp 68-9). Dredging and the removal of coral for construction is one factor; terrestrial erosion is another, often resulting in the smothering of reefs (Kanehoe Bay, Hawaii, is a classic case). Equally destructive is pollution caused by sewage, industrial waste, thermal and desalination effluents, and oil. Other factors – including tourism, with collection of corals, shells, and puffer fish for souvenir trade; mining and blasting; oil and gas production; over-fishing; even nuclear testing – all contribute to the death of reefs. The most complete and widespread destruction of coral reefs is now conducted by the US military. Diego Garcia in the Indian Ocean, once the fertile home of 2,000 people, is now virtually completely covered in concrete, its biological productivity destroyed.

of rising human numbers and demand for resources are eradicating these irreplaceable ecosystems at an alarming rate.

The tragedy of the commons

"Whales have become newly symbolic of real values in a world environment of which man is newly aware. Whales live in families, they play in the moonlight, they talk to one another, and they care for one another in distress. They are awesome and mysterious. In their cold, wet, and forbidding world they are complete and successful. They deserve to be saved, not as potential meatballs, but as a source of encouragement to mankind."

VICTOR SCHEFFER, AMERICAN WHALE EXPERT

The ocean has given rise to all manner of unhappy conflicts, from disputes over resources to depletion of marine species, whether seen as competitors or as prey. This is not so much due to the aggressive nature of fishing peoples and others "who go down to the seas in ships". Rather the problem lies with the nature of the marine realm as a "commons environment".

Apart from coastal zones, all people enjoy open access to the ocean, for them to exploit as they see fit. Each individual or nation has viewed the resources as free for the taking; each has sought to outdo the rest. This self-defeating process is compounded as the competition for a diminishing resource grows more severe, until all too often, the exploiters drive the resource to extinction.

The tragic futility of this outcome is symbolized by the record of the great whales. A whale swimming in the ocean belongs to nobody; when it is killed, it becomes the private property of the whaler, and the profit accrues to the whaler alone. As the

Whose ocean?

Conflict is the theme song of the oceans – a commons environment where largely uncontrolled exploitation and competition for resources makes the original inhabitants inevitably the losers. Of these, none has suffered more heavily than the great whales.

A non-sustainable harvest

The first species to attract unwanted human attention was the right whale – so dubbed because it was the "right" one to pursue – easy to catch and a rich source of oil and whalebone. The Victorian desire to cut a fine figure led to a huge demand for whalebone corsets, and the right whale was ultimately driven almost to extinction. Another early victim was the Western grey whale, hunted for oil, meat, and blubber. These disasters were only the beginning of the story. In this century, with improved technology, soaring demand for whale products, and no common management strategy, we have over-hunted species after species. As each stock fails, we turn our attention to the next.

Since 1900, whaling has focused on the Antarctic, where the whales congregate in summer to feed. The first Antarctic whale to be hunted commercially was the humpback: around 7,000 a year were taken in the 1900s. Once humpback stocks dropped, the blue whale, our largest living mammal, became the target; again, about 7,000 a year were harvested in the 1930s. Next in line was the fin whale – but now, more efficient catcher boats produced a massive rise in exploitation levels. More than 26,000 fin whales were taken in 1940 – the peak catch for this species. After the fin, the sei whale became the main quarry, with a peak catch of 20,000 in 1965. Finally, the whaling industry switched to the smaller minke whale, of which 8,000 were taken in 1970.

In 1946 the International Whaling Commission was established to regulate the industry but was an abject failure as quotas lacked a sound scientific base, and even then they were often ignored. By the late 1970s, as world concern for whales mounted and more non-whaling countries joined the IWC, stricter limits were applied and finally a moratorium on commercial whaling was passed, beginning in 1986. But the IWC can only make rules; it cannot enforce them. More than 14,000 whales have been killed since 1986. Japan, Norway, and Iceland have exploited a loophole in IWC regulations allowing whales to be taken for "scientific purposes". Under this guise these nations continued the hunt, Japan killing 327 minkes in its 1991/2 harvest of Antarctic waters.

Blue whale
Weight 100 tonnes
Length 26 metres

Sei whale
Weight 14 tonnes
Length 16 metres

Humpback whale
Weight 35 tonnes
Length 15 metres

400,000

200,000

100,000

100,000

3000 – 4000

1900   1920   1940   1960   1980

Defeat of the giants
One after another, the Antarctic populations of the great whales have plummeted. Hunting of these species is no longer viable, their future unsure. Because of slow maturation and calving years, it will take many years for the populations to recover, if at all. The population curves, left, are based on estimates of Southern Ocean populations for 1900-90.

Fin whale
Weight 40 tonnes
Length 22 metres

2000 – 100,000

500 – 11,000

24,000

hunting intensifies, whalers have to spend extra effort in finding their next target. Worse still, all people (including the whalers) face an aggravated risk of a whale species becoming extinct.

Concern for the whales has led to a moratorium on commercial whaling, although certain countries continue the hunt. Any return to full-scale commercial whaling is probably a long way off, but a newly-proposed "Management Procedure" provides, in theory, a workable quota system for harvesting whales. When this was unveiled at the 1991 International Whaling Commission meeting, Iceland stated that it would not accept any management procedure that did not meet the wants of its whalers. In 1992 it resigned from the Commission.

Whaling industries are pressurized into recovering their investment as soon as possible. Such short-term economic pressure militates against any long-term sustainable harvest of a resource with such a lengthy self-renewal period. A whale population may grow by only 5 percent a year – a rate that is way below a typical rate of discount on capital investment.

The present-day ocean crisis can be likened to the parable of the "grazing commons" in medieval England. Traditionally, herdkeepers grazed their cattle on common pasture. One area might be used by 10 herdkeepers each with one cow. If one keeper brought along an extra cow, this would result in slightly less grazing per cow, although for the enterprising keeper, the loss would be offset by receiving two shares of the grazing. If, however, all the other keepers followed suit, they would all find themselves poorer and poorer: while each user works out a plan that appears rational, the collective result is that the commons are wrecked.

The troubled oceans

Competition for ocean space and resources takes many forms. The map below highlights typical incidences of several kinds: territorial disputes, resource disputes, and conflicts between human fishing and the needs of marine mammals. As competition for fish stocks intensifies, territorial disputes between fishing nations become commonplace. The Cod Wars between Iceland and Britain in the 1960s and '70s were a dramatic example, receiving much media coverage. Britain eventually had to back down, in the face of Iceland's adoption of the Exclusive Economic Zone (EEZ) – a principle which Britain itself was soon to reinforce.

A rather different confrontation arises when a marine resource serves several needs. In the North Atlantic, capelin is taken for use as fishmeal, but is also eaten by cod – itself a valuable food fish.

Conflict between humans and marine mammals has always been intense. Apart from the hunt for whales, local fisheries regularly cull dolphins, porpoises, and seals to safeguard their interests. Marine mammals are also killed incidentally in massive numbers. Drift nets of monofilament nylon are used on a huge scale. In the South Pacific, for example, each boat in fleets fishing for tuna sets out each night over 60 km of nets, 60 m in depth. A comparable operation for salmon takes place in the North Pacific. Whales, dolphins, seals, and turtles all fall victim in addition to the intended catch. Despite a UN ban on such fishing from 1992 it may continue as the ban is not binding.

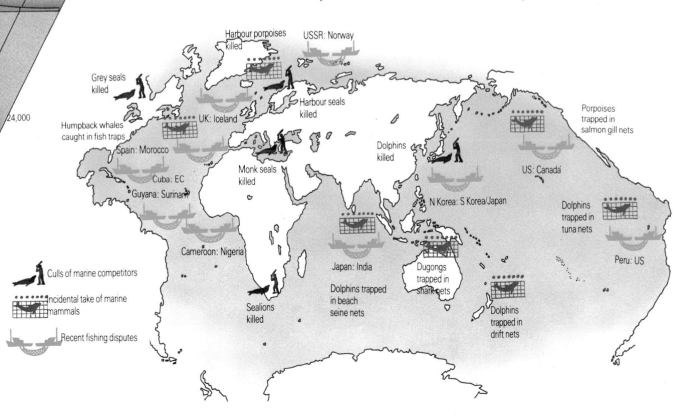

Harbour porpoises killed

USSR: Norway

Grey seals killed

Harbour seals killed

Humpback whales caught in fish traps

UK: Iceland

Dolphins killed

Porpoises trapped in salmon gill nets

Spain: Morocco

Monk seals killed

US: Canada

Cuba: EC

Guyana: Surinam

N Korea: S Korea/Japan

Dolphins trapped in tuna nets

Cameroon: Nigeria

Japan: India

Dugongs trapped in shark nets

Peru: US

Culls of marine competitors

Sealions killed

Dolphins trapped in beach seine nets

Dolphins trapped in drift nets

Incidental take of marine mammals

Recent fishing disputes

MANAGING THE OCEAN

Despite much research, we lack the understanding to exploit marine resources in a sustainable manner. The ocean is beset with crises, ecological and political. And although attempts at management have a long history, there are relatively few success stories. Fortunately, there are signs of a new approach. Fisheries planning is becoming more precise, and fresh moves are under way to control pollution and protect marine environments. A series of conventions and treaties are coming into force,

and the concept of common heritage is still alive, if less robust than was initially hoped, in the new Law of the Sea.

A new fishing strategy

How shall we do a better job of taking a sustainable harvest from the sea? Hitherto we have employed an approach resembling that of hunter-gatherers toward agriculture; we have exploited wild creatures in a wild manner, with predictable results.

We can take a huge step forward right away by applying measures we have long known to be valid and yet persistently ignored to our cost. We can agree on realistic catch quotas, for instance, based on a true understanding of the population dynamics of fisheries, and enforce them strictly. We can impose moratoria on fishing stocks, before they crash, rather than too late as with the Peruvian anchovy and North Sea herring. We can set minimum net-mesh sizes (or caught fish sizes) to

Fishery commissions and advisory bodies
1 Regional Fisheries Advisory Commission for SW Atlantic (CARPAS)
2 Fishery Committee for the EC Atlantic (CECAF)
3 General Fisheries Council for the Mediterranean (GFCM)
4 Indian Ocean Fishery Commission (IOFC)
5 Indo-Pacific Fishery Commission (IPFC)
6 WC Atlantic Fishery Commission (WECAFC)
7 International Baltic Sea Fishery Commission (IBSFC)
8 NW Atlantic Fishery Organization (NAFO)
9 International Commission for the SE Atlantic Fisheries Commission (ICSEAF)
10 International N Pacific Fisheries Commission (INPFC)

Harvesting the sea

With ever-rising demand for the ocean's living protein, and increasing depletion of fisheries and marine mammals, we can no longer afford to ignore ecosystem interactions in our fishing policies. Despite a welter of fishery bodies, quotas for both fish and mammals have generally been too high, reflecting political compromise rather than scientific advice, and protection has been ineffectual. In the future we shall need more realistic quotas and tougher enforcement, while fishery management will increasingly involve choices – where to tap a food chain, which species to harvest and how much, and which to protect. The effects of human intervention are already apparent in some ecosystems as shown below. Before we interfere even more seriously, as in the Antarctic, we must study the consequences.

Fishery management

Since the 1930s there has been a proliferation of fishery commissions and advisory bodies dealing with fishing areas and quotas of fish and mammal species. Some work in a consultative capacity, others set quotas and have their own back-up research programmes. The North Pacific Fur Seal Commission was the first of these bodies to be established in 1911. Within the new Exclusive Economic Zones (EEZs), coastal states rather than fishery commissions have had more direct control over quotas and catches.

Shifting the balance

Since the mid-1960s, the North Sea harvest has remained level. But the nets are no longer filled with mackerel and herring, once so abundant. Instead we haul in a mixture of smaller species, with some cod and haddock. True, our fishing patterns have changed. But human exploitation has apparently shifted the balance of North Sea ecosystems too. Out of a fairly constant fishing stock, herring and mackerel once comprised two-thirds, but now make up less than one-third; while pout, sand-eels, sprats, and larger gadoids have multiplied, perhaps because their larval forms are less likely to be preyed upon since depletion of the herring and mackerel.

Cod, haddock, sand-eel, sprat, and pouting

Herring and mackerel

All weights in million tonnes

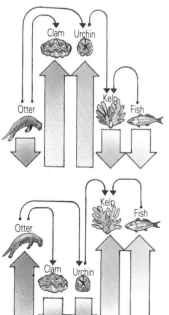

Clam Urchin

Otter Kelp Fish

Otter Kelp Fish

Clam Urchin

Otter vs clam

Protection of the Californian sea otter in 1972 was welcomed by conservationists, but had some unforeseen consequences. The upper diagram shows the pre-protection ratio of the sea otter to other coastal species, the lower one the situation today. An increase in sea otters has been matched by a decrease in clams and sea-urchins, on which the otter feeds. Reduction of sea-urchins has resulted in vigorous growth of kelp (seaweed), which in turn provides shelter for fish. Traditional clam gathering and harvesting of sea-urchins once provided income for the coastal towns. These have now been partially replaced by tourism.

allow immature fish to live and grow – a basic piece of common sense. We can institute laws to protect ocean communities, and cease to persecute the dolphins, seals, and other marine predators, or to entrap them carelessly in our nets.

Probably the two most important measures we can take, however, are to reduce our pressure on ocean resources overall, and adopt a fundamentally modified strategy for fishing. Thus far we have tended to concentrate on individual fish species, invoking a concept of maximum sustainable yield. For the future, we must take account of ecological interactions within an ocean community, and apply a multi-species management strategy.

In the Antarctic Ocean, for example, we find many species of whale and seal, together with vast colonies of seabirds, all supported by abundant stocks of krill. Whereas there used to be 100,000 blue whales, they have now been reduced to just a few thousand; and all other krill-feeding whale species, with the exception of the minke, have been grossly depleted. As a result, their consumption of krill has been reduced from about 150 million tonnes a year to only about 40 million tonnes. So far as we can tell, this has released a large amount of krill for the seabirds and seals.

These feeding links have major significance for our plans to harvest krill. Would the deficit, if we create one, hit the seabirds and seals, which are under no threat – or would it fall on the whale species, at least six of which are struggling to move away from the brink of extinction?

Human hunters have long posed a threat to marine mammals. Not only whales, but turtles, dolphins, manatees, dugongs, Californian sea otters, sealions, certain seals, and polar bears have suffered widespread loss. Most of the 100-plus species need some form of protection. But safeguard measures are not enough, unless we also take account of prey stocks, many of which are increasingly harvested for

Krill – the vital link
Antarctic waters are among the most productive in the world, the main link in the food chain being small shrimp-like krill. During the summer they feed on phyto- and zooplankton, growing from about 45 mm to 150 mm. Summer swarms of krill have been estimated at 650 million tonnes, a huge amount of living matter for only one species.

Changing demands on krill
Prior to their exploitation, the great whales consumed a large proportion of the annual crop of krill which also supported abundant seabird and seal colonies. Since the great whales' decline, seals and seabirds have been the major krill consumers. Human fisheries take only 0.4% of the total krill stock (1991). Future human demand must be taken into account if ever the baleen whales are to recover.

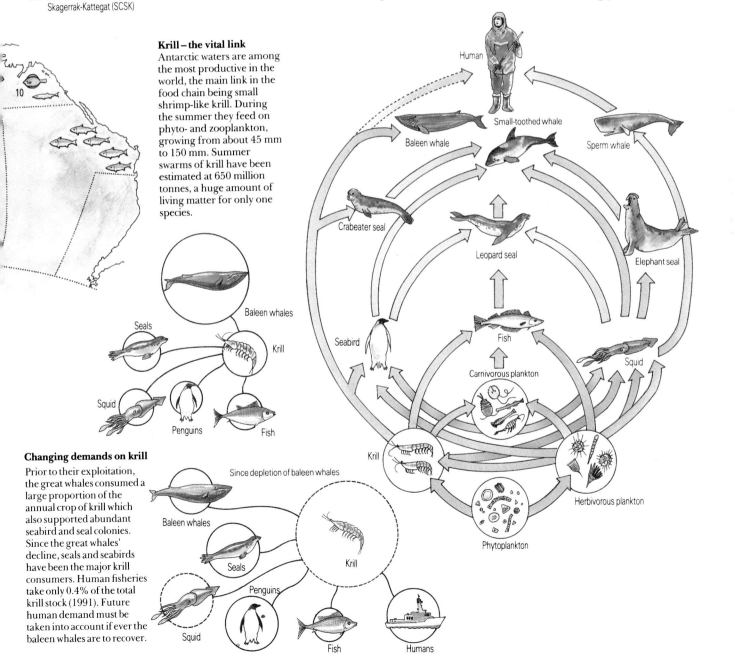

human consumption. An ecosystem-level approach will not only help the disappearing mammals; it will support the long-term welfare of the whole ocean community, including humankind.

Pollution control

The ocean offers some scope for us to dispose of waste; but it requires vast care. Common sense demands that we should control and reduce pollution at source, and clean up its past legacy. In dealing with pollution problems which cross regional boundaries, the international convention is a useful tool. Under the International Convention for the Prevention of Pollution from Ships (known as MARPOL), limits are set on the amounts of oil, noxious substances, sewage, and garbage that ships can discharge. In certain sensitive zones, such as the Baltic and Mediterranean Seas, oil discharge is prohibited altogether. What is more, oil cargoes can be "fingerprinted" through additives, and culprits of pollution identified. As a result, the amount of oil released at sea has been considerably reduced, even though oil tanker traffic has increased. Also of great importance is the London Dumping Convention of 1972 which is concerned with ocean dumping and has banned the dumping of radioactive waste at sea.

Among other actions, international conventions have established legally binding agreements and rules for dumping: a grey list of substances permitted to be dumped in trace amounts; a black list for those requiring special authorization.

Matching the international approach are important conventions dealing with specific areas. The Bonn Agreement of 1969 focused on the control of oil pollution in the North Sea – a heavily polluted and congested shipping area. The agreement safeguards vulnerable coastal areas from spillage, with member countries agreeing to co-operate in any clean-up operations. The Helsinki Convention (1974) has been the first to cover not only marine pollution, but also the more serious issue of land-based pollution. This convention highlights the need for an overall strategy at regional level, and served as a prototype for the United Nations Environment Programme (UNEP).

To its credit, UNEP picked up the environmental football, and has run far with it. Its Regional Seas Programme involves 130 states, 16 UN agencies, and more than 40 international and regional organizations, all working with UNEP to improve the marine environment and make better use of its resources. In some regions, mutually hostile nations sit down at the same table to resolve common problems through common solutions.

This strategy represents a quantum leap in "environmental diplomacy". The problems of ocean pollution have seemed intractable precisely because they have been international problems. The

Clean-up for the ocean

Spearheading the clean-up of the oceans is the regional approach exemplified in UNEP's Regional Seas Programme, backed by international conventions and improved technology. Since the 1975 Mediterranean Action Plan, nine more plans have followed, most recently in South Asia. Long-term success, however, depends on finance. UNEP sees its role as a catalyst, providing or finding sufficient seed money to bring a programme into existence – $8 million, for example, in the first 5 years of the Mediterranean Plan. The main funds must be raised by the countries themselves.

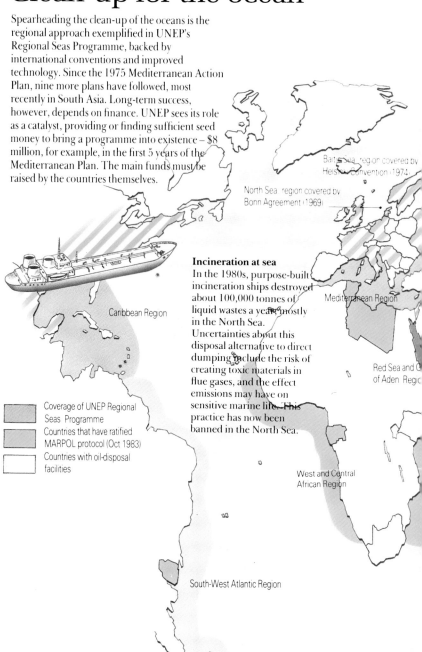

Baltic Sea region covered by Helsinki Convention (1974)

North Sea region covered by Bonn Agreement (1969)

Mediterranean Region

Red Sea and Gulf of Aden Region

Caribbean Region

Incineration at sea
In the 1980s, purpose-built incineration ships destroyed about 100,000 tonnes of liquid wastes a year, mostly in the North Sea. Uncertainties about this disposal alternative to direct dumping include the risk of creating toxic materials in flue gases, and the effect emissions may have on sensitive marine life. This practice has now been banned in the North Sea.

West and Central African Region

South-West Atlantic Region

Coverage of UNEP Regional Seas Programme
Countries that have ratified MARPOL protocol (Oct 1983)
Countries with oil-disposal facilities

MARPOL
In 1973 MARPOL was formulated to control all forms of pollution from ships. It sets minimum distances from land for the discharge of treated and untreated sewage, garbage, and toxic waste. Oil pollution provisions oblige ships over 400 tonnes to carry tanks for the retention of oil residues, and ports handling oil to have proper disposal facilities. The Convention also prohibits the discharge of toxic waste in the Baltic and the Black Sea, and the discharge of oil in the Baltic, Black Sea, Mediterranean, Persian Gulf, and Red Sea. By 1989, MARPOL was signed by 52 states.

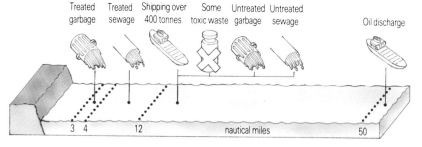

Treated garbage	Treated sewage	Shipping over 400 tonnes	Some toxic waste	Untreated garbage	Untreated sewage		Oil discharge
						nautical miles	
3 4		12					50

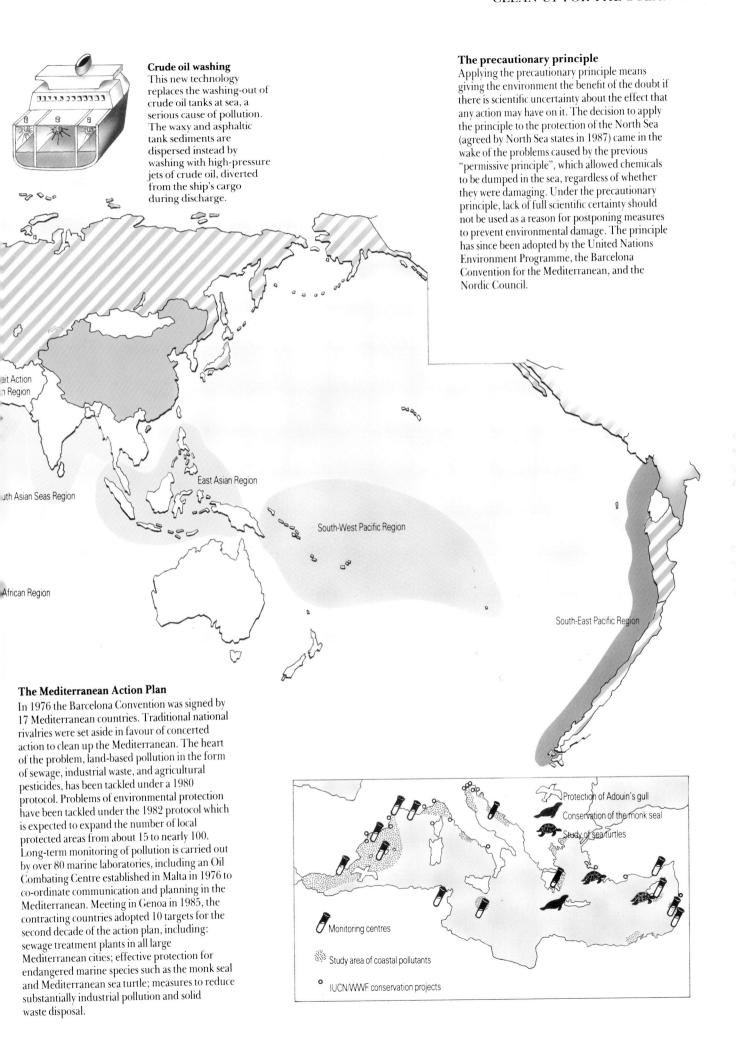

Crude oil washing
This new technology replaces the washing-out of crude oil tanks at sea, a serious cause of pollution. The waxy and asphaltic tank sediments are dispersed instead by washing with high-pressure jets of crude oil, diverted from the ship's cargo during discharge.

The precautionary principle
Applying the precautionary principle means giving the environment the benefit of the doubt if there is scientific uncertainty about the effect that any action may have on it. The decision to apply the principle to the protection of the North Sea (agreed by North Sea states in 1987) came in the wake of the problems caused by the previous "permissive principle", which allowed chemicals to be dumped in the sea, regardless of whether they were damaging. Under the precautionary principle, lack of full scientific certainty should not be used as a reason for postponing measures to prevent environmental damage. The principle has since been adopted by the United Nations Environment Programme, the Barcelona Convention for the Mediterranean, and the Nordic Council.

ait Action
n Region

East Asian Region

uth Asian Seas Region

South-West Pacific Region

African Region

South-East Pacific Region

The Mediterranean Action Plan
In 1976 the Barcelona Convention was signed by 17 Mediterranean countries. Traditional national rivalries were set aside in favour of concerted action to clean up the Mediterranean. The heart of the problem, land-based pollution in the form of sewage, industrial waste, and agricultural pesticides, has been tackled under a 1980 protocol. Problems of environmental protection have been tackled under the 1982 protocol which is expected to expand the number of local protected areas from about 15 to nearly 100. Long-term monitoring of pollution is carried out by over 80 marine laboratories, including an Oil Combating Centre established in Malta in 1976 to co-ordinate communication and planning in the Mediterranean. Meeting in Genoa in 1985, the contracting countries adopted 10 targets for the second decade of the action plan, including: sewage treatment plants in all large Mediterranean cities; effective protection for endangered marine species such as the monk seal and Mediterranean sea turtle; measures to reduce substantially industrial pollution and solid waste disposal.

Protection of Adouin's gull

Conservation of the monk seal

Study of sea turtles

Monitoring centres

Study area of coastal pollutants

IUCN/WWF conservation projects

Regional Seas Programme demonstrates that the community of nations can prove itself equal to some unusually difficult challenges.

The Antarctic heritage

Antarctica, a common heritage of all humanity if ever there was one, remains under the aegis of the Antarctic Treaty powers. Some 26 nations (1992), seven of whom have territorial stakes in the continent, hold sole decision-making rights over all activities in the region. True, the Treaty has proved a successful experiment in international co-operation. Nations as different as the US, the former USSR, the UK, and Argentina have agreed to keep one-tenth of the Earth's land surface demilitarized, nuclear-free, and devoted to research. But membership is small, the parties meet in secret, and until recently it was absurdly difficult for a nation to accede. Hence the charge that it has been an "exclusive" club dominated by developed nations, an association of the world's largest real-estate operators, and a last stand of colonialism. When the first major world oil crisis in the early 1970s coincided with the first scientific research findings that oil was very likely to be present, attitudes to the continent changed; the claims of Treaty parties took on a real economic significance.

Between 1982 and 1988 the Treaty nations debated whether to adopt a Minerals Convention to permit

Managing Antarctica

Explorers laid the first claims to Antarctica in the early part of this century, and by 1943 the continent was staked out by 7 nations: Argentina, Australia, Chile, France, New Zealand, Norway, and the UK. It was another 30 years before an approach to international management was formulated in the Antarctic Treaty, signed by the claimants and 5 additional nations. Despite territorial disputes, the Treaty powers have since 1959 managed the Antarctic as an area devoted to peaceful purposes. Research is co-ordinated by the Scientific Committee on Antarctic Research (SCAR); regulatory measures include the 1980 Convention on the Conservation of Antarctic Marine Living Resources (CCAMLR), designed to manage krill. The 1988 Minerals Convention (CRAMRA) has not been adopted, but has been supplemented by the 1991 Madrid Protocol for the Protection of the Antarctic Environment, banning mining activities for 50 years. The map shows the scope of the Treaty and CCAMLR, and the potential effect of Exclusive Economic Zones (EEZs).

Legend:
- Antarctic Treaty Limit
- CCAMLR limit
- Research stations
- Potential island Exclusive Economic Zones (EEZs)
- Antarctic Convergence
- Krill distribution
- Krill concentration

Flags:
South Korea '90, Ecuador '90, Uruguay '85, France '59, Japan '59, New Zealand '59, Italy '87, Peru '89, Chile '59, China '85, Sweden '88, Norway '59, Belgium '59, S Africa '59, Australia '59, UK '59, Netherlands '90, Argentina '59, USA '59, Finland '90, Poland '77, Spain '88, USSR/CIS '59, Germany (E '87 W '81), Brazil '83, India '83, Czech '62, Denmark '65, Romania '71, Bulgaria '78, PNG '81, Hungary '84, Cuba '84

Map labels: Poland, Chile, USSR, Argentina, Chile, Chile, Argentina, Argentina, USA, UK, Atlantic Ocean, S Georgia (Arg and UK), S Sandwich (Arg and UK), Norway, Bo, Falklands/Malvinas (UK and Arg), Argentina, former USSR, Germany, S Africa, UK, Weddell Sea, Argentina, Chile, UK, Argentina, Bellinghausen Sea, USA, Pacific Ocean, Peter (Nor), USA, Ross Ice Shelf, New Zealand, Ross Sea, USA, former USSR, Unclaimed, Ball, Scott (NZ), New Zealand

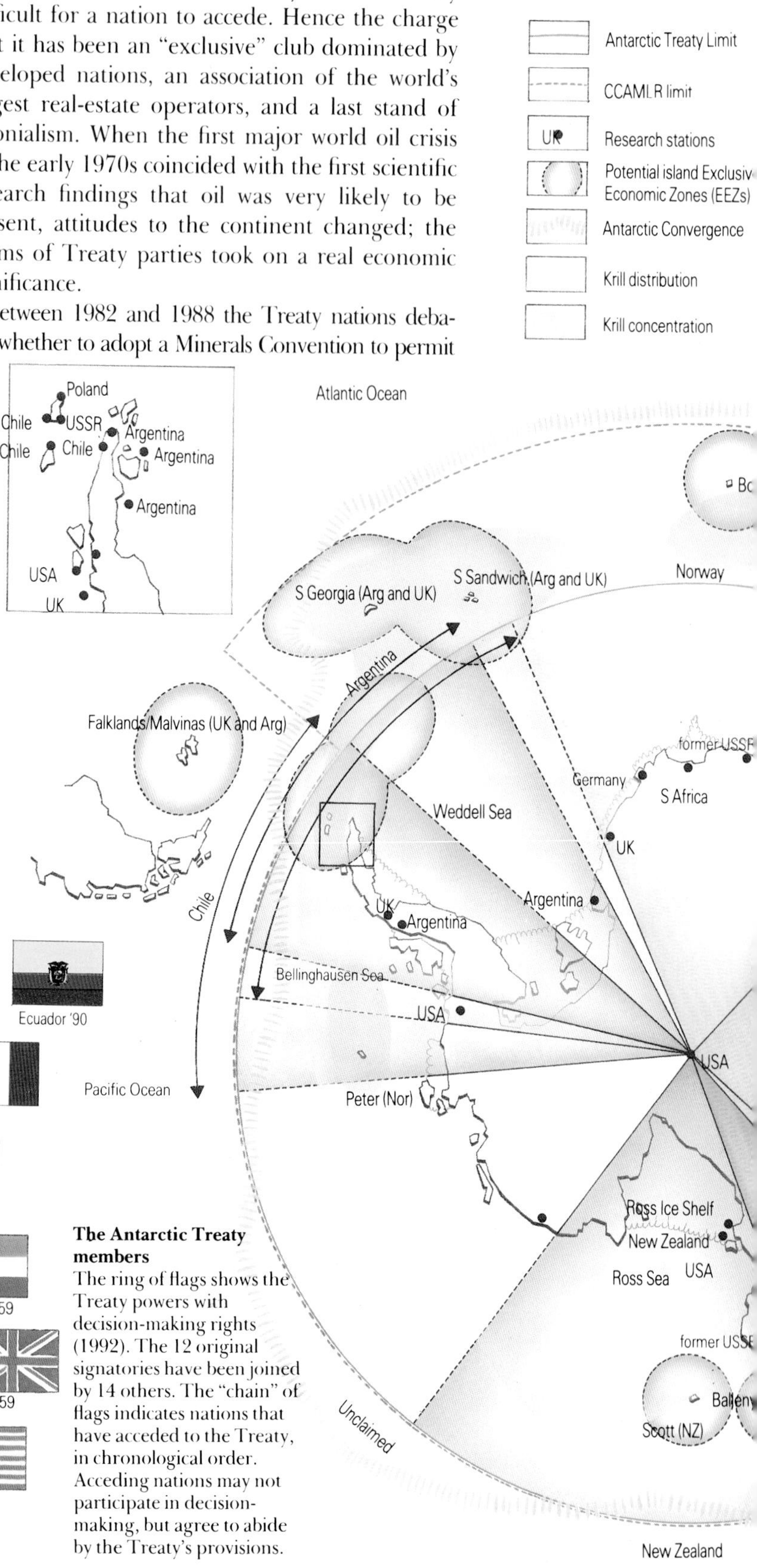

The Antarctic Treaty members

The ring of flags shows the Treaty powers with decision-making rights (1992). The 12 original signatories have been joined by 14 others. The "chain" of flags indicates nations that have acceded to the Treaty, in chronological order. Acceding nations may not participate in decision-making, but agree to abide by the Treaty's provisions.

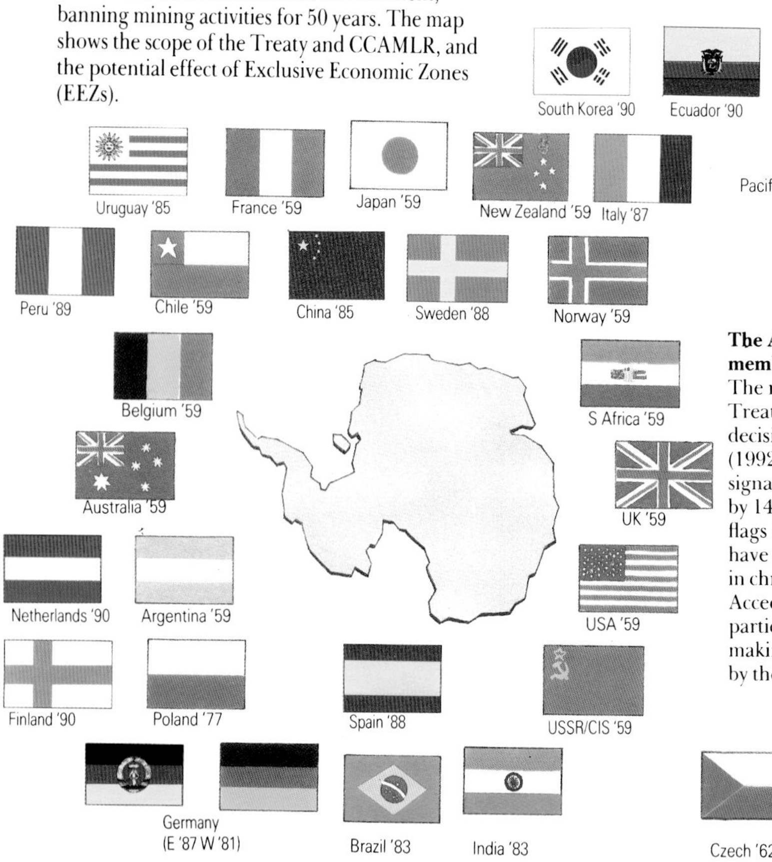

CAMLR
The Convention on the Conservation of Antarctic Marine Living Resources came into force in April 1982. CCAMLR is primarily concerned with controlling Antarctic fisheries, particularly krill, to ensure that catch levels neither jeopardize krill populations nor impede the recovery of the great whales. Despite this new "ecosystem" approach, conservationists maintain that CCAMLR's powers are inadequate.

mining for fossil fuels and other resources. Supporters of controlled mining argued that there was a "duty to develop"; with ever-increasing pressures on the Earth's natural resources it would not be in humanity's interest to deny access to Antarctica's mineral resources.

Conservationists became concerned about the fate of Antarctica in the 1970s and '80s and pressed for the establishment of the continent as a "World Park", dedicated to science and the preservation of its wilderness status. Mining, however well controlled, they argued, would inevitably harm the continent. As the Minerals Convention moved toward adoption, domestic political expediency moved Australia and France to refuse to sign. They were

followed by several other Treaty nations until the hardline pro-mining USA, Japan, and the UK changed their minds with the signing of the Madrid Environment Protocol in October 1991. According to the Protocol "any activity relating to minerals resources, other than scientific research, shall be prohibited" for 50 years. The protocol may then be reviewed and overturned if 75 percent of the nations with voting rights change their minds.

The agreement makes a firm foundation for the continent's future, further measures are vital to regulate pollution, control over-exploitation of fish and krill stocks, and establish a regional sanctuary for whales and seals. The Treaty system has demonstrated an ability to survive

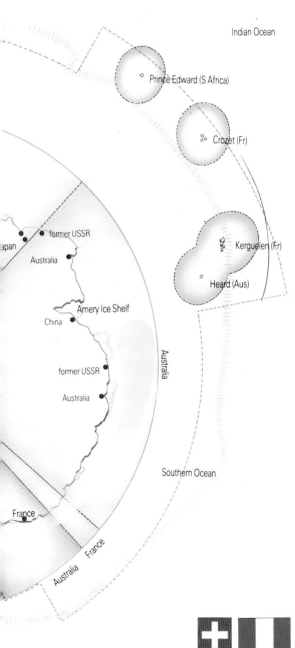

Indian Ocean

Prince Edward (S Africa)

Crozet (Fr)

Kerguelen (Fr)

Heard (Aus)

Antarctica

former USSR

Japan

Australia

Amery Ice Shelf

China

former USSR

Australia

Australia

France

Australia

France

Australia

France

Southern Ocean

1 billion barrels of oil

North Sea

Alaska

Antarctica's hidden wealth
The first direct evidence of Antarctica's huge oil potential came in 1973, when drilling on the continental shelf of the Ross Sea discovered traces of gaseous hydrocarbons, an initial indicator of the possible presence of oil and gas. A secretly-commissioned US government report later estimated that the combined potential of the Ross, Weddel, and Bellinghausen

Seas were some 45 billion barrels of oil (around 15 billion exploitable barrels) and 30 trillion cubic metres of natural gas. Compared to proven oil reserves elsewhere, such as the 8 billion barrels of the North Slope oilfield of Alaska, this is a considerable store. Despite the half-century mining ban, the continent will still be looked upon by some as a source of 21st century minerals.

Antarctic science
Scientists from all over the globe brave the harsh conditions of Antarctica for the chance to work in a unique, pristine environment. Antarctic ice contains the history of the atmosphere, essential for improving our knowledge of global climate. In 1985, British Antarctic scientist Joe Farman was the first to detect the "hole" in the ozone layer.

Global regime
Even before the Treaty, developing countries were pressing for a global regime for Antarctica. A proposal in the 1950s by India was withdrawn after opposition from future Treaty nations; but the matter has been repeatedly raised. In 1982-3, Malaysia asked the General Assembly to put Antarctica on the agenda. To counter criticism, the Treaty powers responded by inviting acceding states to observe their meetings, and agreeing to release more of their internal documents.

Switzerland '90 Guatemala '91

e '87   North Korea '87   Austria '87   Canada '88   Colombia '89

and contribute to change and may be a useful model for international co-operation in a wider context.

From high seas to managed seas

Laws work pretty well for over 40 percent of the oceans. For the rest – the waters beyond either the Exclusive Economic Zones (EEZs) or the continental shelves – the law of the jungle reigns, at least for those countries that are not party to the Law of the Sea Convention.

How have we reached this situation? Nations used to exercise exclusive control only over their "territorial sea", being a narrow coastal zone about three nautical miles wide (roughly the sector that could be covered by cannon fire from land). All the rest were subject to the doctrine of the "freedom of the high seas". This tradition persisted, with minor amendments such as an extension of the territorial sea to 12 miles (and occasionally to 200 miles), until the 1960s, when it became plain that advancing technology was allowing certain nations to take an undue share of the ocean's wealth.

In the 1960s, a philosophy was enunciated that the ocean (or at the very least its deep sea beds) should become a common heritage of humankind, to be managed by an international body such as the United Nations. This "universalist" spirit stimulated the convening of the Third Law of the Sea Conference (UNCLOS), which eventually ran from 1973 to 1982. The aim was to develop a single all-embracing treaty, covering all issues, including fisheries, navigation, continental shelves, the deep sea bed, scientific research, and pollution of the marine environment. The result was to be a "package deal" that would establish a world order for whatever uses were to be made of the oceans.

Alas for high hopes. The 160-plus participating nations have failed to agree on anything near a comprehensive treaty backed by international law and enforced by marine tribunals. The concept of a common heritage needing international management now survives only for the deep sea bed. Worse still, while 161 participating nations have signed the final document with its 320 articles, a number of others have declined, including the United States. Without all the leading maritime nations the Treaty will remain a broken-backed affair. In any case, of the many nations that appear to accept the Treaty in principle, too few have ratified it to bring it into force.

But not all is lost. We still have a large number of international treaties dealing with separate marine issues. The International Maritime Organisation, for example, does a fair job with navigation and the control of pollution from ships, while UNEP's Regional Seas Programme has made a sound start on the clean-up of specific areas.

UNCLOS has enabled us to progress a long way beyond the hopelessly confused situation of the 1960s. Moreover, it is inspiring us to draw upon a large body of customary laws and established

The laws of the sea

Our view of the legal status of the ocean is turning full circle. Once it was a boundless, two-dimensional expanse, belonging to no-one. Now we perceive it as a finite, three-dimensional resource, which should belong to all. Over the centuries, this traditional "freedom of the seas" has been gradually encroached on by national claims, fishing agreements, and a growing body of customary and international laws. In the last 50 years, however, a new principle has emerged to dominate conference tables. Like a beacon, the idea of "common heritage" steered the long deliberations of UNCLOS III. They fell short of the goal, but still achieved a new "written constitution" for the ocean.

Mare Liberum 1609
The Dutchman, Grotius, first proclaimed the "freedom of the seas" in keeping with the exploratory spirit of the age. Territorial waters were limited to about 3 nautical miles from land.

19th century agreements
In 1839 Belgium ignored an Anglo-French Oyster Beds treaty, demonstrating the need for multilateral support. The first international convention on the Policing of the North Sea Fisheries beyond Territorial Waters was signed in 1882.

Regulating exploitation
In 1893, the US tried to control exploitation of fur seals, but an international tribunal ruled this illegal. The Convention on North Pacific Fur Seals (1911) was the first international agreement on stock regulation.

The League of Nations Conference 1930
The first major international conference on the law of the sea raised two issues destined to dominate conferences for 50 years: territorial sea limits and "common heritage". An enlightened delegate, Snr Suarez, proposed that living ocean resources should be viewed as a common patrimony.

Fishery commissions 1930s and 40s
Commissions set up to regulate fisheries proved politically ineffective and unable to implement scientific advice. The NE and NW Atlantic Commissions, for example did not cover all states fishing the area, and most such bodies were slow to appreciate the importance of quotas.

The Achilles heel

The UNCLOS III Convention is presented to the world nations as an "all or nothing" package, a feature which is proving to be the "Achilles heel" of the new Law of the Sea. When a country like the US strongly objects to one UNCLOS recommendation, such as an international regime for the sea beds, it is obliged to reject the entire treaty. Though many states have now signed the treaty, most will spend a long time deliberating before ratifying – it needs 60 ratifications, but as yet has 49. Because full implementation of the Law of the Sea treaty is still some way off, a number of states press for interim measures, and some of these contradict the spirit of the new constitution. Certain industrialized countries have enacted unilateral legislation to enable national licensing of sea-bed mining, though this directly conflicts with the UNCLOS recommendations for international development of sea-bed resources.

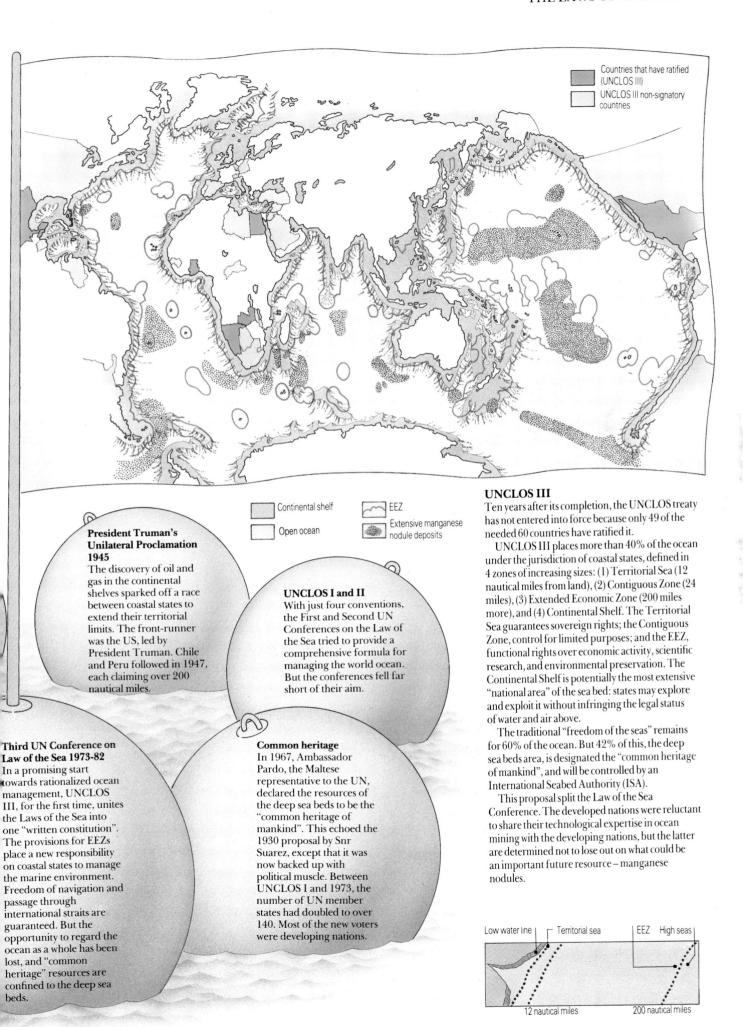

Countries that have ratified (UNCLOS III)

UNCLOS III non-signatory countries

Continental shelf

Open ocean

EEZ

Extensive manganese nodule deposits

President Truman's Unilateral Proclamation 1945

The discovery of oil and gas in the continental shelves sparked off a race between coastal states to extend their territorial limits. The front-runner was the US, led by President Truman. Chile and Peru followed in 1947, each claiming over 200 nautical miles.

UNCLOS I and II

With just four conventions, the First and Second UN Conferences on the Law of the Sea tried to provide a comprehensive formula for managing the world ocean. But the conferences fell far short of their aim.

Third UN Conference on Law of the Sea 1973-82

In a promising start towards rationalized ocean management, UNCLOS III, for the first time, unites the Laws of the Sea into one "written constitution". The provisions for EEZs place a new responsibility on coastal states to manage the marine environment. Freedom of navigation and passage through international straits are guaranteed. But the opportunity to regard the ocean as a whole has been lost, and "common heritage" resources are confined to the deep sea beds.

Common heritage

In 1967, Ambassador Pardo, the Maltese representative to the UN, declared the resources of the deep sea beds to be the "common heritage of mankind". This echoed the 1930 proposal by Snr Suarez, except that it was now backed up with political muscle. Between UNCLOS I and 1973, the number of UN member states had doubled to over 140. Most of the new voters were developing nations.

UNCLOS III

Ten years after its completion, the UNCLOS treaty has not entered into force because only 49 of the needed 60 countries have ratified it.

UNCLOS III places more than 40% of the ocean under the jurisdiction of coastal states, defined in 4 zones of increasing sizes: (1) Territorial Sea (12 nautical miles from land), (2) Contiguous Zone (24 miles), (3) Extended Economic Zone (200 miles more), and (4) Continental Shelf. The Territorial Sea guarantees sovereign rights; the Contiguous Zone, control for limited purposes; and the EEZ, functional rights over economic activity, scientific research, and environmental preservation. The Continental Shelf is potentially the most extensive "national area" of the sea bed: states may explore and exploit it without infringing the legal status of water and air above.

The traditional "freedom of the seas" remains for 60% of the ocean. But 42% of this, the deep sea beds area, is designated the "common heritage of mankind", and will be controlled by an International Seabed Authority (ISA).

This proposal split the Law of the Sea Conference. The developed nations were reluctant to share their technological expertise in ocean mining with the developing nations, but the latter are determined not to lose out on what could be an important future resource – manganese nodules.

Low water line | Territorial sea | EEZ | High seas

12 nautical miles | 200 nautical miles

treaties. It serves to reinforce many existing conventions and organizations. Finally, it provides a starting point from which we can develop new rules, regulations, and practices. We are much better off with UNCLOS than without it.

A testing ground for planet management

The state of the world ocean is a sensitive litmus test of our developing skills in planet management. Focus down almost anywhere in the 360 million square kilometres of the world ocean and you will find some sign of humanity's presence and impact, however slight. Move in close to the major centres of population, and the symptoms are all too apparent.

Year by year, our influence becomes more pervasive. New technologies enable us to reach ever further into the ocean, while industrial pollutants highlight its complexity as a resource by turning up thousands of kilometres from human habitation. As we come to grips with the multiple dimensions of the ocean ecosystem, it is increasingly clear that, while it may sometimes prove more robust than ecosystems on land, our errors in ocean management can be even harder to remedy.

Clearly, we need an agreed strategy for managing our ocean resources, and for developing them as appropriate. The new Law of the Sea Convention reserves three-fifths of the ocean as a common resource, where the traditional freedoms of the seas still apply. It also recognizes the wealth of the ocean floor, accounting for 42 percent of Earth's surface, as a common planetary inheritance of all humankind. Simultaneously it places two-fifths of the ocean under the control of individual nations, together with the relevant sea-bed resources. This may fly in the face of common-heritage values, but it is true that national ownership rights can promote a sense of responsibility.

There are also real dangers in this approach, however. The ocean cannot readily be sliced up like some gigantic pie: it is a continuous ecosystem and many of its components, like the major currents, flow to the horizons with sublime indifference to human politics. Better, say some scientists, to view the ocean as a whole.

The ocean, in short, is an indivisible resource, whereas the nation state tends to be a divisive force at a global level. However well it performs as a management tool to serve local needs, the nation state, virtually by definition, is incapable of matching up to the collective needs of the community of nations, let alone to the broader needs and interests of present and future generations.

How, then, can we re-establish a concept of the ocean as the common heritage of humankind, to be administered as a necessarily shared resource? How can governments be persuaded to approach joint problems in a spirit of joint endeavour? We cannot refuse the challenge. For, without doubt, the world ocean will be a critical proving-ground if we are to succeed in our new role as planet caretakers.

Future ocean

The migratory routes of the Arctic tern, humpback whale, and the Pacific salmon highlight the way marine life links far-flung corners of the world ocean. The map, right, also shows some early building blocks for ocean management. Intense shipping activity in the English Channel, for example, has resulted in traffic-management schemes designed to cut shipping losses and pollution. Nation states which are at odds ideologically have come together to work on management schemes for the Caribbean and for such vulnerable, enclosed seas as the Baltic and Mediterranean. Pressures on the world ocean will grow as we seek to harness both its renewable and non-renewable resources, from mackerel to manganese nodules. Some marine ecosystems, like the Great Barrier Reef, can be designated as international conservation areas. Others, like the Antarctic Ocean, will demand new, multi-species management strategies if they are to produce a sustainable yield.

Cleaning up the Caribbean

The Caribbean may not yet be as polluted as some other semi-enclosed seas, but the prevailing east-to-west currents trap pollutants against the coasts. The Mississippi River injects pollution from US industries, cities, and farms. Pesticide residues wash down from banana, cotton, and sugar plantations. Industry in Puerto Rico and Trinidad discharges effluent direct into the sea, while only 10% of the sewage produced in the region is treated. There are clear implications for Caribbean tourism and fisheries. While the Caribbean Action Plan, focusing on 66 environment and development projects, has been slow to take off, the threat to the Caribbean has induced 28 nations, many with divergent political views, to adopt new treaties to protect and develop the marine environment.

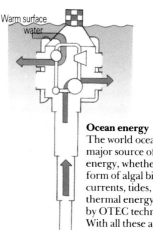

Ocean energy
The world ocean will be a major source of renewable energy, whether in the form of algal biomass, currents, tides, waves or the thermal energy exploited by OTEC technology, left. With all these activities, as with the exploitation of manganese nodules (in the Pacific) and metalliferous muds (in the Red Sea), or with offshore oil production, environmental impact assessment will be a key management tool.

Tern migratory route

Humback whale migratory route

Atlantic salmon

Pacific salmon

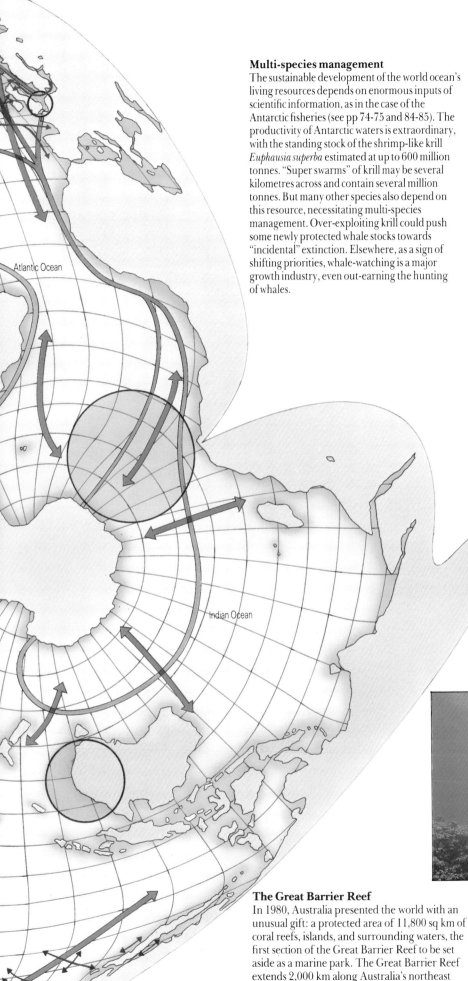

Multi-species management

The sustainable development of the world ocean's living resources depends on enormous inputs of scientific information, as in the case of the Antarctic fisheries (see pp 74-75 and 84-85). The productivity of Antarctic waters is extraordinary, with the standing stock of the shrimp-like krill *Euphausia superba* estimated at up to 600 million tonnes. "Super swarms" of krill may be several kilometres across and contain several million tonnes. But many other species also depend on this resource, necessitating multi-species management. Over-exploiting krill could push some newly protected whale stocks towards "incidental" extinction. Elsewhere, as a sign of shifting priorities, whale-watching is a major growth industry, even out-earning the hunting of whales.

Policing the sea-lanes

In 1990, more than 60 ships were involved in collisions in UK waters alone. Worldwide, 175 ships were lost that year. Traffic management schemes have been introduced in accident-prone spots such as the English Channel (see radar picture, above) but accidents still happen. Oil spillages at sea are a chronic problem, but EC satellite experts have developed methods of detecting slicks from the air, before the offending tankers have time to sail away. Similarly, satellite data helps to show which countries are to blame for the toxic chemicals, radioactive pollution, and untreated sewage that join oil slicks in the North Sea and other polluted regions. As well as policing, satellite research is yielding data essential to better understanding of the marine environment. Europe's ERS-1 satellite, launched in 1991, surveys the oceans and sea ice, providing data for climate modelling and to help answer questions about ocean-atmosphere interaction, such as El Niño, and ocean circulation, as well as detecting pollution.

Atlantic Ocean

Indian Ocean

The Great Barrier Reef

In 1980, Australia presented the world with an unusual gift: a protected area of 11,800 sq km of coral reefs, islands, and surrounding waters, the first section of the Great Barrier Reef to be set aside as a marine park. The Great Barrier Reef extends 2,000 km along Australia's northeast coast and covers 207,000 sq km. The reef supports some 400 species of coral and an estimated 1,500 fish species. Much of the reef has been under pressure for some time, starting with phosphate mining and turtle-meat industries, and followed by sea-bed mining, oil drilling, and tourism. Deforestation and intensive farming on the mainland also impact on the reef, but the embryonic marine park is a portent of better things to come.

ELEMENTS

Introduced by James E. Lovelock

Originator of the Gaia hypothesis

To my mind, the outstanding spin-off from space research is not new technology, but that for the first time we have been able to look at Earth from the outside - and have been stimulated to ask new questions.

Having worked on the Martian atmosphere, looking for signs of life, I later switched back to Earth and concentrated on the nature of our own atmosphere. This work resulted in the Gaia hypothesis, which suggests that the planet is, in a real sense, alive – a superorganism comprising all life on Earth and its environment, and one capable of manipulating conditions to suit its own needs.

It was a long way from devising plausible life-detection experiments to the hypothesis that Earth's atmosphere, and the cycling of elements within it, is actively maintained and regulated by the biosphere. But we found that the chemistry of the atmosphere violates the rules of steady state chemistry. Disequilibria of the scale that we observed (p. 11) suggest that the atmosphere is not merely a biological product, as oxygen often is, but more probably a biological construction – like a cat's fur, or a bird's feathers; an extension of a living system, designed to maintain a chosen environment.

We defined Gaia as a complex entity involving the Earth's biosphere, atmosphere, oceans, and soil, the totality constituting a "feedback" or "cybernetic" system which seeks an optimal physical and chemical environment for life on this planet. Gaia remains a hypothesis, but much evidence suggests that many elements of this system act as the hypothesis predicts.

Meanwhile, the human species, aided by the industries at its command, has significantly altered some of the planet's major chemical cycles. We have increased the carbon cycle by 20 percent, the nitrogen cycle by 50 percent, and the sulphur cycle by over 100 percent. We have increased the flow of toxins into air, water, and food chains. We have reduced the planet's green cover, while our factory outpourings reach the upper atmosphere, and far into the oceans. And as our numbers grow, so will these perturbations.

If the biosphere does control the atmosphere, the control system is unlikely to be easily disturbed. Nevertheless, we shall have to tread warily to avoid the cybernetic disasters of runaway positive feedback or of sustained oscillation between two or more undesirable states. We could wake one morning to find that we have landed ourselves with the lifelong task of planetary maintenance engineering. Then, at last, we should be riding in that strange contraption, "Spaceship Earth".

I hope and believe that we shall achieve a sensible and economic technology which is more in harmony with Gaia. We are more likely to achieve this goal by retaining and modifying technology than by a reactionary "back to nature" campaign. A high level of technology is by no means always energy-dependent. Think of the bicycle, the hang glider, modern sailing craft, or a mini-computer performing in minutes human-years of calculation, yet using less electricity than a light-bulb. The elemental resources of Gaia – energy, water, air, and climate – are so abundant and self-renewing as to make us potential millionaires. And potentially, at least, we have the intelligence to learn how to work with Gaia, rather than undermining her.

James E. Lovelock.

THE ELEMENTAL POTENTIAL

We all know that energy supplies us with heat and light. It carries us about. It makes our machines function. Indeed, it sustains our entire economic system. So much is obvious. But energy also fuels our lifestyles in less apparent ways. It not only cooks our food, it grows it. Without massive energy subsidies, in the form of fertilizers and pesticides, our agriculture would be much less productive. When we sit down to a meal, we are in effect, eating oil and coal.

By exploiting the planet's fossil fuels, which represent the biologically stored solar energy of millennia, we have been able to build up and power an industrial civilization which is radically different, both in nature and scale, from earlier civilizations. A single tonne of oil generates energy equivalent to the energy output of 660 horses over 24 hours.

But this new energy wealth is not shared equitably: an average American consumes 280 times as much energy as the average Ethiopian. Without sufficient, affordable energy, the world's developing nations will find that critical development programmes are stillborn. Given the constraints on fossil fuels and the problems associated with nuclear power (see p. 122), there has been growing interest in harnessing the largest nuclear reactor in our solar system: the sun.

The sun radiates more energy into space than 200,000 million million of our largest existing commercial nuclear reactors – although the Earth receives only one part in a billion of this vast output. Even so, our annual solar energy budget is roughly equivalent to 500,000 billion barrels of oil, or to perhaps a thousand times the world's proven oil reserves in the late 1980s. At any given moment, incoming solar energy striking the Earth's atmosphere is equivalent to some 40,000 one-bar electric fires burning constantly for every man, woman, and child of the Earth's human population.

By the late 1980s, however, all the world's solar collectors were yielding energy equivalent to a mere 0.014 percent of the total oil consumption. Exploitation of the energy trapped by plants (biomass energy) was much more significant, supplying at least 15 percent of the world's energy budget. The potential of biomass is, as yet, hardly tapped, though many possibilities exist for its future exploitation (see pp 128-9).

No-one doubts that the non-renewable energy sources, predominantly the fossil fuels and nuclear power, will continue to make a solid contribution to

The global powerhouse

The sun's energy is the mainspring of all life on Earth. Without it the oceans would freeze. Temperatures on the planet's surface would drop almost to absolute zero (-273°C). Solar energy drives the great geophysical and geochemical cycles that sustain life, among them the water cycle, the oxygen cycle, the carbon cycle, and the climate. The sun provides our food, by photosynthesis, and most of our fuel. Fossil fuels are simply stored solar energy – the product of photosynthesis millions of years in the past. Over 99% of the energy flow in and out of the Earth's surface results from solar radiation. Heat from the Earth's core and the gravitational forces of sun and moon supply the rest. Solar radiation striking the Earth is equivalent to all the energy from 173 million large power stations going full blast all day, every day. But 30% of this energy is reflected away back into space. Most of the rest either warms the air, sea, and land (47%) – or fuels evaporation and the water cycle (23%).

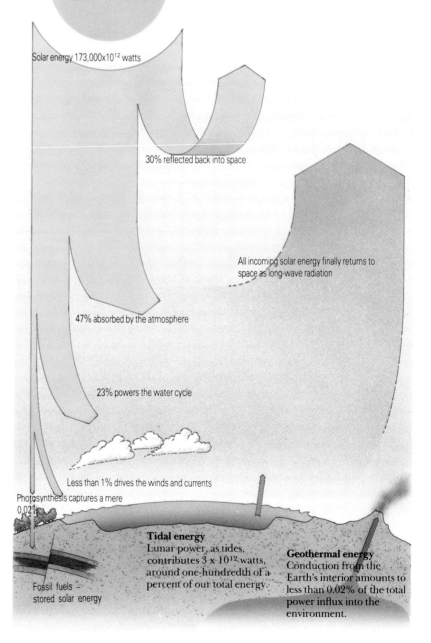

Solar energy 173,000x10^{12} watts

30% reflected back into space

All incoming solar energy finally returns to space as long-wave radiation

47% absorbed by the atmosphere

23% powers the water cycle

Less than 1% drives the winds and currents

Photosynthesis captures a mere 0.02%

Tidal energy
Lunar power, as tides, contributes 3 x 10^{12} watts, around one-hundredth of a percent of our total energy.

Geothermal energy
Conduction from the Earth's interior amounts to less than 0.02% of the total power influx into the environment.

Fossil fuels – stored solar energy

The human energy budget

Prior to the Industrial Revolution, the sun was the only source of energy widely available to humankind. It provided food for muscles – a fit person generates in a day's work the equivalent of a single-bar electric fire used for an hour. Wood has been used since prehistory. Sails to use sun-created wind were first raised 5,000 years ago, windmills 2,000 years later, and water-wheels, which use water raised by the sun, 2,000 years after that. Coal came into general use just 300 years ago, and oil and gas only in the last 100 years. Not until the 20th century did non-solar energy arrive in the forms of geothermal and nuclear power. The natural flows of energy that have been used for millenia are known as *renewable* sources. The amount of energy fossil fuels can supply is ultimately limited by geology. These are known as *non-renewable* sources. The pie-chart, below, shows the current contribution of each energy source and how this might change by the year 2000. As global energy demand grows and non-renewable sources begin to run out, so attention is turning back to the renewables.

Non-renewables

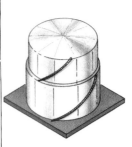

Oil
Oil is the world's largest energy source – supplying 40% of our energy. Oil consumption fell in the early 1980s, and has since been growing by only 2% annually; consumption of other fuels is increasing much faster.

Coal
Coal is the most plentiful fossil fuel. Just three regions, China, the former USSR, and the US, own 66% of the world's reserves. Coal use has been growing by some 3% per year, but concerns about the consequent problems of acid rain and greenhouse gas emissions could constrain its growth.

Natural gas
Natural gas accounts for 20% of the world's current annual energy budget. It is expected to be the fastest growing energy source, as many countries plan to use it to reduce dependence on oil and lessen the environmental problems associated with fossil fuels. Combustion of natural gas emits only about half the carbon dioxide of an equivalent amount of coal.

Renewables

2000 1990   World energy supplies

Biomass
Biomass is plant or animal matter that can be converted into fuel. It currently provides about 15% of the world's energy, and it is the main fuel for half the world's population, mostly in the form of wood. In some poor countries it accounts for 90% of all fuel burnt.

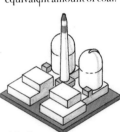

Nuclear power
Nuclear power, once hailed as the answer to all the world's energy problems, now supplies 5% of global energy production. There are currently 426 plants operating in 30 countries and perhaps less than 50 under construction. Sharply rising costs and the loss of public confidence have resulted in few new reactors being ordered.

Hydropower
Falling water generates 20% of the world's electricity, about 7% of total energy demand. It is still greatly underexploited (only 7% of the developing world's potential is harnessed), although large dams often carry heavy environmental and social costs.

Solar
The sun already contributes significantly to the energy needs of buildings through windows and walls. Features to maximize this energy have been incorporated in countless new buildings. Solar collectors are widely used for domestic hot water in many countries: including two-thirds of homes in Israel. Cells to convert sunlight to electricity are increasingly viable as an energy source as manufacturing becomes cheaper.

Power from the sea
Ocean power comes in four main forms: wave power, tidal power, current power, and ocean thermal energy conversion – which exploits temperature differences between the surface and depths. The ultimate energy potential is massive, but only a small fraction is likely to be harnessed.

Geothermal
The Earth's temperature rises 1°C every 30 m down, more in geologically active areas. Geothermal power makes use of this heat, either directly as hot water or to produce electricity. There are geothermal plants in 27 countries, currently producing the equivalent of 50 million MW of electricity per year.

Wind
Winds are caused by uneven heating of the Earth's surface. The power of the wind is proportional to the cube of windspeed, so doubling the speed increases the power 8 times. Windmills can be used either to generate electricity or to do mechanical work.

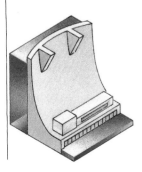
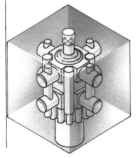

our energy needs. But we shall also need to develop many of the more promising renewable energy options shown here if we are to ensure that the world has enough energy to take us through the 21st century.

Energy-rich, energy-poor

The grossly unequal endowments of non-renewable and renewable energy inherited by virtue of geology, biological history, or geographical location mean that there can be no single energy solution for the problems which have emerged since the first oil shock engineered by the Organization of Petroleum Exporting Countries (OPEC). Coal cannot be *the* alternative to oil, any more than nuclear fission or renewable energy can.

Those who have rehearsed the transition away from oil, in whichever direction, have found that many of the glib formulas which surfaced after the OPEC breakthrough simply do not work. Big may not always be best, but small is certainly not uniformly beautiful. The growing recognition that the only way forward is to consider all potential sources of energy, together with all the component needs which go to make up the total energy picture, has led some countries to focus more on what page energy they

Energy units key
Energy (measured in joules and tonnes of oil equivalent) is the capacity to do work. Power (measured in watts) is the rate of energy delivered. 1 joule is work done when 1 kg is moved 1 m with an acceleration of 1 m per second per second. 1 watt = 1 joule/second. 1 kilowatt-hour (kWh) = 1,000 watts for 1 hour. 1 mtoe (million tonnes of oil equivalent) is roughly equal to 12 billion kWh, or 1.5 million tonnes of coal.

The energy store

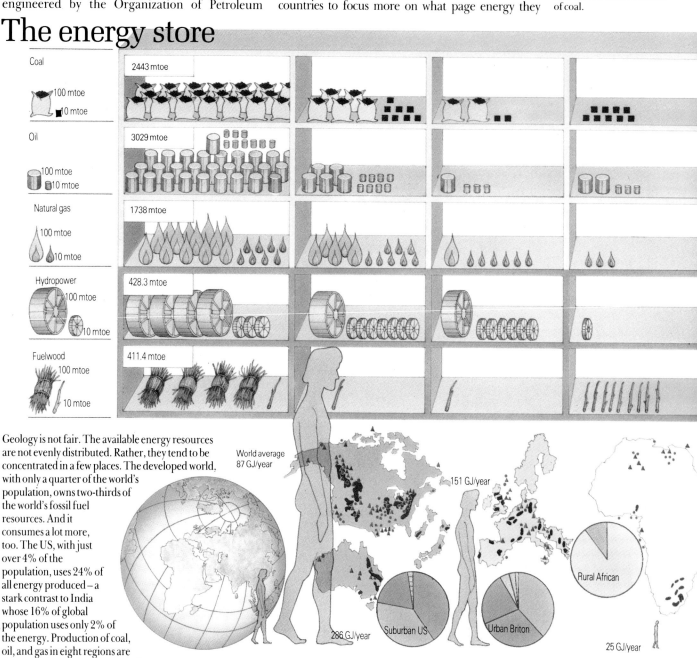

Coal
100 mtoe
10 mtoe
2443 mtoe

Oil
100 mtoe
10 mtoe
3029 mtoe

Natural gas
100 mtoe
10 mtoe
1738 mtoe

Hydropower
100 mtoe
10 mtoe
428.3 mtoe

Fuelwood
100 mtoe
10 mtoe
411.4 mtoe

World average 87 GJ/year

151 GJ/year

286 GJ/year

Suburban US

Urban Briton

Rural African

25 GJ/year

Geology is not fair. The available energy resources are not evenly distributed. Rather, they tend to be concentrated in a few places. The developed world, with only a quarter of the world's population, owns two-thirds of the world's fossil fuel resources. And it consumes a lot more, too. The US, with just over 4% of the population, uses 24% of all energy produced – a stark contrast to India whose 16% of global population uses only 2% of the energy. Production of coal, oil, and gas in eight regions are shown on the top three shelves; the two renewables, hydropower and wood (bottom two shelves), are shown in terms of consumption only. The regional maps, right, show the location of fossil fuel resources. Areas such as Africa, India, and Latin America, with few fossil fuel reserves, must rely on fuelwood and expensive oil imports. The height of each figure reflects regional per capita energy consumption. The adjacent pie-charts compare energy sources.

N America, Oceania, Japan
This region has the highest per capita consumption, using twice as much as W Europe and 17 times that of S Asia. But, while the US has huge reserves, Japan must import more than 80% of its needs.

Western Europe
Even though W Europe has reduced its oil dependence since 1973, oil is still the main source of energy. North Sea reserves are unable to meet demands – W Europe imports more than half its supplies.

Africa
Apart from Libyan and Nigerian oil and South African coal, Africa is ill-supplied with fossil fuels. Biomass supplies 80% of the energy needs of an African villager. Often wood is burnt inefficiently.

are likely to need – and on the whole range of sources which might help meet that need.

Increased energy efficiency will help close the gap between supply and demand, but many countries will continue to show striking mismatches between energy they consume, let alone need, and energy they produce – or can potentially produce. Many developed nations owe their prosperity to domestic fossil fuel stocks which they are close to exhausting, if they have not already done so. Thanks to world trade, however, developed nations, with only one-quarter of the global population, continue to consume four-fifths of the global energy budget.

The leading fossil fuel, meanwhile, is still oil, which accounts for 42 percent of commercial energy worldwide – although its contribution could fall to 35 percent (of a larger total energy budget) by the year 2000. Thereafter, its contribution could tail away as it comes to be used more as a chemical feedstock than as a fuel. Natural gas could last one and a half times as long as oil. Coal, by contrast, is available in abundance; total reserves are estimated to be 220 times the amount we consume each year. But environmental problems, notably global warming, will prevent any easy transition back to coal. The need remains for cleaner energy sources.

Coal
Oil
Natural gas
Hydropower
Biomass
Commercial fuels
Draught animals
Human labour
Nuclear energy

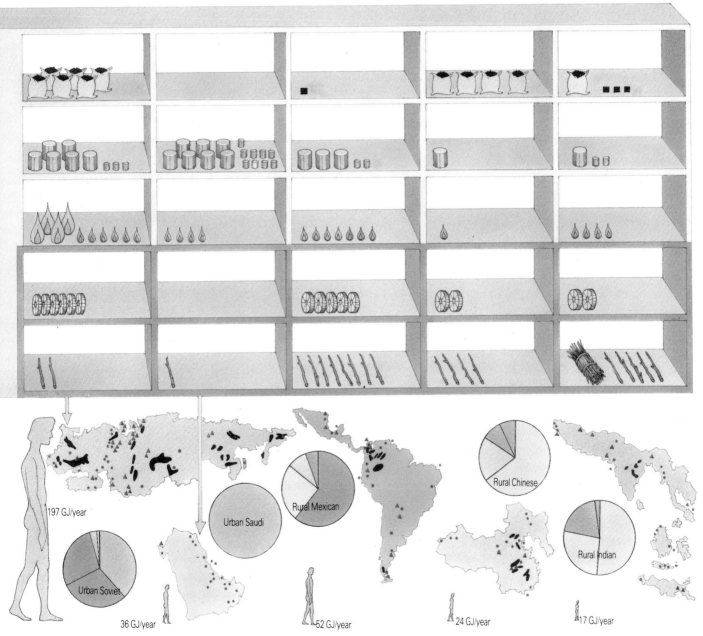

197 GJ/year

Urban Soviet

36 GJ/year

Urban Saudi

Rural Mexican

52 GJ/year

Rural Chinese

Rural Indian

Rural Chinese

24 GJ/year

17 GJ/year

Eastern Europe, former USSR
The former USSR is rich in energy and its resources are by no means fully explored. The world's largest oil producer, it has recently begun to export large quantities of natural gas.

Middle East
Over 65% of the world's known reserves of oil are located in the Middle East. Producing 10% of the world's energy, it consumes only 2% – although demand is likely to grow.

Latin America
Although poorly provided with reserves, Latin America is a net exporter of energy, thanks mainly to the Mexican and Venezualan oilfields. Fuelwood shortages can be acute in some countries.

China
China is the world's largest producer of coal. But so large is its population that its per capita consumption is still low. Animal and human power provide nearly a third of the energy used.

South Asia
This energy-poor region contains several of the bottom countries in terms of energy consumption per capita. Almost all the energy used is directed to the production and cooking of food.

Among the renewables, biomass and hydropower make the most significant contributions to the world energy budget. As with food and water, we have enough energy in total – but supply and distribution are out of phase with developing demand.

Our changing climate

Climate is an expression of the great interacting realms of atmosphere, land, and ocean. A region may enjoy a climate of hot dry summers and cold dry winters, with intervening seasons of warm moist conditions. Or it may suffer a climate of warm damp summers and mild wet winters, with nothing much better – as viewed by some people at least – in between. By contrast, weather is what we experi-

ence day to day, in the form of cloudy or blue skies, rain and humidity, wind and the like. So whereas weather can change from hour to hour and week to week, climate represents an average pattern over years, decades, and centuries.

Over lengthy periods, moreover, climate can change. A long-term shift of only 1°C is enough to trigger profound shifts, as witness the phenomenon of the Little Ice Age in Western Europe, which peaked in the late 17th century. A drop of 4°C is enough to bring on a full ice age, with swift results for the planetary habitat: at the onset of the last glaciation almost 11,000 years ago, advancing ice sheets eliminated huge forests of the northern hemisphere in just a century. As for more recent change,

The maritime effect
The winds, the Earth's rotation, and the placement of the continents cause the great ocean currents. Because they move huge masses of water, sometimes cold, sometimes warm, from region to region, they also influence the climate. Thus the UK, which is warmed by the North Atlantic Drift originating off Florida, has a milder climate than Labrador, which is on the same latitude, but is cooled by a current from the Arctic Ocean.

The climate asset

Five major factors shape climate: the sun's energy, the atmosphere, the Earth's shape and position in space, its rotation, and the oceans. Because the Earth is a sphere, air is warmed more at the Equator than at the poles. As the warm, moist air rises, it flows polewards, cooling and drying as it goes. On reaching latitude 30° it begins to sink, warm up, and flow back towards the Equator. These vertical flows of air on either side of the Equator, called "Hadley cells", are responsible for the belts of desert and arid land found around the Tropics of Cancer and Capricorn. Air, as wind, flows from areas of high pressure, associated with higher temperatures, towards areas of low pressure. The energy of the Earth's rotation, the "Coriolis force", bends the winds to give the characteristic patterns shown on the map, right. The diagram, below, illustrates the vegetational impact of rainfall and temperature. Where cold polar winds meet the warm westerlies of the mid-latitudes, the changeable weather typical of temperate regions is found. The movement of large masses of ocean water also influences climate since small differences in temperature absorb, or release, enormous quantities of heat. The map shows the seven main climatic types, the size of the circles indicating annual mean precipitation.

Latitude and temperature
The subsolar point, where the sun appears directly overhead, varies with the seasons between 23°N and 23°S. Here the sun's rays are perpendicular to the Earth's surface. With increasing latitudes, less sunshine is intercepted and average temperatures fall.

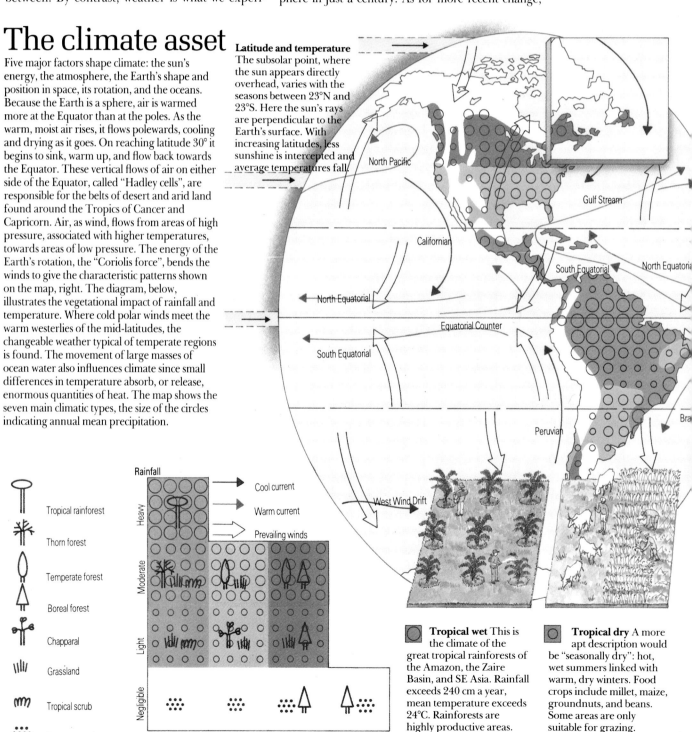

North Pacific

Gulf Stream

Californian

North Equatorial

South Equatorial

North Equatorial

Equatorial Counter

North Equatorial

South Equatorial

Peruvian

Bra

West Wind Drift

Rainfall

Cool current

Warm current

Prevailing winds

Heavy

Moderate

Light

Negligible

Tropical rainforest

Thorn forest

Temperate forest

Boreal forest

Chapparal

Grassland

Tropical scrub

Desert or tundra

Hot   Warm   Cool   Cold

Tropical wet This is the climate of the great tropical rainforests of the Amazon, the Zaire Basin, and SE Asia. Rainfall exceeds 240 cm a year, mean temperature exceeds 24°C. Rainforests are highly productive areas.

Tropical dry A more apt description would be "seasonally dry": hot, wet summers linked with warm, dry winters. Food crops include millet, maize, groundnuts, and beans. Some areas are only suitable for grazing.

what we now assume is "normal" is actually one of the warmest phases of the past 1,000 years. This warm phase should, instead, be regarded as approaching an extreme for sustained "good weather". In the northern-hemisphere land areas between 1881 and 1983, the warmest year was 1981 – while three of the winters since 1978 rank among the six warmest.

Climatic change, in short, is the norm. The one thing that we can be sure about with regard to future climate is that it will feature deep-seated shifts, even if we leave climate to get on with its own natural course – without disrupting it by, for example, contributing to carbon dioxide build-up in the atmosphere (see p. 110).

All this has major implications for our capacity to keep producing food. Throughout the world, climate is a critical factor in agriculture. During the late 1960s, the Sahel drought brought disaster to entire nations. In 1972, another drought inflicted such damage on the Soviet wheat crop that it helped to quadruple world prices within two years. In 1974, a delayed monsoon in India wrought havoc for millions of people. In 1975, pulses of cold air ravaged Brazil's coffee crop, causing inflationary upheavals in coffee prices around the world.

Conversely, of course, a stable climate, or rather a climate with slow and predictable changes, can be a tremendous asset. Just as much as energy or water, climate represents an "elemental" dimension to our

How land shapes climate
The rain-shadow effect of coastal mountains is well known. Warm, moist winds rise on the windward side of mountains. As they rise they expand and cool, causing the water vapour they carry to condense and fall as rain. The dry, cool air descends on the other side of the mountains. As it does so it is compressed and warmed. The resultant dry, warm wind creates a markedly different climate and vegetation, as in California's Sierra Nevada.

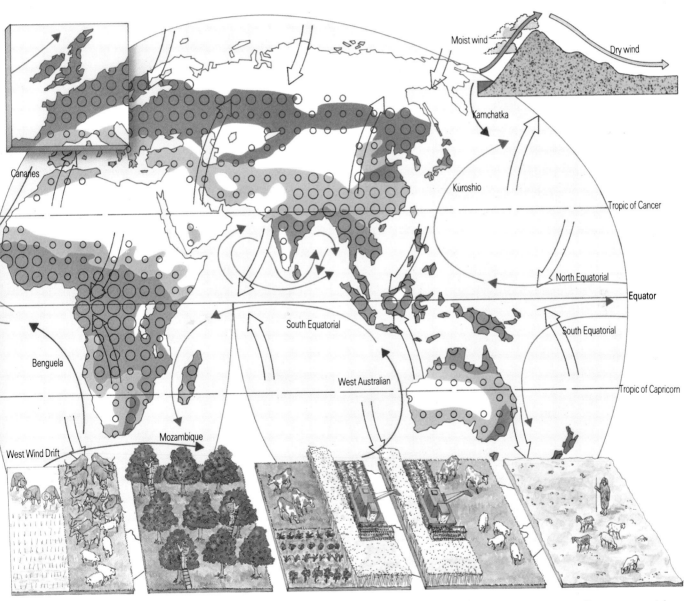

Warm humid The climate of southeastern US and much of China is warm and humid. It is a ideally suited to multiple cropping – the warm temperatures and regular rainfall enable crops to be grown all year round.

Warm dry The long, dry summers and mild, wet winters of the Mediterranean fostered Western civilization. With irrigation, many crops can be grown. But it can be an unforgiving climate if badly managed.

Cool humid The dominant role of north Europe had its foundations in the cool, humid climate. The constraints on agriculture are the low levels of sunlight, the cold winters, and short-term summer dryness.

Cool dry The great grasslands of the world, the Steppes of Russia, the Plains and Prairies of N America, the Pampas of S America, are cool and dry. These are the lands of corn and cattle. Rainfall is low and irregular in summer.

Desert or mountains Mountainous regions, with poor soil cover and low temperatures, are unsuited for most crops. High temperatures, low rainfall, and scant vegetation also make deserts unsuitable. Nomads roam here.

Earth that can basically enrich our lives. Provided we have time to adapt to long-term fluctuations of climate, we can surely hope to cope with a future of ever-changing climate.

The liquid of life

Water, water everywhere – but it is astonishing how little of it is directly usable. Only a small fraction is fresh, and 77.5 percent of this freshwater is "locked up" in ice-caps and glaciers. And of that tiny fraction of available freshwater, only 3 percent occurs in the atmosphere, rivers, and lakes, the rest being held in underground aquifers.

Even so, the "usable hydrosphere" contains more water than we are likely to need in the foreseeable future. The problem, as is so often the case with natural resources, is that water is not evenly distributed around the globe. Many people spend their time fighting floods, others go thirsty.

Perhaps more than we care to recognize, our lifestyles depend on the availability of freshwater. If, for whatever reason, our taps were to run dry, our household routines would collapse, our health would be at risk, factories would grind to a halt, and agriculture would be in dire straits. The entire fabric of our societies could begin to unravel. We may take freshwater for granted, in short, but we do so at our peril.

Certain sectors of our economies are particularly thirsty, with agricultural irrigation among the top consumers. The demand for water for irrigation in Africa and Latin America is likely to account for 30 percent of the growth in water consumption by the year 2000. In global terms, however, there is enough usable freshwater to meet the domestic, industrial, and irrigation needs of at least twice the current world population.

Although the sun evaporates almost half a million cubic kilometres of water from the seas each year, we should think of usable water as that proportion of evaporated water which ends up on land – and runs off into rivers and lakes. Here we are talking of less than 40,000 cubic kilometres of water, or less than one-tenth of the total originally evaporated from the seas.

This figure, furthermore, is only a year-round average, taking no account of seasonal and other fluctuations. The average low-water flow of inhabited continents is only about one-third of the total figure, with the remainder of the water disappearing in the form of floods before we can harness it.

So the "stable runoff" available for use is more like 14,000 cubic kilometres. This is still a phenomenal amount of water. Of course, as we become more "advanced", so we use ever-increasing amounts of water: whereas the absolute minimum a person needs for domestic use is 5 litres a day, with a more realistic figure around 20 litres, a developed world citizen consumes well over 100. When we add in industry, this total can jump to 500 litres, especially in northern cities. Nonetheless, improving

The freshwater reservoir

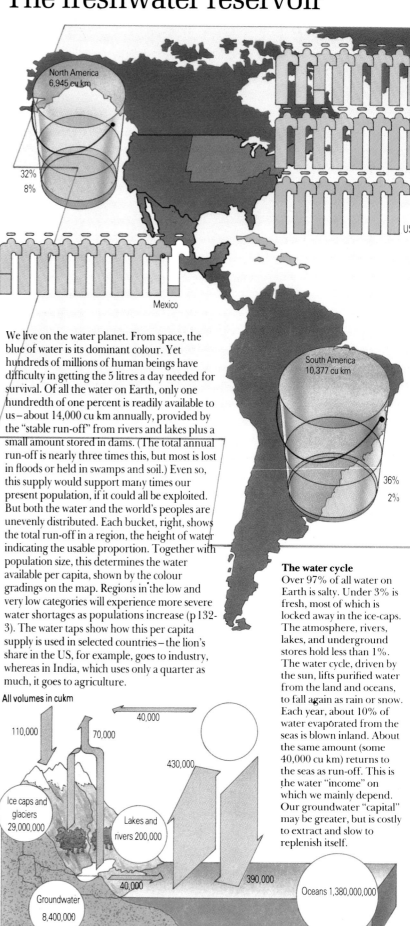

We live on the water planet. From space, the blue of water is its dominant colour. Yet hundreds of millions of human beings have difficulty in getting the 5 litres a day needed for survival. Of all the water on Earth, only one hundredth of one percent is readily available to us – about 14,000 cu km annually, provided by the "stable run-off" from rivers and lakes plus a small amount stored in dams. (The total annual run-off is nearly three times this, but most is lost in floods or held in swamps and soil.) Even so, this supply would support many times our present population, if it could all be exploited. But both the water and the world's peoples are unevenly distributed. Each bucket, right, shows the total run-off in a region, the height of water indicating the usable proportion. Together with population size, this determines the water available per capita, shown by the colour gradings on the map. Regions in the low and very low categories will experience more severe water shortages as populations increase (p 132-3). The water taps show how this per capita supply is used in selected countries – the lion's share in the US, for example, goes to industry, whereas in India, which uses only a quarter as much, it goes to agriculture.

All volumes in cukm

The water cycle

Over 97% of all water on Earth is salty. Under 3% is fresh, most of which is locked away in the ice-caps. The atmosphere, rivers, lakes, and underground stores hold less than 1%. The water cycle, driven by the sun, lifts purified water from the land and oceans, to fall again as rain or snow. Each year, about 10% of water evaporated from the seas is blown inland. About the same amount (some 40,000 cu km) returns to the seas as run-off. This is the water "income" on which we mainly depend. Our groundwater "capital" may be greater, but is costly to extract and slow to replenish itself.

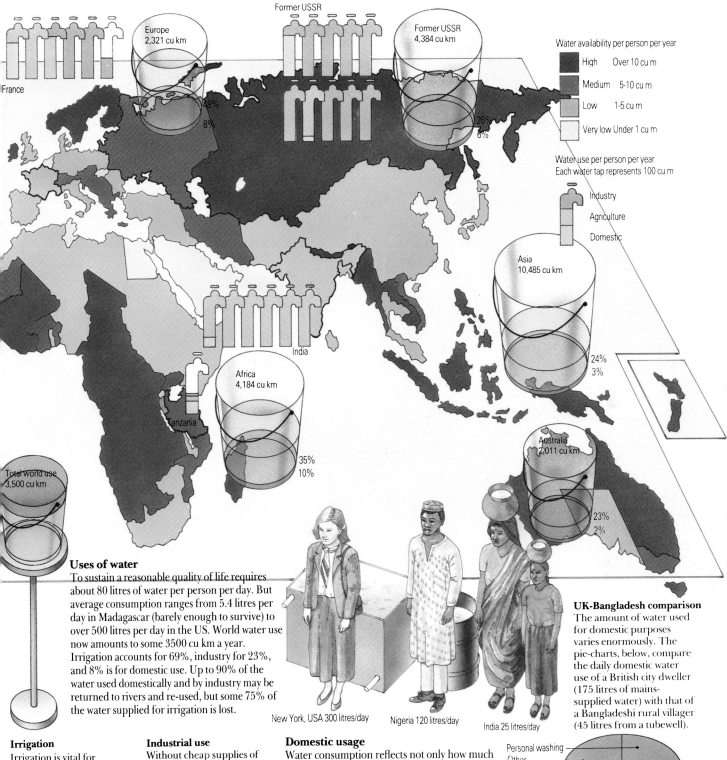

France

Europe
2,321 cu km

Former USSR

Former USSR
4,384 cu km

Water availability per person per year

■	High	Over 10 cu m
■	Medium	5-10 cu m
▨	Low	1-5 cu m
□	Very low	Under 1 cu m

Water use per person per year
Each water tap represents 100 cu m

Industry
Agriculture
Domestic

48%
8%

26%
6%

Asia
10,485 cu km

24%
3%

India

Africa
4,184 cu km

35%
10%

Total world use
3,500 cu km

Australia
2,011 cu km

23%
2%

Tanzania

Uses of water

To sustain a reasonable quality of life requires about 80 litres of water per person per day. But average consumption ranges from 5.4 litres per day in Madagascar (barely enough to survive) to over 500 litres per day in the US. World water use now amounts to some 3500 cu km a year. Irrigation accounts for 69%, industry for 23%, and 8% is for domestic use. Up to 90% of the water used domestically and by industry may be returned to rivers and re-used, but some 75% of the water supplied for irrigation is lost.

New York, USA 300 litres/day

Nigeria 120 litres/day

India 25 litres/day

UK-Bangladesh comparison

The amount of water used for domestic purposes varies enormously. The pie-charts, below, compare the daily domestic water use of a British city dweller (175 litres of mains-supplied water) with that of a Bangladeshi rural villager (45 litres from a tubewell).

Irrigation

Irrigation is vital for agriculture. 17% of the world's cultivated land is irrigated. But, since irrigated land is cropped more than once a year, its contribution to the world's harvests is much greater than 17%. It may contribute as much as 35% of the global harvest.

Industrial use

Without cheap supplies of water, industry would grind to a halt. Water is used as a coolant, a solvent, in washing applications, and for dilution of pollution. As shown, below, the amount of water needed for different products varies greatly. The production of plastics is particularly thirsty.

Domestic usage

Water consumption reflects not only how much is available per head, but also how difficult it is to fetch or how expensive to buy. In countries with piped water, consumption is much higher than in developing countries, where a 2-km walk to find water is not unusual. One cubic metre (1,000 litres) of poor-quality water in the Third World may cost $20, against 10 cents for the same volume of high-quality water in a rich country. An urban Third World household with one tap may use 40% more than its rural equivalent. The figures, above, show domestic use per capita for three "average" families.

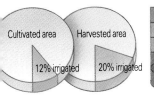

Cultivated area   Harvested area

12% irrigated   20% irrigated

Steel	150 cu m	
Paper (processing)	250 cu m	
Wheat		500 cu m
Plastic		up to 2,000 cu m

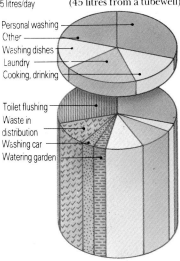

Personal washing
Other
Washing dishes
Laundry
Cooking, drinking

Toilet flushing
Waste in distribution
Washing car
Watering garden

water catchment, storage, supply, use, and recycling techniques should enable us to stretch our global water resource almost indefinitely – provided that we start soon.

Mining our mineral wealth

The planet's mineral wealth has been tapped since pre-history, its surface dotted with the workings and spoil-heaps of earlier mining activities. Where once we picked shallow excavations with antlers and other crude tools, today we use dynamite and massive machines to extract the minerals which underpin almost every aspect of our lives.

Fortunately, we still have sufficient stocks of most minerals to last us well beyond the year 2000. The "mineral showdowns" which were predicted in the wake of the first OPEC oil crisis now seem a rather more distant prospect – although the potential for such showdowns undoubtedly exists.

Industry depends heavily on some 80 minerals, including a number which, like aluminium and iron, are in relatively plentiful supply. True, a few countries will soon exhaust their domestic supplies of such materials, but there is enough elsewhere to go round, without the risk of cartelized cutbacks on production and supply.

A small number of minerals, however, qualify as strategic commodities. That is, they are critically important to industry, while being in relatively short supply. Chromium, for example, does more than put a shine on bumpers: it also contributes to irreplaceable alloys used in tool steel and in jet engines. Manganese is essential for high-grade steels. Platinum is used for catalytic converters, which are at the heart of many auto-emissim control systems, and for advanced communications equipment. Cobalt is crucial for high-strength, high-temperature alloys used in aerospace.

The US, the EC, and Japan import most of their supplies of these key commodities, primarily from politically volatile regions of central and southern Africa. Fortunately, however, new sources are becoming available as new discoveries are made – although some of these potential sources will be expensive to exploit.

Overall, three-quarters of the 80 materials are abundant enough to meet all our anticipated needs, or, where they are not, ready substitutes exist. But at least 18 minerals represent a rather thornier problem, even when greater recovery and recycling are taken into account: for example, lead, sulphur, tin, tungsten, and zinc.

Inevitably, the major mineral consumers will face rising prices for some key materials. Supply restrictions are less likely, however, since producer countries in the Third World are generally dependent on steady exports for foreign exchange. Zambia, for example, relies on minerals for over half its national income. Meanwhile, countries like the US have stockpiled enough reserves of the more critical minerals to last them for many years.

The mineral reserve

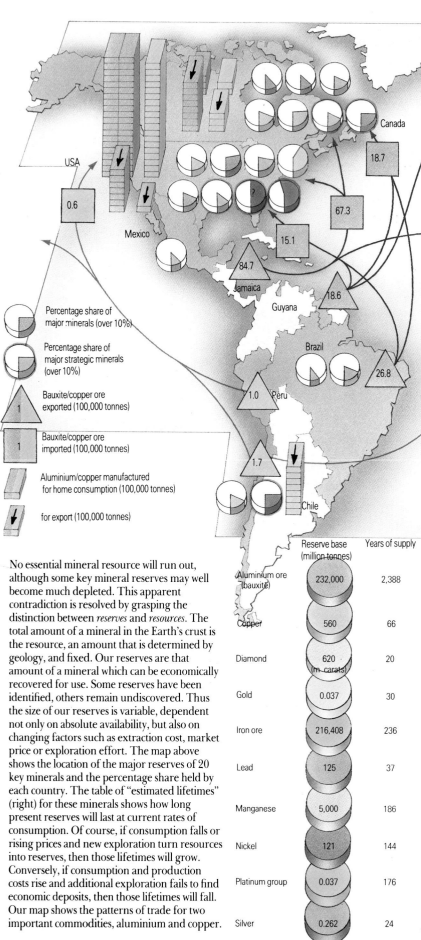

No essential mineral resource will run out, although some key mineral reserves may well become much depleted. This apparent contradiction is resolved by grasping the distinction between *reserves* and *resources*. The total amount of a mineral in the Earth's crust is the resource, an amount that is determined by geology, and fixed. Our reserves are that amount of a mineral which can be economically recovered for use. Some reserves have been identified, others remain undiscovered. Thus the size of our reserves is variable, dependent not only on absolute availability, but also on changing factors such as extraction cost, market price or exploration effort. The map above shows the location of the major reserves of 20 key minerals and the percentage share held by each country. The table of "estimated lifetimes" (right) for these minerals shows how long present reserves will last at current rates of consumption. Of course, if consumption falls or rising prices and new exploration turn resources into reserves, then those lifetimes will grow. Conversely, if consumption and production costs rise and additional exploration fails to find economic deposits, then those lifetimes will fall. Our map shows the patterns of trade for two important commodities, aluminium and copper.

	Reserve base (million tonnes)	Years of supply
Aluminium ore (bauxite)	232,000	2,388
Copper	560	66
Diamond	620 (m. carats)	20
Gold	0.037	30
Iron ore	216,408	236
Lead	125	37
Manganese	5,000	186
Nickel	121	144
Platinum group	0.037	176
Silver	0.262	24

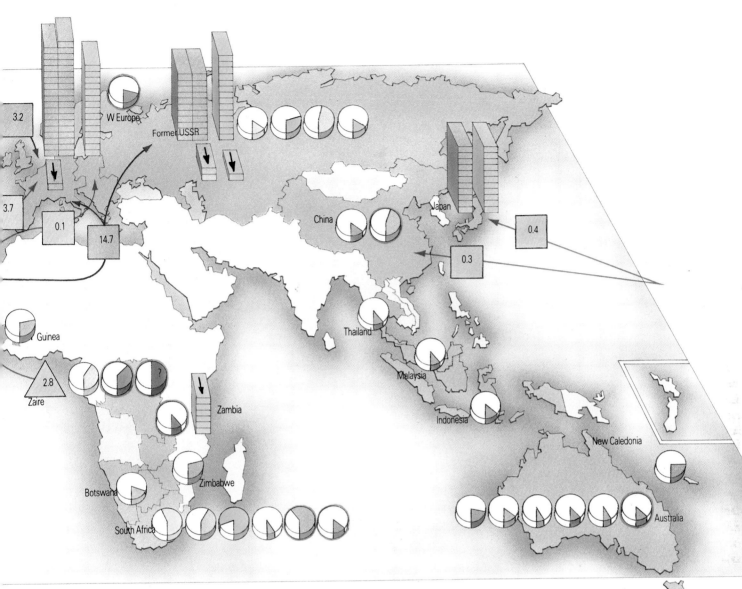

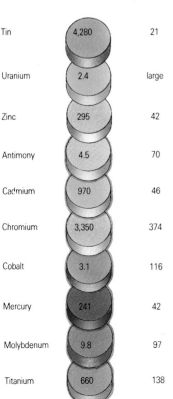

Tin	4,280	21
Uranium	2.4	large
Zinc	295	42
Antimony	4.5	70
Cadmium	970	46
Chromium	3,350	374
Cobalt	3.1	116
Mercury	241	42
Molybdenum	9.8	97
Titanium	660	138

Stockpiling

Minerals are an integral part of diplomacy. Each country tries to ensure access to those minerals which affect its vital national interests, especially defence. Strategically important minerals are vulnerable to interruptions of supply, often for political reasons. Lacking indigenous supplies, many Western countries have been building up stockpiles of such minerals as manganese, chromium, cobalt, and platinum. One unintended effect of stockpiles is to even out the sharp fluctuations in mineral prices. Substitution and recycling can also be strategic options, boosting supply.

American stockpiles

The US has stockpiled over half a tonne of bauxite (the ore from which aluminium is made) for each American citizen. As the world's largest consumer of strategic minerals, the US is particularly vulnerable to interruptions to its supplies, importing 100% of its titanium, 97% of its manganese, and over 90% of its chromium, largely from developing countries.

Recycling

Almost half the iron needed for steel-making now comes from scrap and nearly a third of the aluminium. Recycling can bring major energy savings. For example, the energy required to produce one tonne of secondary aluminium from scrap is only 5% of the energy used to extract and process primary metal from ore. Scrap is now a vital source of supply for metals.

Substitution

Metals that are easy to substitute include antimony, cadmium, selenium, tellurium, and tin. Tin has been losing out to glass, plastics, steel, and aluminium in the can-making and packaging industries: aluminium now accounts for over 90% of all US drinks cans. But substitution is no panacea: platinum is an unrivalled catalyst and stainless steel depends on chromium.

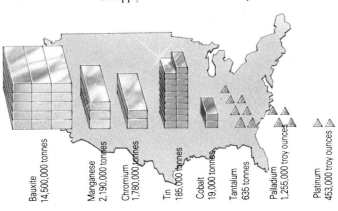

THE ELEMENTS CRISIS

Over the past 20 years the world has woken up to a new crisis: the elements crisis. Some elements are seen to be in increasingly short supply, while other elements, in the form of pollution, are turning up in the wrong places – and causing havoc in the process. Nothing woke up more people to the new realities than the first oil shock of 1973-4.

Oil – the end of an era?

Until 1973, the world's consumers forgot – if, indeed, they ever knew – that they were spending energy capital accumulated over many millions of years. They forgot that they were exploiting Nature's literally "unrepeatable offer". OPEC had its own reasons for shocking its clients out of their complacency, but in doing so it did the world a long-term service.

The first reaction to skyrocketing oil prices was a sharp reversal of consumption through increased energy efficiency and efforts to shift to other fuels. Even so, the impact of rising oil prices was dramatic. West Germany, for example, managed to cut its oil imports between 1973 and 1980, but its total energy import bill rose almost ninefold, reaching DM75 billion in 1981.

The success of energy conservation is amply demonstrated by the reduction in energy intensity – the amount of energy used to produce a unit of gross domestic product (GDP) – in OECD countries. Between 1973 and 1985, energy consumption within these countries grew by only 5 percent, while their GDP grew by almost one-third. But after the collapse of world oil prices in 1986, the energy intensity in the US and EC began to increase again.

The most pronounced effects of the energy crisis, however, are to be found in the developing countries – where commodity exports buy ever-less oil. In 1975, for example, a tonne of copper bought 115 barrels of oil. Six years later it bought only half as much oil. Furthermore, the poorer nations have been less successful in reducing their oil dependence, leaving them acutely vulnerable to further oil shocks. During the 1991 Gulf crisis the economies of developing countries were particularly hard hit by the unprecedented daily oil price fluctuations that rocked the global economy.

The problem is compounded by the declining oil output from the former USSR and US (the world's first and second greatest producers) and from the North Sea. With the Middle East's share of the oil

The oil crisis

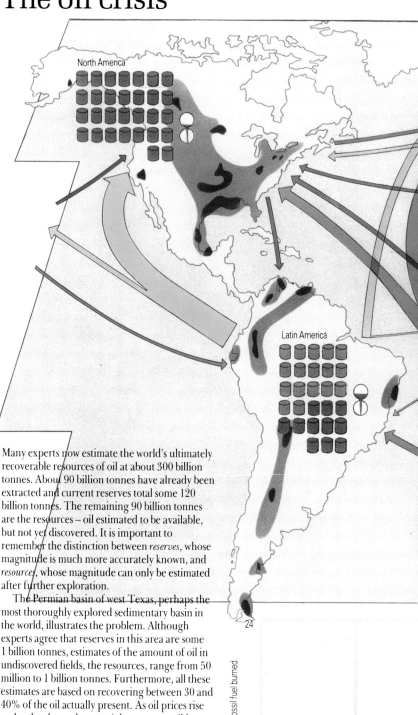

Many experts now estimate the world's ultimately recoverable resources of oil at about 300 billion tonnes. About 90 billion tonnes have already been extracted and current reserves total some 120 billion tonnes. The remaining 90 billion tonnes are the resources – oil estimated to be available, but not yet discovered. It is important to remember the distinction between *reserves*, whose magnitude is much more accurately known, and *resources*, whose magnitude can only be estimated after further exploration.

The Permian basin of west Texas, perhaps the most thoroughly explored sedimentary basin in the world, illustrates the problem. Although experts agree that reserves in this area are some 1 billion tonnes, estimates of the amount of oil in undiscovered fields, the resources, range from 50 million to 1 billion tonnes. Furthermore, all these estimates are based on recovering between 30 and 40% of the oil actually present. As oil prices rise and technology advances, it becomes possible to recover more of the available oil. How much oil there is for use is thus a function not only of the absolute amount present, but also of its price and the level of technology.

The availability of oil is also a function of politics since the resources are very unevenly distributed. About 30% of the world's remaining resources are in the Middle East, about 25% in Eastern Europe, China, and the former USSR, and about 20% in North and South America. The map summarizes the world oil situation in 1990, showing how much oil each region has produced, the size of its reserves, and the lifetime of those reserves at current consumption and price levels. The arrows plot the trade in oil between nations, not detailed shipping movements.

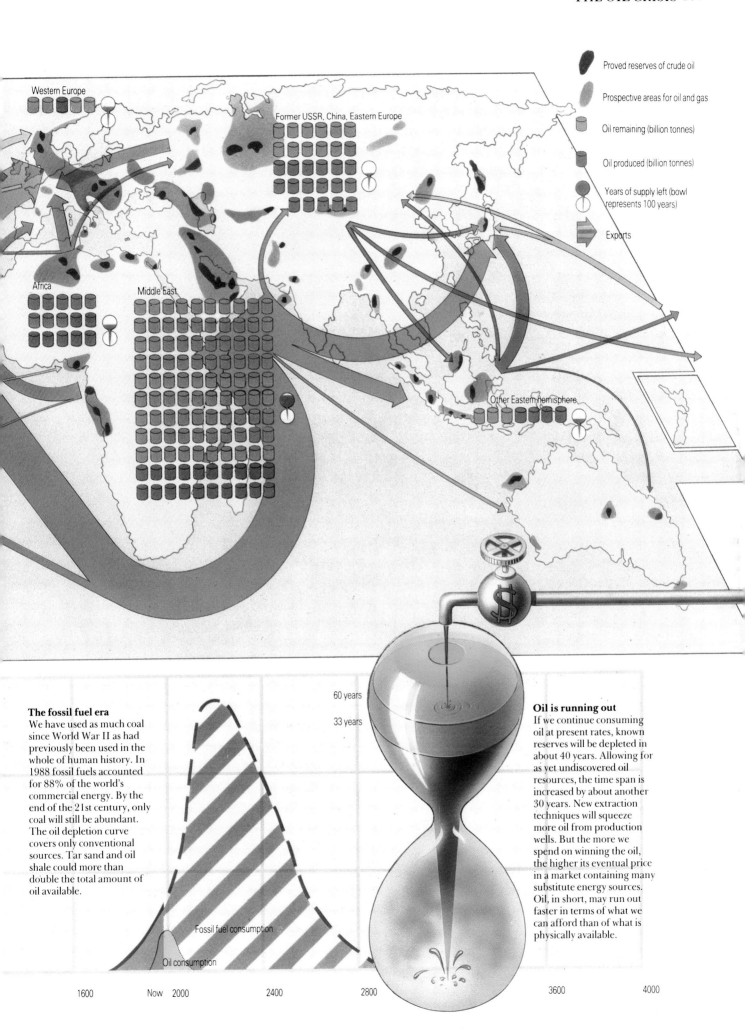

Proved reserves of crude oil

Prospective areas for oil and gas

Oil remaining (billion tonnes)

Oil produced (billion tonnes)

Years of supply left (bowl represents 100 years)

Exports

Western Europe

Former USSR, China, Eastern Europe

Africa

Middle East

Other Eastern hemisphere

60 years

33 years

The fossil fuel era
We have used as much coal since World War II as had previously been used in the whole of human history. In 1988 fossil fuels accounted for 88% of the world's commercial energy. By the end of the 21st century, only coal will still be abundant. The oil depletion curve covers only conventional sources. Tar sand and oil shale could more than double the total amount of oil available.

Fossil fuel consumption

Oil consumption

Oil is running out
If we continue consuming oil at present rates, known reserves will be depleted in about 40 years. Allowing for as yet undiscovered oil resources, the time span is increased by about another 30 years. New extraction techniques will squeeze more oil from production wells. But the more we spend on winning the oil, the higher its eventual price in a market containing many substitute energy sources. Oil, in short, may run out faster in terms of what we can afford than of what is physically available.

1600   Now 2000   2400   2800   3600   4000

market expected to increase from 27 percent in 1990 to nearly 40 percent by the end of the decade, the market will be as reliant on this region as it was in the mid-seventies. Oil is no longer the stable, cheap resource it once was.

The other energy crisis
More than two billion people, or the majority in the developing world, rely on wood as fuel to cook their food. More than half of them are overcutting available trees. For the poorest third of humankind, the real energy crisis is fuelwood – and it was so long before OPEC appeared on our horizons. Fuelwood dominates daily life, as increasing amounts of time and labour are spent in finding it and carrying it home. If it has to be bought, the price can account for two-fifths of a family's cash income. And the problem is growing worse.

Half of all wood cut worldwide each year is used as fuel for cooking and heating, at least four-fifths of it in the developing world. A minimum of 1.5 billion people encounter daily difficulty in finding enough fuelwood, even though they average only about 3 kilograms a day, little more than a few sticks. They cut trees faster than the timber stock replenishes itself. Moreover, their growing numbers and the increasing human densities spread the problem further afield. As a result, a potentially sustainable wood-gathering activity becomes destructive and, ultimately, unsustainable. The fuelwood issue, in fact, is a classic instance of how impoverished people in the developing world can find themselves obliged to destroy tomorrow's livelihood in order to secure today's essentials. They do it not out of ignorance. They do it out of tragic compulsion.

Worse still, at least 125 million people simply cannot lay hands on sufficient fuelwood to meet even their minimal needs. For many families, it now costs as much to heat the supper bowl as to fill it.

For many in the developing world, then, the energy crisis does not devolve into a debate on how to limit the electricity consumed in a gadget-orientated home. It strikes directly at the struggle to keep body and soul together.

We have already seen (pp 38-41) how deforestation brings a tide of troubles in its wake. In the case of fuelwood deficits, there are some added woes. Millions of families seek substitutes in materials such as cattle dung. In Asia and Africa alone, at least 400 million tonnes of dung are burned as fuel each year. If applied to fields, this dung could help produce 20 million tonnes of grain – enough food to sustain tens of millions of people.

Unless we devise a response to this great and growing energy problem, we can expect that within just another three decades the number of people overcutting an already thin fuelwood resource will more than double, while the number facing acute shortages

The fuelwood crisis

Of the 2 billion people who rely on wood as their primary energy source, some 70% do not have access to secure supplies. By the year 2000, some 2.7 billion people, half the population of the developing world, will find their minimum fuel needs insecure or unmet. The map below indicates the severity of the crisis in developing countries (except China). It is worst in the arid

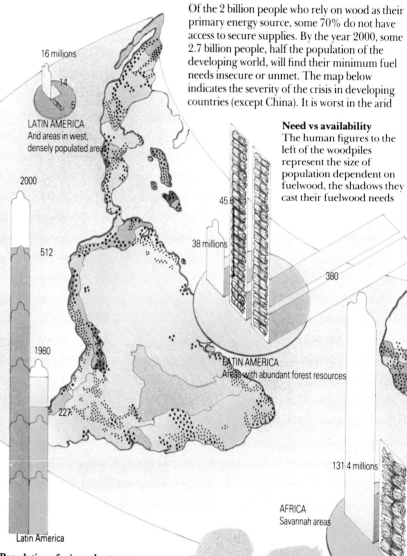

Need vs availability
The human figures to the left of the woodpiles represent the size of population dependent on fuelwood, the shadows they cast their fuelwood needs

Populations facing shortages
The numbers of people facing shortages are shown in columns for 4 regions. Each figure in a column represents 100 million people – those in orange (1980) and red (2000) face acute scarcity or deficit, those in yellow are depleting their resources faster than they are replenished.

A vicious spiral
Virtually all of the trees within a radius of 70 km of Ouagadougou in the Sahel have been consumed for fuelwood. Scarcity occurs most in the ecologically fragile drylands and highlands where loss of tree cover leads to flooding, soil erosion, and the silting up of river beds or dams. The search for alternative fuels leads to the burning of dung that would otherwise have been returned to the soil. This reduces crop yields, forcing farmers to clear more forest to maintain food supplies. This, in turn, reduces the availability of fuelwood and results in another twist down the vicious spiral illustrated on the right.

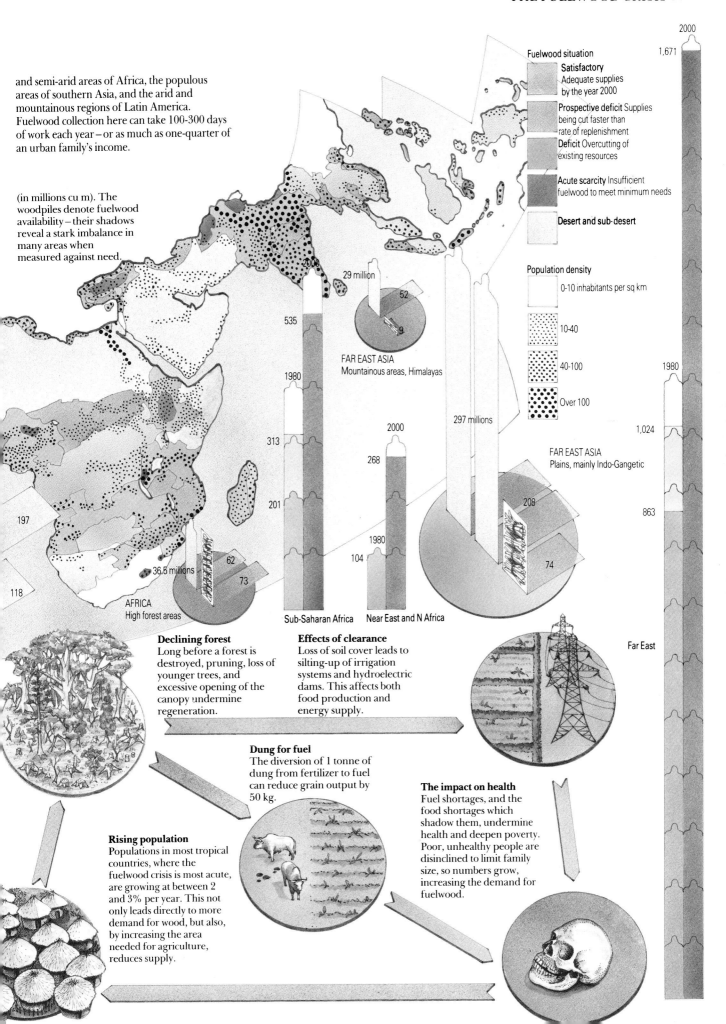

and semi-arid areas of Africa, the populous areas of southern Asia, and the arid and mountainous regions of Latin America. Fuelwood collection here can take 100-300 days of work each year – or as much as one-quarter of an urban family's income.

(in millions cu m). The woodpiles denote fuelwood availability – their shadows reveal a stark imbalance in many areas when measured against need.

Fuelwood situation

Satisfactory Adequate supplies by the year 2000

Prospective deficit Supplies being cut faster than rate of replenishment

Deficit Overcutting of existing resources

Acute scarcity Insufficient fuelwood to meet minimum needs

Desert and sub-desert

Population density

0-10 inhabitants per sq km

10-40

40-100

Over 100

29 million
52
9

FAR EAST ASIA
Mountainous areas, Himalayas

535
1980
313
201
62
73
36.5 millions

AFRICA
High forest areas

197
118

2000
268
104
1980

Sub-Saharan Africa

Near East and N Africa

297 millions

FAR EAST ASIA
Plains, mainly Indo-Gangetic

208
74

2000
1,671
1980
1,024
863

Far East

Declining forest
Long before a forest is destroyed, pruning, loss of younger trees, and excessive opening of the canopy undermine regeneration.

Effects of clearance
Loss of soil cover leads to silting-up of irrigation systems and hydroelectric dams. This affects both food production and energy supply.

Dung for fuel
The diversion of 1 tonne of dung from fertilizer to fuel can reduce grain output by 50 kg.

The impact on health
Fuel shortages, and the food shortages which shadow them, undermine health and deepen poverty. Poor, unhealthy people are disinclined to limit family size, so numbers grow, increasing the demand for fuelwood.

Rising population
Populations in most tropical countries, where the fuelwood crisis is most acute, are growing at between 2 and 3% per year. This not only leads directly to more demand for wood, but also, by increasing the area needed for agriculture, reduces supply.

could exceed one billion. The implications of these shortages will be tremendous. And the longer the damage continues, the harder it will be to find a viable long-term solution.

In the hothouse

The Earth is naturally protected by radiation-absorbing gases, notably carbon dioxide and water vapour. These serve to retain some of the sun's warmth in what is commonly referred to as the "greenhouse effect". Without these gases, the Earth's surface temperature would be far cooler, and considerably more hostile to life.

In the early Earth, carbon dioxide accounted for over 70 percent of the atmosphere. The sun, however, was about a quarter less powerful: it was carbon dioxide and other "greenhouse gases" that kept the temperature warm enough for life. Over the aeons, the sun has grown hotter, but temperatures have remained comfortable because carbon dioxide concentrations have steadily declined. So the human-induced increase in atmospheric carbon since the onset of the Industrial Revolution is essentially the reversal of a natural trend. In 1750 the carbon dioxide concentration was about 280 parts per million by volume (ppmv). At present the concentration is around 360 ppmv, and is projected to be 450 ppmv by 2025. This points to a steady warming of the planet's climate, with the potential of large-scale disruption to the world's

The greenhouse effect

Incoming radiation
Of the incoming short-wave solar radiation, only 24% hits the Earth's surface directly – 3% of which is promptly reflected back into space. The rest of the incoming short-wave radiation gets caught up in the Earth's atmosphere, where it is either scattered back to space (25%), deflected to the surface (26%), or simply absorbed (25%).

2025 450 ppmv
1990 360 ppmv
1750 280 ppmv

Total outflow of energy
3 + 25 + 67 + 5 = 100

Incoming solar radiation

25%

100%

3%

Outgoing terrestrial radiation

67%

5%

Carbon dioxide levels in the atmosphere have increased by about 30% from 1750 to 1990. Before this century, carbon dioxide levels had not risen above 300 ppmv at any time in the past 100,000 years.

25%

24%

26%

Carbon dioxide plays a key role in determining the Earth's climate. It lets through most of the incoming short-wave radiation from the sun, but traps and retains much of the long-wave radiation that the Earth radiates out towards space. The net effect is to keep the Earth's surface at a higher temperature than if carbon dioxide were not present in the atmosphere. This is the "greenhouse effect".

The Earth has remained at a relatively stable temperature, and one suitable for life to flourish, for billions of years. Gaian scientists argue that life on the planet has moulded the environment, just as the changing environment influenced the evolution of life. The carbon dioxide cycle is one of the most important aspects of this interaction. Carbon is released to the atmosphere by volcanic activity, respiration, and decay, and now by human activity. Plant life removes carbon from the air and "fixes" it in living cells. Burial of dead organisms removes carbon from the system. Two major carbon dioxide "sinks" are the oceans and the tropical rainforests. The latter can fix between one and two tonnes of carbon for each square kilometre of land area each year. Hence there is a further reason to reduce tropical deforestation and accelerate reforestation.

Energy absorbed by atmosphere
25 + 109 + 29 = 163
Energy re-radiated from atmosphere
96 + 67 = 163

29%

Outgoing radiation
The Earth, being cooler than the sun, emits its energy in longer wavelengths (infra-red radiation). Because the atmosphere is largely opaque to these wavelengths, they bounce back and forth between the surface and atmosphere. So the equivalent of 109% of incoming solar radiation is absorbed – and 67% is re-radiated into space.

109%

96%

114%

Absorbed at surface
24 + 26 – 3 = 47

Lost from surface
114 + 29 – 96 = 47

Latent and sensible heat

Latent heat (24%) and sensible heat (5%) are two significant flows of energy which are also absorbed by the atmosphere.

biota, agriculture, and sea levels. The official scientific estimates, produced by the Intergovernmental Panel on Climate Change (IPCC), suggest a range of global warming from 1.5°C to 4.5°C by 2100 – with a best guess at around 2.5°C. Over the last 100,000 years of glacial advance and retreat, the global temperature does not appear to have fluctuated by more than 2°C.

About two-thirds of the 7 billion tonnes of carbon dioxide added to the atmosphere each year are due to our burning of fossil fuels, and one-third is due to deforestation, land use changes, and fuelwood burning. The main carbon dioxide emitters are the northern fossil fuel-dependent industrial countries. Yet in the course of time China and India could become major emitters, depending on their energy supply-demand relationships, and the role of alternative energy technology.

Carbon dioxide is singled out as the main problem pollutant because it is so ubiquitous, it is least susceptible to major reduction in a politically and economically acceptable manner, and it is the largest single contributor to greenhouse-related warming. Other human-influenced "greenhouse gases" include methane, accounting for around 15 percent of global warming, the chlorofluorocarbons, or CFCs (22 percent), and nitrous oxide (6 percent). It is worth noting that though carbon dioxide is regarded as the most potent source, methane absorbs radiation 25 times more

Global emissions

The distribution of greenhouse gas emissions by country is a matter of sharp controversy. The US is clearly the biggest polluter, accounting for nearly 25% of all greenhouse gases, followed by Europe and the former USSR. More difficult is the analysis of deforestation sources, and rice production where methane is the major pollutant. The Earth has a capacity to absorb greenhouse gases and many in the developing world are questioning how this capacity should be shared between countries. Their view is that the rich North has usurped this capacity in its industrialization, and that the countries of the South are emitting gases well within their share, and need to continue to emit before they get their development paths on a sustainable track.

The greenhouse gases

Carbon dioxide's main sources are fossil fuel combustion, wood burning, deforestation and land use change generally. **Methane** is increasing at about 1% annually and is very difficult to control because it is associated with the emissions from cattle and other ruminants, from rice paddies and other wetlands, and from leaks in the gas and coal industries. The **CFCs** (and other halogenated compounds) are mostly linked to refrigerants, foam production, and fire retardants. **Nitrous oxides** are associated with fossil fuel combustion, the use of fertilizers, land use change, and forest destruction. Low-altitude **ozone** is created by the action of sunlight on aerial pollutants, notably from car exhausts and industry. See pie chart for the relative contributions to the warming.

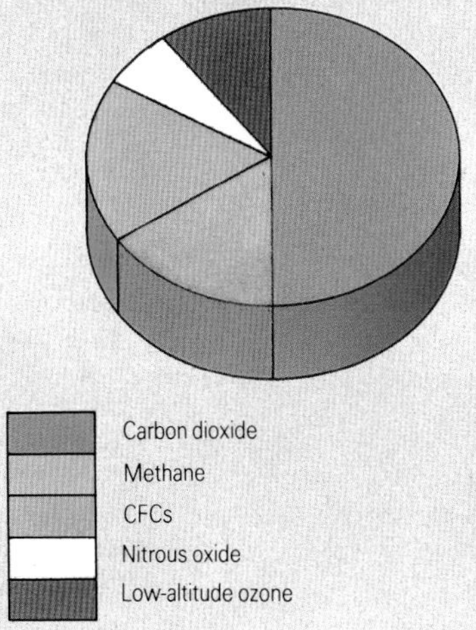

- Carbon dioxide
- Methane
- CFCs
- Nitrous oxide
- Low-altitude ozone

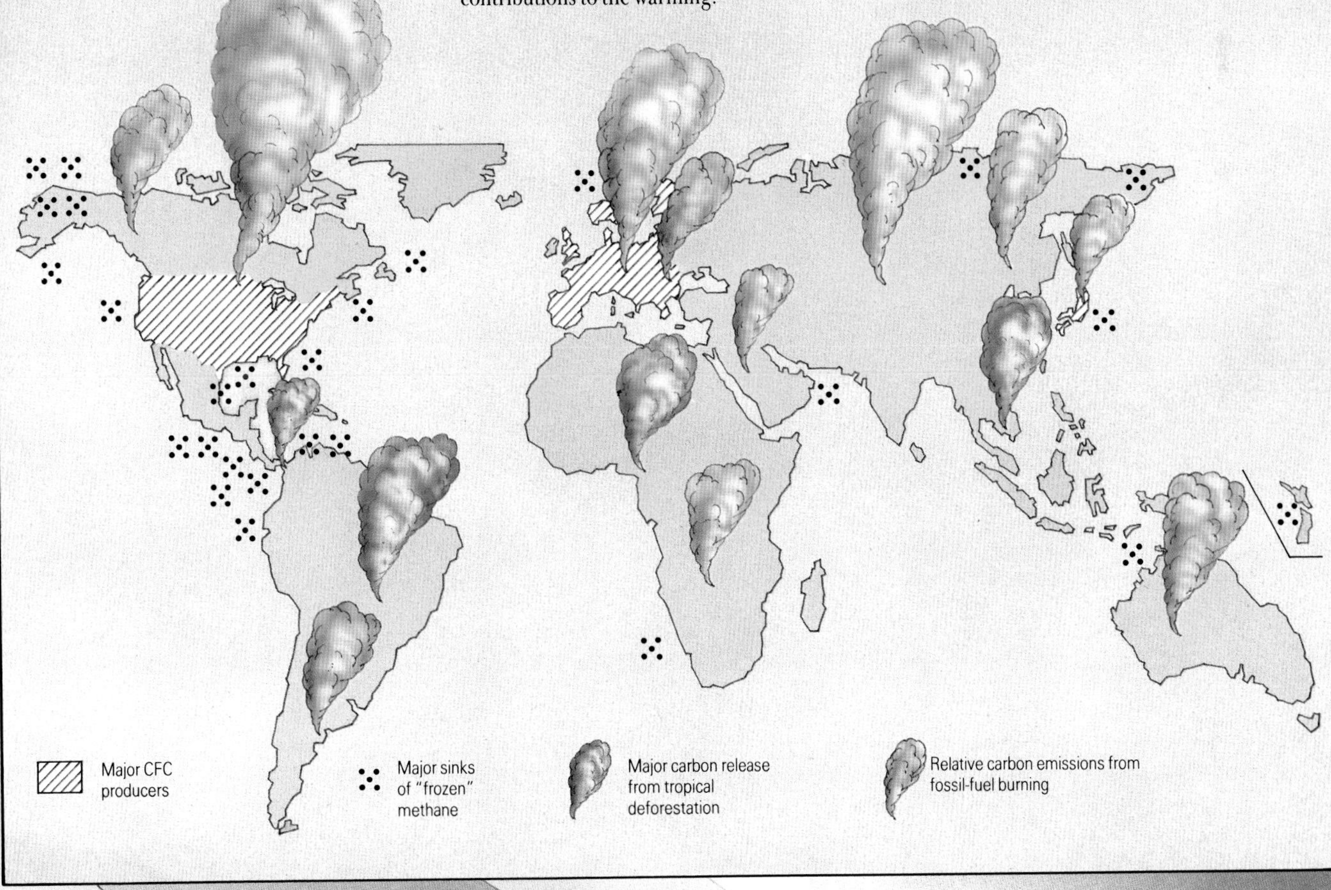

Major CFC producers

Major sinks of "frozen" methane

Major carbon release from tropical deforestation

Relative carbon emissions from fossil-fuel burning

effectively, and CFCs do so 15-16,000 times better. So, what are the effects on the world's biota, agriculture, and sea levels likely to be?

The greenhouse effects

The temperature record shows a global increase of 0.6°C since 1940. In addition, weather patterns are changing, and the decade of the 1980s saw more hotter years than any previous decade. However, it would be unwise to pin all this down to global warming alone. Calculations of the warming depend on extremely sophisticated computer models of the atmosphere and ocean circulation, which are still in the process of being refined. Only by the end of the decade should we know a little more accurately just what to expect next century.

For sea level, the best guess is an increase globally of around 40 cms by 2100. This will obviously affect densely populated coastal areas, and may, in time, lead to the inundation of whole islands, such as in the Maldives and Pacific atolls. Particularly vulnerable are large coastal cities and crowded agricultural regions, such as river deltas, in the South. Salt contamination of natural water supplies will also be a problem. It has been estimated, for example, that some 13 million people in Bangladesh and over 20 million in coastal China would be at risk.

Regarding agriculture, there will be gainers and losers, depending on the location of national boundaries and micro-regional variations in rainfall. Because average temperatures are expected to increase more near the poles than near the equator, the shift in climate and agricultural zones will be more pronounced in higher latitudes. The geography of migrating food production could result in very different trading patterns and possible territorial conflict. More northerly and southerly regions, presently rather poor in agricultural output, could become more productive. Higher yields in some areas may compensate for decreases in others – but then again they may not, particularly if today's major food exporters, such as the US, suffer serious losses.

Human health may prove to be a more significant issue. Change in temperature and humidity will affect the location of disease-carrying insects and parasites which may affect people with no natural defences. It is likely that the costs of health care will rise and impoverished countries with poorly-developed health provision could suffer greatly as a result.

Although most scientists agree that human activities are changing the climate, they do not agree on the rate at which it will occur, nor on its specific impacts. This makes it difficult to put a "price tag" on either climate change or on policies to prevent it. However, if the Earth's surface warms as predicted, millions of people are likely to become more vulnerable to the effects of famine, drought, flooding, and more. The money spent

Climate chaos

Most scientists agree that human activities are changing the climate; the possible consequences range from the moderate to the destructive. Sea level rise would endanger coastal cities and productive lowlands; weather patterns might become more extreme; climate zones could shift, altering agricultural and biodiversity patterns. With the altered distribution of food and water, conflicts might ensue and mass migrations increase, from rural to urban areas, and from South to North. It is likely that the poor would suffer most as they have fewer options for response. The workings of Gaia's life-support systems remain largely mysterious. What we do know inspires admiration for an organism that has survived enormous changes in chemistry and species in the past (including volcanism and meteorite impact). Similar catastrophic effects could possibly be associated with climate change, notably the rapid melt of the West Antarctic ice sheet and the release of vast stores of frozen methane in the tundra. Should such events synchronize – an improbable scenario – the short-term consequences for humanity would be severe. But the planet would most certainly cope. The human remedy is to act with prudence and in good time, to avoid pressing the world to the limits of human tolerance and adaptability.

Sea level rise
The best guess is a global increase of around 40 cm by 2100, affecting densely-populated coastal cities and agricultural regions, as well as island nations. Salinization of natural water supplies would be serious.

Refugees
The impact will be most noticeable in regions already vulnerable to variations in rainfall and temperature and will probably create a new burst of refugees amongst the most destitute, adding to the burden and social tension in recipient nations.

Doomsday scenarios
A rapid melting of the West Antarctic ice sheet and the release of vast stores of frozen methane (a potent greenhouse gas) in the tundra of the North represent "worst case", devastating scenarios.

Biodiversity
Climatic zones could shift several hundred kilometres towards the poles – flora and fauna could lag behind. Climate change would favour the common, most opportunistic species at the expense of the rare and least adaptable – such as those in protected areas.

Unpredictable weather
Direct association between atmosphere warming and greater extremes of weather is difficult to prove but it is possible that more extreme weather patterns and storm surges should be planned for.

Agricultural changes
The impact on agriculture depends crucially on the speed of the onset of climate change, and the scope for adaptation in crop and livestock genetics and water management. There may be severe effects in some regions, especially those least able to adjust.

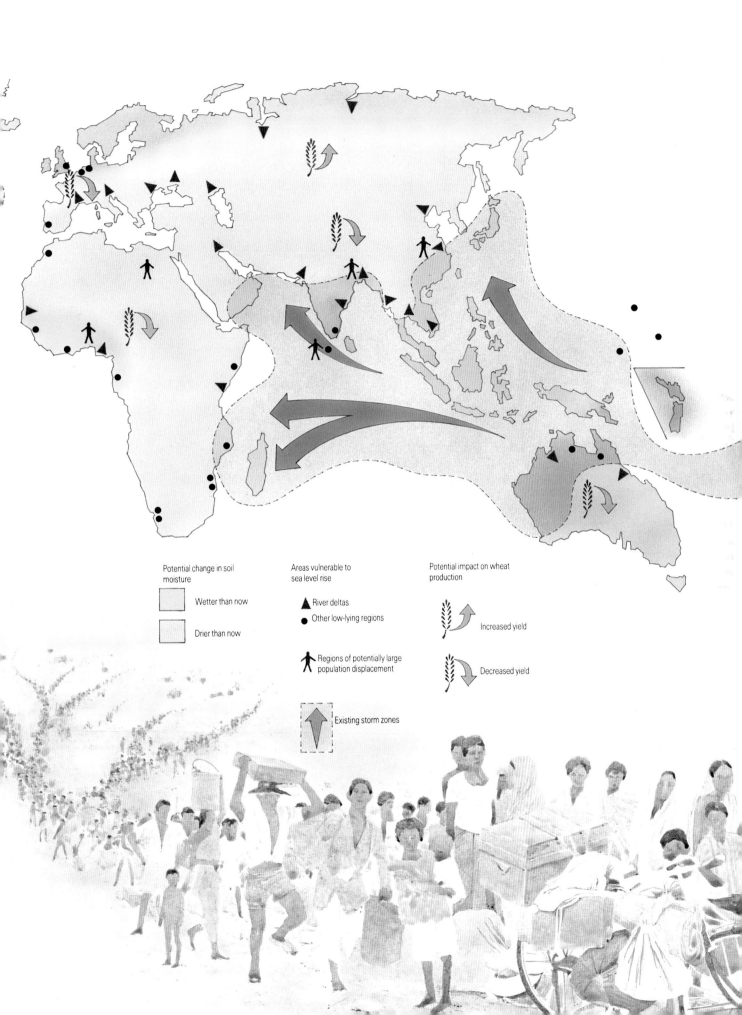

Potential change in soil
moisture

Wetter than now

Drier than now

Areas vulnerable to
sea level rise

▲ River deltas

● Other low-lying regions

🚶 Regions of potentially large
population displacement

⬆ Existing storm zones

Potential impact on wheat
production

Increased yield

Decreased yield

on taking action now could be viewed as an insurance premium for protection against a hard-to-measure but potentially devastating risk.

Ozone depletion

Climatic change and ozone depletion are chemically connected. Both are truly global in extent: there is no hiding place from their effects. One of the main greenhouse gases is the CFC group, and the formulations CFC 11 and CFC 12 are also the most active as ozone destroyers.

The ozone layer is found in the stratosphere 15-50 kms above the Earth's surface. It acts as a shield, absorbing high energy ultraviolet (UV-B) radiation which is hazardous to life, and also heats the outer skin of the atmosphere.

In the early 1970s it was James Lovelock who first identified the build-up of CFCs in the atmosphere, these wholly man-made chemicals whose active life is over 70 years. Concern for the effect of CFCs on stratospheric ozone mounted but the first clear evidence of ozone depletion came in 1984 when British scientists in Antarctica found an area as big as the United States contained no ozone at all. This is the notorious "ozone hole", the result of complex chemical reactions in a slow-moving, circumpolar vortex where the interaction of the chemicals, spring sunshine, and ice crystals become very effective ozone destroyers. Since 1979 about 15 percent of Antarctic ozone has been destroyed, although locally and episodically up to 95 percent of ozone disappears.

In recent years loss of ozone has been recorded over the Arctic, though the extent of destruction is much less and there is no "hole" as such. But the Arctic atmosphere is "primed" for further ozone loss, and with the natural chemical activity associated with the plume from the 1991 eruption of Mount Pinatubo in the Philippines, ozone losses in the Arctic for the spring of 1992 measured as much as 15-18 percent.

It is estimated that "background" rates of ozone depletion could be 2 percent annually, but CFCs are not the only industrial pollutants to blame. Various halons, carbon tetrachloride, and methyl chloroform, used as fire retardants and solvents, also deplete stratospheric ozone. Recently methylbromide, used as a fungicide in many parts of the world, has been identified as a powerful ozone destroyer.

Ironically, because ozone heats the outer atmosphere, so its loss should contribute to a reduction in the global warming effects of CFCs. The relationship is by no means a cancelling out of opposing effects, however, and the loss of stratospheric ozone as an ultraviolet shield is altogether seen as a more serious threat than its contribution to atmospheric cooling.

Increased exposure to UV-B will potentially give rise to increased numbers of human skin diseases and cataracts, may damage agriculture and aquatic ecosystems, and even accelerate climate change.

Holes in the ozone map

A thin veil of ozone high in the Earth's atmosphere protects life below from the portion of the sun's ultraviolet radiation that would otherwise place stress on many forms of life. Yet this ozone veil is being damaged by chemicals released at the Earth's surface. The global depletion of the ozone layer is around 2% annually. Each 1% loss of ozone is likely to cause an increase of about 2% in the hazardous ultraviolet radiation. The consequences of increased exposure to radiation represent threats to human health, as well as animals, plants, and the environment. Already there are reports of cataract damage to sheep in southern Chile, where the ozone hole is regularly found each southern spring. The tropics are so far the only areas which are not affected by ozone depletion. The map, right, shows ozone trends as percentage change per decade for northern and southern springs.

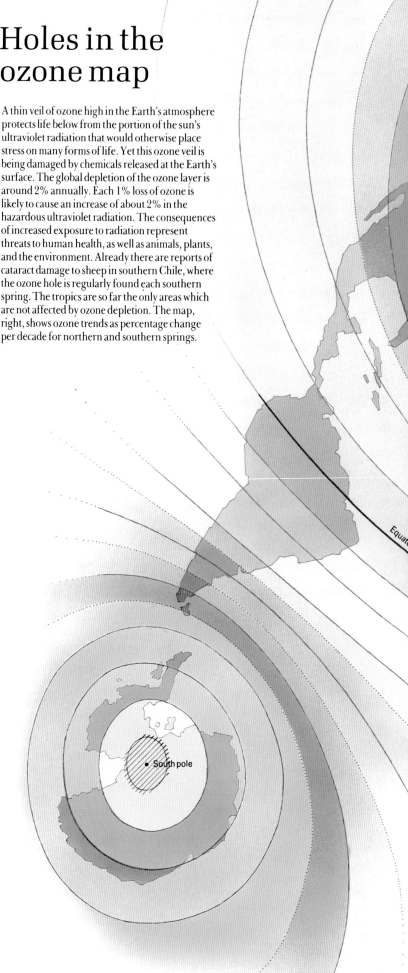

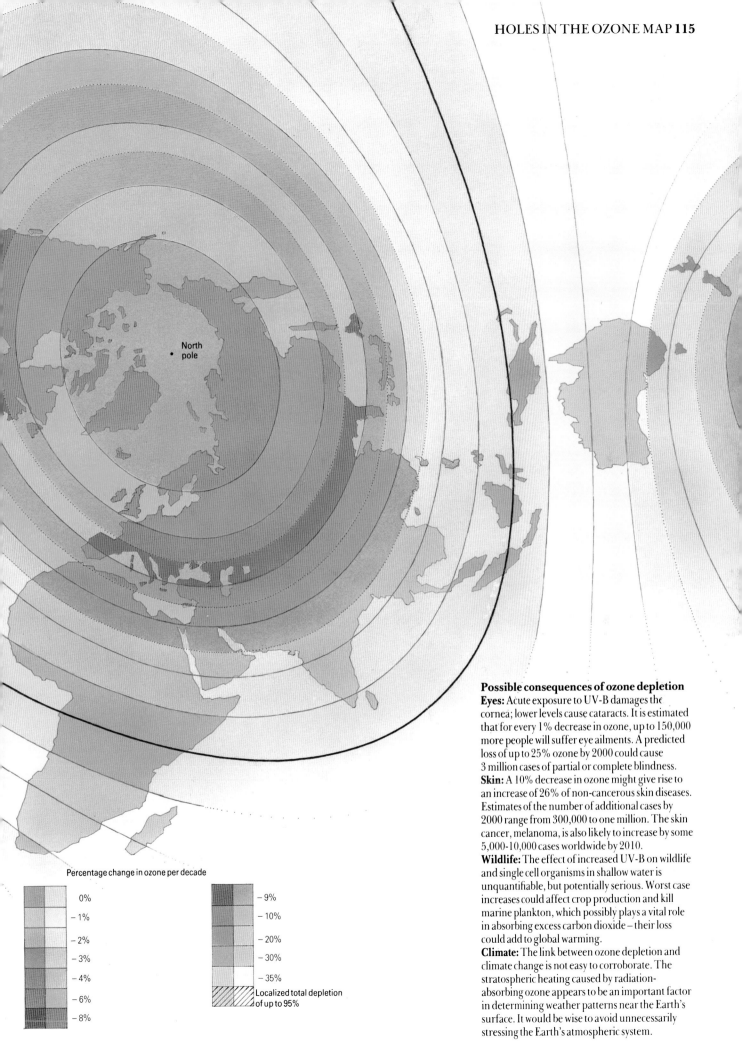

North
pole

Percentage change in ozone per decade

	0%
	−1%
	−2%
	−3%
	−4%
	−6%
	−8%

	−9%
	−10%
	−20%
	−30%
	−35%
	Localized total depletion of up to 95%

Possible consequences of ozone depletion
Eyes: Acute exposure to UV-B damages the
cornea; lower levels cause cataracts. It is estimated
that for every 1% decrease in ozone, up to 150,000
more people will suffer eye ailments. A predicted
loss of up to 25% ozone by 2000 could cause
3 million cases of partial or complete blindness.
Skin: A 10% decrease in ozone might give rise to
an increase of 26% of non-cancerous skin diseases.
Estimates of the number of additional cases by
2000 range from 300,000 to one million. The skin
cancer, melanoma, is also likely to increase by some
5,000-10,000 cases worldwide by 2010.
Wildlife: The effect of increased UV-B on wildlife
and single cell organisms in shallow water is
unquantifiable, but potentially serious. Worst case
increases could affect crop production and kill
marine plankton, which possibly plays a vital role
in absorbing excess carbon dioxide – their loss
could add to global warming.
Climate: The link between ozone depletion and
climate change is not easy to corroborate. The
stratospheric heating caused by radiation-
absorbing ozone appears to be an important factor
in determining weather patterns near the Earth's
surface. It would be wise to avoid unnecessarily
stressing the Earth's atmospheric system.

Whereas ozone in the stratosphere is beneficial, ozone near the Earth's surface is a pollutant, one of many invisible threats to our air.

Atmospheric pollution

In both developed and developing countries, air pollution is a common hazard. In many countries in the North, acid rain has become a major problem while the cities of the developing world are experiencing serious air pollution.

Millions of tonnes of sulphur dioxide and nitrogen oxides are released into the atmosphere by industry and vehicles. These gases react with water in the air to form acid rain, mist and snow, which falls to the ground, far removed from the point of origin, often in another country. To date, this phenomenon hits worst at parts of Canada and the US, and central and northern Europe – though China, Australia, and Brazil are noticing early signs of this silent scourge. Thousands of lakes are now lifeless, unable to support fish of many sorts, and numerous species of plants, birds, and insects are declining in acidified areas. Europe's forests are in the grip of a deadly epidemica 1990 UN survey found defoliation worst in Byelorussia, Czechoslovakia, and the UK. Another UN study estimated that the economic loss of European forests due to air pollution is running at $29 billion annually. If other impacts are added, such as the effects on human health and buildings, the cost could be almost ten times as much.

International measures have been agreed to begin to check the problem of acid rain, but some of the most polluting nations have persistently refused to join in some of the agreements. A common argument is that pollution-reduction targets are not scientifically based. Yet a basic dilemma with acid rain – as with many environmental problems – is that the onslaught apparently arrives so fast and its workings are so complex that we cannot afford to wait until we have a 100% correct answer.

Acid rain has become something of an umbrella term; and other air pollutants are included under its heading. Some of the most important are volatile organic compounds (VOCs), associated with vehicle emissions and the chemical industry, ammonia, and ozone near the Earth's surface, formed by the action of sunlight on nitrogen oxides and VOCs. Ozone pollution can increase human susceptibility to infection and respiratory disease, reduce the yield of certain crops, and damage natural vegetation. Cities worldwide are afflicted. In 1992 ozone pollution reached record levels in Mexico City, prompting officials to adopt stringent emergency measures to cut vehicle use and industrial activity, and to fund a pilot study of a plan to use 100 giant industrial fans to "suck" polluted air from the city. Such Jules-Verne responses illustrate how far the problem has got out of hand.

Cars are also the source of three-quarters of the 330,000 tonnes of lead that are ejected as fine

The invisible threat

Acidification ranks among the most serious threats to the environment in the northern hemisphere. Heavily industrialized areas pump some 90 million tonnes of sulphur dioxide into the air each year. Hardest hit are southern Sweden, Norway, parts of central Europe, and the eastern part of North America. Some 18,000 lakes in Sweden alone are now so acidified that fish stocks have been severely reduced. In Bavaria and other areas of central Europe, whole forests are dying. The map shows areas that are particularly sensitive to acid rain, typically with thin, rocky topsoils.

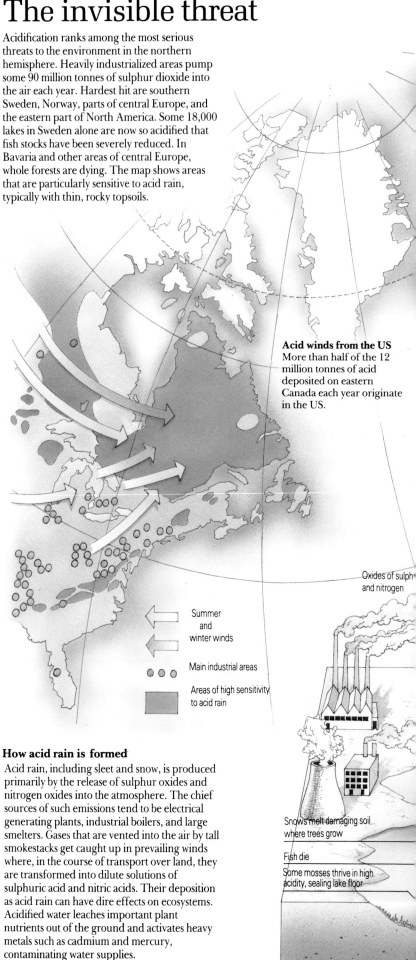

Acid winds from the US
More than half of the 12 million tonnes of acid deposited on eastern Canada each year originate in the US.

Summer and winter winds

Main industrial areas

Areas of high sensitivity to acid rain

Oxides of sulphur and nitrogen

Snows melt damaging soil where trees grow

Fish die

Some mosses thrive in high acidity, sealing lake floor

How acid rain is formed

Acid rain, including sleet and snow, is produced primarily by the release of sulphur oxides and nitrogen oxides into the atmosphere. The chief sources of such emissions tend to be electrical generating plants, industrial boilers, and large smelters. Gases that are vented into the air by tall smokestacks get caught up in prevailing winds where, in the course of transport over land, they are transformed into dilute solutions of sulphuric acid and nitric acids. Their deposition as acid rain can have dire effects on ecosystems. Acidified water leaches important plant nutrients out of the ground and activates heavy metals such as cadmium and mercury, contaminating water supplies.

Lead emissions from cars

Each year 450,000 tonnes of lead are released into the air by humans, compared with 3,500 tonnes from natural sources. Lead from vehicle exhaust represents more than half of this pollution. Lead is added to petrol to improve its combustion properties, and is released to the air as fine particles. The illustration (below right) demonstrates the various pathways by which lead finds its way to children living in areas of high traffic density.

Third World worst

Alerted to the dangers of lead, pressure groups in the North have successfully campaigned for the reduction of lead in petrol. Many Third World countries, however, still have intolerably high levels of lead in the atmosphere because oil companies persist in selling heavily leaded petrol. A Motor Gasoline Quality Survey in 1983 revealed that levels of lead in petrol in the Third World are almost consistently double those in the developed world.

Blood lead levels (μg/dl)

35

30
Blacks

25
Hispanics

20

15

10
70　71　72　73　74　75　76　77
Year

The maximum lead intake

for an average 70-kg human is about 6 μg/kg body weight/day. Children absorb lead more easily, so a maximum for them would be 1.2 μg/kg/day. There is a strong correlation, left, between leaded petrol sales and blood lead in New York children.

Air 54%
(direct inhalation 5%, deposited on food 22%, deposited on fingers 27%)

Average daily lead intake per urban 2-year-old child
45 μg

Leaded petrol sales

Water 7%

Food 39%

The rain in Scandinavia

Most of the acid in the rain falling on Scandinavia comes from sources in Europe, particularly the UK. Yet Gaian scientists argue that marine algae contribute as much as 40% of the sulphur, probably exacerbated by agricultural run-off into the North Sea causing algal blooms.

Oxides combine with water to become acid rain

Acid snow at high altitudes

Smog in Los Angeles

Air pollution comes in guises other than acid rain. Photochemical smog, of the kind formed in the Los Angeles basin, has had a dramatic impact on the ecosystems of southern California, causing extensive damage to forests as well as posing a serious health hazard. The smog is composed of a number of chemicals, notably ozone and peroxyacetyl nitrate (PAN), both of which are extremely harmful to plants. These substances are formed by the action of strong sunlight on a mixture of nitrogen oxides and hydrocarbons exhausted to the air by vehicles, combustion, and industrial processes.

The geographical features of Los Angeles serve to promote the formation of such smogs. Wind patterns and the surrounding mountains create conditions for the formation of temperature inversions, trapping air pollution close to the ground. Controls designed to reduce emissions from vehicles and other sources have been introduced in a successful attempt to combat the smog.

A temperature inversion occurs when a layer of warm air overlies a layer of cooler air, preventing air pollution from escaping. The smog in Los Angeles is seen as a brownish orange haze between May and October.

Sunlight

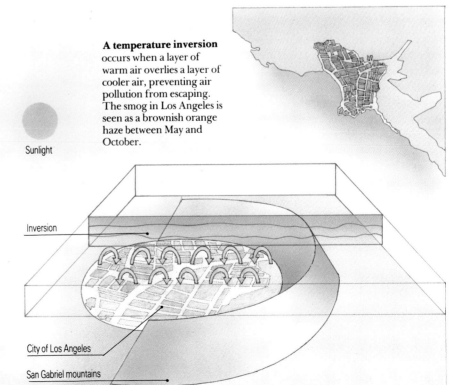

Inversion

City of Los Angeles

San Gabriel mountains

particles into our skies each year. Equally serious, there is now a build-up of charcoal-caused smoke over Third World cities such as Lagos, Jakarta, and Calcutta – the poor person's smog.

Death from polluted water

Water, a supposed fount of life, kills at least 25 million people in developing nations each year, three-fifths of them being children. Of the world's leading diseases, four out of five depend on water for their impact: they either breed in or are spread by water. All too often, there is not enough water in Third World communities for basic domestic needs, so it tends to be used time and again. It ends up as thoroughly dirty water, an ideal habitat for pathogens, including carriers of disease.

The 1980s witnessed unprecedented efforts which brought water and sanitation to many hundreds of millions. Yet there are still an estimated 1.23 billion people in the developing world without access to adequate and safe water supplies, and 1.74 billion without appropriate sanitation services. That is 31 percent without water, and 43 percent without sanitation. In practical terms, lack of drinking water means no supply within several hundred metres; lack of sanitation means no bucket latrine or pit privy, let alone sewerage system. So ponds and rivers are the main sources of drinking water – and the main sites for impromptu toilets. While the number of hospital beds per 1,000 people is often thought of as a sound criterion of health services, a far better measure is the number of household taps and serviceable toilets.

The significant increase in the number of people with water and sanitation services was offset by the rapidly expanding population. In 1990, in urban regions of the developing world, there were almost 100 million more people without access to the basic facilities than there were in 1981. In developed nations, by contrast, at least 98 percent of citizens have ready access to clean water, as much as they want.

The cost to human health in the Third World is enormous. In one way or another, water is implicated in trachoma blindness (500 million sufferers), malaria (350 million), schistosomiasis (250 million), and elephantiasis (250 million), plus typhoid, cholera, infectious hepatitis, leprosy, yellow fever, and probably worst of all, diarrhoea. Every hour more than 500 children die from diarrhoeal diseases: if this scourge could be eliminated, we would see an end to much of the malnutrition that afflicts the Third World, and that fosters many other diseases. In Africa alone, one million children succumb to malaria each year.

In addition to water-related deaths, many times more sufferers are left grossly debilitated, hardly able to do a decent day's work. In India, waterborne diseases claim 73 million workdays each year, while costs through medical treatment and lost production amount to almost $1 billion.

There are still further costs. If inadequate or

Water that kills

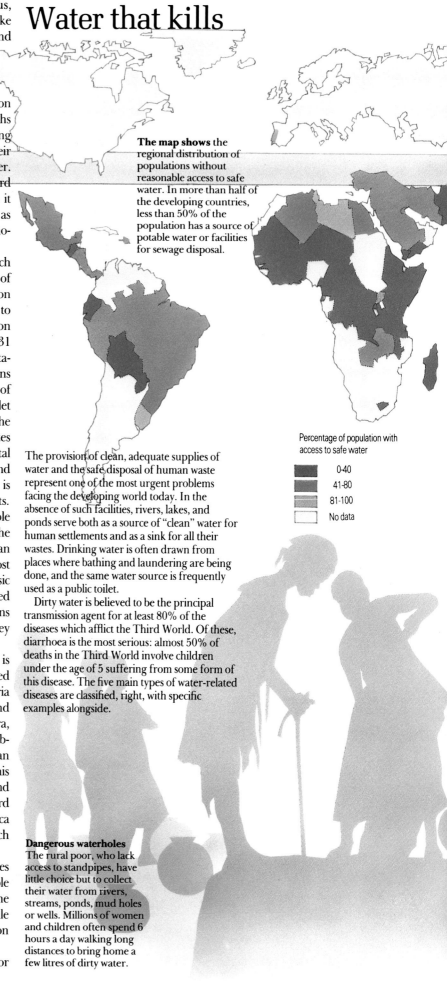

The map shows the regional distribution of populations without reasonable access to safe water. In more than half of the developing countries, less than 50% of the population has a source of potable water or facilities for sewage disposal.

Percentage of population with access to safe water

0-40

41-80

81-100

No data

The provision of clean, adequate supplies of water and the safe disposal of human waste represent one of the most urgent problems facing the developing world today. In the absence of such facilities, rivers, lakes, and ponds serve both as a source of "clean" water for human settlements and as a sink for all their wastes. Drinking water is often drawn from places where bathing and laundering are being done, and the same water source is frequently used as a public toilet.

Dirty water is believed to be the principal transmission agent for at least 80% of the diseases which afflict the Third World. Of these, diarrhoea is the most serious: almost 50% of deaths in the Third World involve children under the age of 5 suffering from some form of this disease. The five main types of water-related diseases are classified, right, with specific examples alongside.

Dangerous waterholes
The rural poor, who lack access to standpipes, have little choice but to collect their water from rivers, streams, ponds, mud holes or wells. Millions of women and children often spend 6 hours a day walking long distances to bring home a few litres of dirty water.

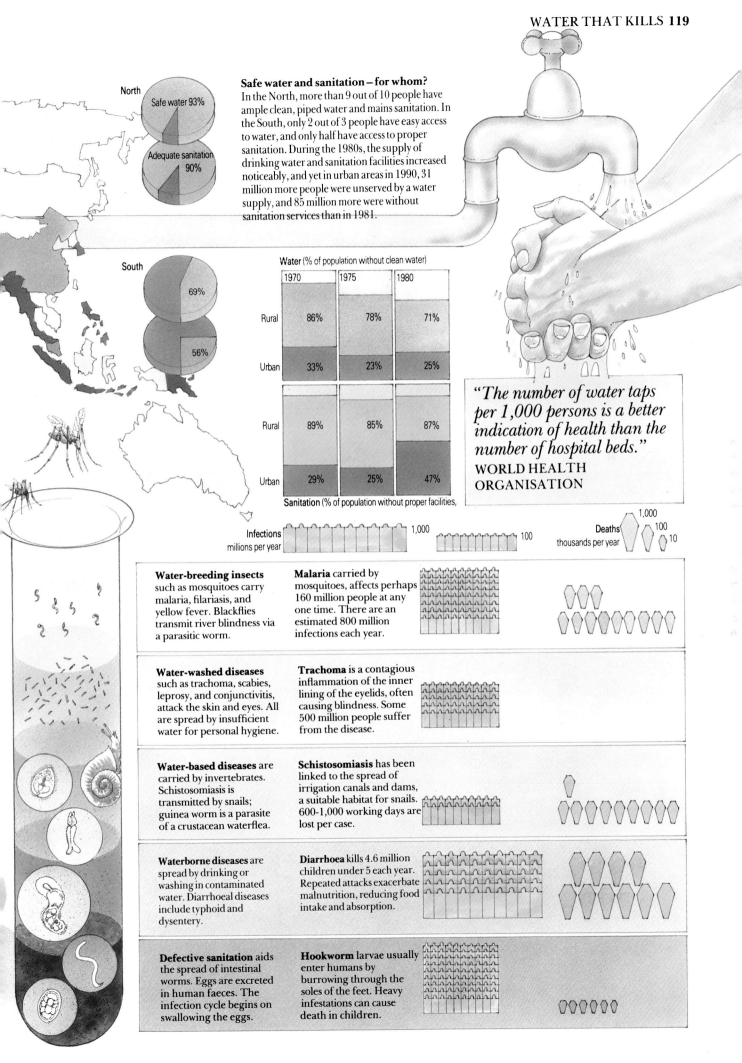

North
Safe water 93%
Adequate sanitation 90%

Safe water and sanitation – for whom?
In the North, more than 9 out of 10 people have ample clean, piped water and mains sanitation. In the South, only 2 out of 3 people have easy access to water, and only half have access to proper sanitation. During the 1980s, the supply of drinking water and sanitation facilities increased noticeably, and yet in urban areas in 1990, 31 million more people were unserved by a water supply, and 85 million more were without sanitation services than in 1981.

South
69%
56%

Water (% of population without clean water)

	1970	1975	1980
Rural	86%	78%	71%
Urban	33%	23%	25%
Rural	89%	85%	87%
Urban	29%	25%	47%

Sanitation (% of population without proper facilities,

"The number of water taps per 1,000 persons is a better indication of health than the number of hospital beds."
WORLD HEALTH ORGANISATION

Infections millions per year — 1,000 · · · 100

Deaths thousands per year — 1,000 · 100 · 10

Water-breeding insects such as mosquitoes carry malaria, filariasis, and yellow fever. Blackflies transmit river blindness via a parasitic worm.

Malaria carried by mosquitoes, affects perhaps 160 million people at any one time. There are an estimated 800 million infections each year.

Water-washed diseases such as trachoma, scabies, leprosy, and conjunctivitis, attack the skin and eyes. All are spread by insufficient water for personal hygiene.

Trachoma is a contagious inflammation of the inner lining of the eyelids, often causing blindness. Some 500 million people suffer from the disease.

Water-based diseases are carried by invertebrates. Schistosomiasis is transmitted by snails; guinea worm is a parasite of a crustacean waterflea.

Schistosomiasis has been linked to the spread of irrigation canals and dams, a suitable habitat for snails. 600-1,000 working days are lost per case.

Waterborne diseases are spread by drinking or washing in contaminated water. Diarrhoeal diseases include typhoid and dysentery.

Diarrhoea kills 4.6 million children under 5 each year. Repeated attacks exacerbate malnutrition, reducing food intake and absorption.

Defective sanitation aids the spread of intestinal worms. Eggs are excreted in human faeces. The infection cycle begins on swallowing the eggs.

Hookworm larvae usually enter humans by burrowing through the soles of the feet. Heavy infestations can cause death in children.

unsafe water causes child deaths on a vast scale, so child deaths stimulate birth rates, which in turn serve to perpetuate population growth.

Hazardous chemicals – a growing problem

Slowly but steadily, we are poisoning our environments. After doing a good job on cutting back the traditional pollutants, notably visible contaminants such as smoke, we are fostering an insidious spread of toxic chemicals and metals in quantities that are very hard to detect. Some of these pollutants are deadlier than anything we have had to face to date. DDT, dieldrin, and other pesticides are suspected of causing cancer and birth defects, among other problems. Polychlorinated biphenyls (PCBs)

constitute a family of chemicals with over 200 types, and are used in plastics, electrical insulators, and hydraulic fluids; they are unusually toxic and persistent, affecting the vital organs. Heavy metals, such as mercury, lead, cadmium, chromium, and nickel, are believed to cause cancer, plus disorders of the lungs, heart, kidneys, and the central nervous system.

Yet we generate vast amounts of these materials. In the US, industry spews forth at least 265 million tonnes of hazardous waste each year, or about one tonne per citizen. Worse, much is disposed of in a manner that does not meet basic standards of environmental safety. Until recently, the main mode of disposal has involved simple dumping in landfill tips, more than 1,100 of which are health hazards.

Widening circle of poison

Our planetary environment is a closed system. Persistent materials pumped into the air, dumped in rivers and seas, or hidden from view in landfills do not simply melt away. The waste by-products of our industrial processes are a growing threat – sometimes an international threat. The developed world regularly exports to the Third World hundreds of millions of kilograms of potentially lethal chemicals – chemicals classified as too dangerous for use in their country of origin. The irony is that these same chemicals (especially pesticides) return to the developed world on bananas, coffee beans, tomatoes, and other food stuffs imported from the developing nations. In 1981 the US revoked export restrictions on these substances on the grounds that US exports were being put at a "competitive disadvantage".

Metallic and non-metallic pollutants
Since the 1930s, when arsenic was the only metal known to be a carcinogen, beryllium, cadmium, cobalt, chromium, iron, lead, nickel, selenium, titanium, and zinc have been added to the list. PCBs, which are totally synthetic and have no natural counterparts, are classified as carcinogens. Once PCBs enter the environment, a process of bio-amplification occurs. Samples taken in the North Sea indicate a PCB concentration of 0.000002 parts per million (ppm) in sea water, but 160 ppm in marine mammals. Cancer-forming chemicals, including PCBs, are found in the tissues of 99% of all Americans. The US National Research Council has estimated that up to 20,000 Americans may die each year from relatively low levels of pesticides in domestically-grown food.

Chemical time-bombs
Abandoned hazardous waste dumps can be chemical time-bombs. In 1978 the community of Love Canal, New York, was itself abandoned when an unusual number of cancers and birth defects came to light. The Hooker Chemical Company had used the site from the late 1940s to dump dioxin, lindane, and mirex. It had taken just 30 years for these chemicals to work their way to the surface. Often such materials contaminate underground water.

1967

Toxic waste

Contamination of water supplies

Exporting the problem
Faced with tightening environmental controls, some waste disposal companies are looking south for a solution to their problems. Millions of tonnes of hazardous wastes are shipped to developing world countries for "disposal" each year.

Bogotá – river of death
Only about one-quarter of Colombia's population has access to pollution-free water; along the banks of the Bogotá River this figure is almost nil. Colombia's largest industrial complex and over 5 million people use the Bogotá as an open sewer. Tocaima, a small town totally dependent on the river for domestic and agricultural water supply, has the country's highest infant mortality rate; between 1974 and 1979 this rate increased by nearly 70%, while cases of gastro-enteritis and diarrhoea increased by one-third. Tocaima is not alone in suffering ill-health.

67% growth in synthetics
US production of organic chemicals grew from 4.75 million tonnes in 1967 to 7.9 million tonnes in 1977 – an increase of 67%.

1977

Pesticides

As for advanced nations, they manufacture 70,000 different chemicals, most of which have not been thoroughly tested to assess their long-term effects on humans. Certain pollutants, especially pesticides and PCBs, accumulate in fatty tissues of organisms. As they pass up food chains – from, say, microscopic creatures in water bodies, to plants, finally to eaters of plant-eaters such as humans – they become concentrated. This process of bio-amplification can magnify their effective dosage by 10-100 times at each stage. As a result, their eventual concentration can increase a million times by the time they reach humans.

Nor do the advanced nations retain these hazardous chemicals for their own use. Of the US output of pesticides, worth $8 billion per year, two-fifths is exported, chiefly to developing countries where safety regulations are not well developed. Of these US exports, one-quarter is made up of chemicals whose use is severely restricted or banned at home. Thus arises a widening "circle of poison".

Ironically, much of the Third World produce treated with developed-world pesticides is grown for developed-world markets. Almost half the coffee imported by the US from Latin America contains chemical residues, while beef shipments are sometimes stopped because they harbour several times more DDT (used as a spray against cattle diseases) than is permitted by US domestic regulations. Similarly, Germany has rejected contaminated

DDT
One of the best-documented examples of how a substance can enter the food chain and become concentrated is DDT. Wholesale use of DDT has resulted in pollution of rivers, and the absorption of tiny amounts by small fish. Instead of being excreted, DDT tends to lodge in the fatty tissues of living organisms. As small fish are eaten by big fish, which, in turn, are eaten by birds (and people), so the ratio of DDT to body weight increases. Although DDT was prohibited for use within the US as long ago as 1972, the US still manufactures over 18 million kg a year for export, largely to the Third World. Ignorance in the Third World of the dangers involved with the use of pesticides such as DDT is a major problem. A sample of rural workers in Central America shows that they have 11 times as much DDT in their bodies as the average American citizen.

Perfect produce?
Extreme importance is attached by Northerners to the appearance of imported food. Some 20% of all agrochemicals used only serve to improve the look of vegetables and fruit.

Body blows
Heavy metal pollutants are responsible for wide-ranging damage to vital organs. Mercury and lead attack the central nervous system (1), nickel and beryllium damage the lungs (2), antimony can lead to heart disease (3), and cadmium causes kidney damage (4).

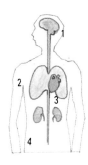

Third World toxics
Chemical companies in the North often sell products to less regulated markets in the South. They also build new factories there, to escape environmental controls. By 1975, El Salvador produced 20% of the world's total output of the pesticide parathion.

Chemical-resistant pests
Indiscriminate pesticide use has produced new strains of pest which are resistant to the commonly used sprays. To counter this, even higher doses are used, further endangering human health and wildlife – and accelerating resistance.

Pesticide poisonings
The United Nations estimates that 40,000 people in developing countries are killed, and two million injured, every year as a result of pesticide poisoning. Developing world countries, accounting for only 20% of world pesticide consumption, suffered 50% of poisonings and 90% of the resulting deaths.

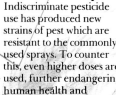

tobacco grown in Thailand, while Scandinavian countries have refused fruit and vegetables from Africa. The issue of hazardous chemicals amounts to one of our greatest environmental challenges.

Nuclear energy – a failed promise?

The vision of cheap nuclear electricity may have evaporated, but many protagonists still believe that nuclear power provides the main hope of the abundant energy they see as critical for future social and ecological stability, notwithstanding the disaster at Chernobyl in 1986. An expanding nuclear industry, they argue, will reduce reliance on fossil-fuel burning and hence reduce the "greenhouse effect" (pp 110-13). There are risks,

they admit, but these must be traded off against potential benefits. During the '60s and '70s, energy utilities in the US, UK, Japan, France, the former USSR, and other countries committed huge financial and scientific resources to their nuclear development programmes. By 1989, some 426 reactors in 27 countries supplied nearly 17 percent of the world's electricity.

But early predictions to the effect that nuclear power would provide 50 percent of the world's electricity by the year 2000 now seem laughable. The world's nuclear power industry looks as though it is driving itself into a cul-de-sac. The industry could possibly live with the disapproval of the environmentalists, but now its most serious

The nuclear dilemma

The nuclear industry is sick. Far from providing a cheap and plentiful supply of energy that would satisfy demand for the foreseeable future, it has provided us with an expensive energy source fraught with intractable technical problems and unacceptable environmental and health risks. Long-lived radioactive wastes cast a shadow which reaches across the generations.

A nuclear dead-end?

In 1974, the International Atomic Energy Authority estimated that by the year 2000 there would be 4450 thousand megawatts of nuclear capacity around the world. By the time of the Chernobyl disaster in 1986, this estimate had been cut ninefold. Nuclear power, for all the promises of its early enthusiasts, has run into a cul-de-sac. Falling energy demand, rising real costs, and public concern have all played their parts in bringing about this state of affairs. Many billions of dollars have been wasted, particularly in the US, on projects that have been abandoned before completion. Electricity companies are discovering that it is cheaper and easier to encourage people to use energy more efficiently, and so reduce demand, than it is to build new capacity to meet growing demand. A dollar invested in energy efficient technology frees up to seven times more energy than a dollar spent on nuclear power. Weapons proliferation is another danger. Even more alarming, this material could fall into the hands of terrorists. The nuclear era may turn out to be unexpectedly short-lived.

	1989	1982	
			10 operating reactors
			1 operating reactor
	△		Cancelled orders
			31-75% of electricity generated by nuclear power
			11-30%
			0-10%
			% unavailable
			Emerging nuclear countries
	①		Nuclear accident site

A catalogue of disasters

1 1957, Windscale, UK A reactor fire led to the contamination of some 800 sq km of land. At least 20 people died from cancer.

2 1958, Urals, USSR An explosion of a nuclear waste site contaminated land and probably killed hundreds of people.

3 1968, Detroit, USA Part of the core of a fast breeder reactor overheated and began to melt.

4 1969, Colorado, USA Spontaneous ignition in a nuclear waste pile caused a release of plutonium dust.

5 1972, New York, USA A plutonium works had to be closed permanently after an explosion.

6 1975, Browns Ferry, USA A workman started a fire with a candle that knocked out emergency systems and nearly destroyed the reactor.

7 1976 Windscale, UK A leak of 2 million litres of radioactive water.

8 1979 Harrisburg, USA Operator error led to the world's most serious core meltdown accident.

9 1981 Windscale, UK Release of iodine 131 into Cumbrian countryside. Local milk supplies contaminated.

10 1986 Chernobyl, USSR Control of the chain reaction lost. Explosion released radioactive cloud that spread right across Europe.

Terrorism

Waste

Warheads

Citizens' movements

Escalating plant costs

Risk of accidents

problem is in the field of economics. Massive cost over-runs, with US plants proving 5-10 times more expensive than had been projected, resulted in the largest municipal bond default in US history, pushing many power utilities to the brink of bankruptcy. No new nuclear reactor has been ordered in the US since 1978, and all orders made in the four preceding years have been cancelled – mainly for economic reasons. The year 1990 was the first in the 35-year history of commercial nuclear power that no new construction began on a power plant.

But those countries which have declared nuclear moratoria have not typically done so because of economic problems. Sweden's reactors were among the safest and most efficient in the world but will be phased out by 2010. The real problems are social and political: the environmental and health risks, the inability of the industry to dispose safely of radioactive waste, and the proliferation of nuclear weaponry have helped to derail nuclear power. But it was the world's worst nuclear accident, at Chernobyl in April 1986, which doubled public opposition to nuclear power in many countries.

Although the nuclear industry is down, it certainly is not out. Despite the well-publicized risks involved, some countries, China included, still want to press ahead with nuclear power. The nuclear industry is not going to disappear, but is seen as a particularly high-risk option – not as the energy panacea originally promised.

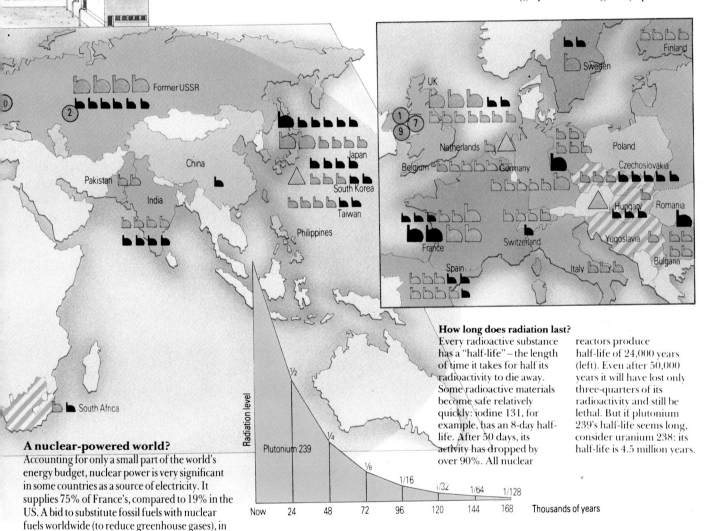

How long does radiation last?
Every radioactive substance has a "half-life" – the length of time it takes for half its radioactivity to die away. Some radioactive materials become safe relatively quickly: iodine 131, for example, has an 8-day half-life. After 50 days, its activity has dropped by over 90%. All nuclear reactors produce half-life of 24,000 years (left). Even after 50,000 years it will have lost only three-quarters of its radioactivity and still be lethal. But if plutonium 239's half-life seems long, consider uranium 238: its half-life is 4.5 million years.

A nuclear-powered world?
Accounting for only a small part of the world's energy budget, nuclear power is very significant in some countries as a source of electricity. It supplies 75% of France's, compared to 19% in the US. A bid to substitute fossil fuels with nuclear fuels worldwide (to reduce greenhouse gases), in meeting an energy demand likely to double by 2025 would require the building of one new plant every two and a half days. This would result in around 18 times as many nuclear plants as at present, posing severe economic and environmental risks, as well as the dangers of nuclear weapons proliferation. For any developing world country wanting to go nuclear, the high capital cost and the small size of their electrical grids are major obstacles. Indeed, so unlikely is nuclear power to be of benefit to such countries as Pakistan or Iran, that the only plausible explanation for nuclear developments seems to be the desire to acquire a nuclear-weapon-making capability.

Dealing with waste
Thirty-five years after the first commercial nuclear reactor began operating, there is still no acceptable solution to the problem of radioactive waste. Used fuel rods can be recycled to recover uranium and plutonium, but this process creates more wastes for disposal. The by-products of nuclear power production are accumulating exponentially, stored underground or in water pools at nuclear plants – a total of 84,000 tonnes in 1990, projected to more than double by the year 2000. Other radioactive waste problems include the tailings from uranium mining, and the emerging question of how to shut down old stations.

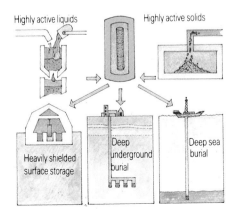

MANAGING THE ELEMENTS

Beset as we are with crises of misuse or overuse of Earth's elements, we can take heart from efforts to get to grips with some. We are using oil more efficiently. For those people who rarely see a drop of oil, relying instead on fuelwood, there is hope that more trees may eventually be planted than are chopped down. We have had a UN Water and Sanitation Decade. In a rare show of accord, the world's governments have put together a convention to safeguard the ozone layer.

Managing energy

The way we manage energy must change. Up to now, policy-makers have typically followed one energy path: increasing energy *supply* to meet the energy *demand*. This has given us ever-more gas and oil fields, together with ever-more consumption of resources and pollution. The sheer scale of our energy use – even greater when the growing needs of the developing world are considered – makes that path increasingly dangerous. Our ability to respond to the threat of global warming will depend on how quickly we are prepared to follow a different, new energy path.

When Thomas Edison founded the first electric utility, his perception was to sell illumination, rather than electricity. People do not want electricity, oil, and gas; they want the services which can be provided by using these fuels – heating, lighting, transport, cooking, and so on. The real demand to be met is for these energy services. The challenge of the new energy path is to meet that demand as efficiently and as cleanly as possible. Two energy options are available that are sustainable, environmentally supportable, and technically feasible: renewable energy and energy efficiency.

Renewable energy resources are available in immense quantities. Contrary to popular belief, renewables already supply about 20 percent of the world's energy – mostly biomass and hydropower – and solar energy is likely to be the cornerstone of the new energy system. Yet whatever measures are taken to reduce our dependence on fossil fuels and uranium, and however fast renewable energy technologies are introduced, there will remain a need to use fossil fuels for the foreseeable future. So we also need to emphasize the introduction of cleaner and more efficient technologies for burning fossil fuels.

Furthermore, there is little point in going to all the expense and trouble of tapping renewable energy with low impact technologies only to find

The new energy path

There are two sides to the energy business: the supply side and the demand side. Up until now, energy management decisions have been dominated by supply-side thinking: building new power stations and digging new oil wells to supply energy. And this has dictated our profligate use. But energy is a means not an end: people do not want electricity, oil, gas – they want the services they provide – heat, light, mobility. The dominance of the energy supply-side is giving way to the demand-side. The energy structure can be redesigned to supply services with reduced energy consumption and pollution. Meeting this challenge involves implementing new policies: on the supply side, reduce the impact of fossil fuels by using cleaner-burning technologies, and introduce renewable energy sources to replace them; on the demand side, use energy far more efficiently than we have in the past.

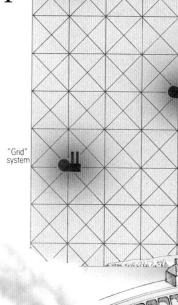

"Grid" system

The shape of the new system
In contrast to the massive, centrally-controlled coal and nuclear plants which can take upwards of ten years to plan and build, renewable sources can be small, local, and quick to develop, allowing cost-effective and flexible planning. Using diverse and decentralized renewable sources can minimize transmission and distribution losses, while strengthening the overall electricity grid. Reliability improves, and dependence on fossil fuel lessens.

High energy/supply-side landscape
The dominant energy policy today (and since World War II) has an energy demand-supply balance tipped in favour of supply, justifying the building of new power stations, drilling of new gas fields, or opening of new oil pipe lines. It bears no regard to people's actual needs, and meets what it considers to be an ever-increasing demand for energy through a complex, centralized distribution system, with highly-polluting technologies.

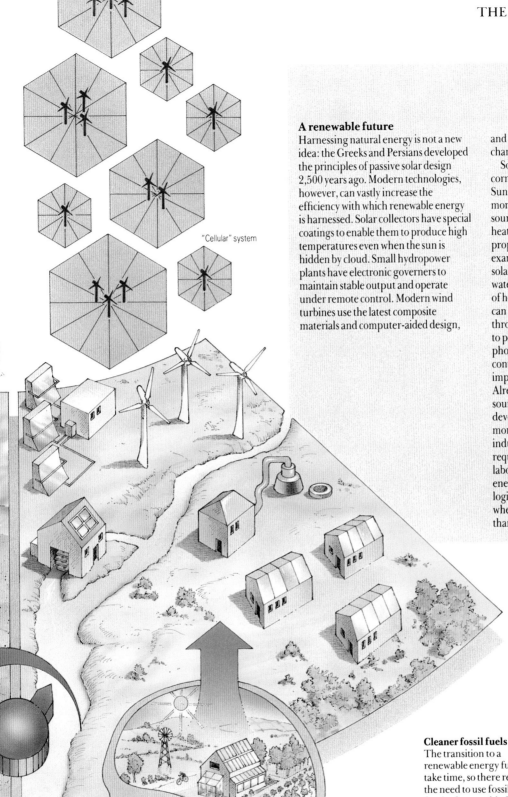

"Cellular" system

A renewable future

Harnessing natural energy is not a new idea: the Greeks and Persians developed the principles of passive solar design 2,500 years ago. Modern technologies, however, can vastly increase the efficiency with which renewable energy is harnessed. Solar collectors have special coatings to enable them to produce high temperatures even when the sun is hidden by cloud. Small hydropower plants have electronic governers to maintain stable output and operate under remote control. Modern wind turbines use the latest composite materials and computer-aided design, and can anticipate and respond to changing wind behaviour.

Solar energy is likely to be the cornerstone of the new energy system. Sunshine is available in great quantity, more widely distributed than any other source, and admirably suitable for water heating, which accounts for a huge proportion of world energy usage. For example, there are more than 1.5 million solar collectors supplying domestic hot water in Tokyo alone; in Israel two-thirds of homes have such heaters. Solar energy can also be turned into electricity – either through heating water to produce steam to power a turbine, or through photovoltaic cells. The price of solar cells continues to fall as technical improvements and demand increase. Already, solar cells are the least expensive source of electricity for much of the rural developing world. Renewables create more jobs than conventional energy industries because their capital requirements are more modest and their labour needs are greater. A renewable energy strategy would therefore be logical for Latin America and Africa where there is more spare labour than capital.

Cleaner fossil fuels

The transition to a renewable energy future will take time, so there remains the need to use fossil fuels for the foreseeable future.

Acid emissions from coal-fired plants have been greatly reduced by the fitting of flue gas desulphurization equipment, although the price paid for this is a reduced efficiency entailing the burning of more coal for the same amount of energy released. For any new power stations there are cleaner, more efficient technologies to reduce the amount of coal needed, but for further reductions in carbon emissions the basic wastage of heat at the power station must be addressed.

Currently, only about 40% of the energy released from burning fossil fuels provides a useful service. The rest is lost as waste heat.

By tapping the "waste" heat and piping it direct to local homes and industry, the efficiency of modern plants can be up to 80%. Such "Combined Heat and Power" (CHP) schemes therefore reduce the amount of fossil fuel needed (and hence carbon emissions).

CHP is well established in many countries, ranging from small systems for single factories and hospitals to large power stations heating extensive residential districts in cities as diverse as Stockholm and Paris.

Low energy/demand-side landscape

The demand for electricity is a reflection of the demand for the services electricity can provide – hot showers, clean clothes, data storage/retrieval, etc. With such a services-orientated perspective, the energy demand-supply balance is tipped in favour of demand. Electricity consumers are then better placed to make suppliers recognize their needs, and meet them in the most appropriate ways.

it leaking from buildings, and inefficiently used in appliances and vehicles. A determined drive to introduce energy efficient technology and behaviour in all sectors of the economy is the most vital component of the new energy path.

The fifth fuel

The shift to renewable energy technologies will take time. Our second energy option, by contrast, we could implement quickly. Energy efficiency means producing the same final energy services – light, heat, transport – but using less energy to do so. This means less economic cost, less conflict over the siting of power plants, and, for many countries, less military or political cost to maintain access to foreign energy resources. Furthermore, energy efficiency offers the most direct, cost-effective method to counter global warming and air pollution.

Consider the humble compact fluorescent lamp. Over its lifetime it will save $30 in electricity, and prevent a coal-fired power station from belching nearly a tonne of carbon dioxide into the air. It saves resources, reduces pollution, and creates wealth. In the US, it is calculated that today's best technologies can save about three-quarters of our electricity for less money than the operating costs of an existing nuclear or coal plant.

The well-being of society is not inexorably tied to its level of energy consumption. True, for the 30 years following World War II, energy consumption did seem proportional to economic growth; but since the rise of OPEC, economic growth in Western Europe and Japan has been three times as great as energy growth. In part, this has been due to the reduced dependency on oil. In part too, it has been due to energy savings: from 1975 to 1985, for example, the energy intensity of the US economy fell by a remarkable 22 percent. Since 1980, the US has got four times as much new energy from efficiency savings as from all increases in supply, saving consumers and industry $150 billion a year. Yet this is just the beginning: insulating superwindows, new motors, and other technologies are being developed that are several times more efficient than the older models. At least ten car companies have built prototype vehicles that drive between 65 and 130 miles to the gallon.

Efficiency technologies, from super-insulation to "smart" motors, are improving so quickly that estimates of energy needs to perform any given task are revised downwards every year. One conservative estimate suggests that the North American economy could remain as active as it is now, but using half as much energy. And this is with the technologies currently available. North America would then be brought to the present efficiency levels of Western Europe and Japan – the most energy-efficient economies in the world. But even these economies could double or even quadruple their efficiency in the foreseeable future.

Some optimistic calculations suggest that, with improved efficiency, the world as a whole could

An energy-efficient future

Current patterns of energy use cannot be sustained. Oil and gas will become too scarce and expensive for all but the most essential uses. Coal will suffer a similar fate within a few more centuries. But the environmental implications of burning fossil fuels are even more pressing. Nuclear power is a long way from fulfilling its early promise, and faces huge public opposition. A key objective in a new energy strategy is to increase the efficiency of energy use in all sectors of the economy; the domestic, industrial, and transport sectors, each of which typically accounts for about a third of an industrial nation's energy consumption. In a true sustainable energy economy, people would live in energy-efficient, solar homes, travel by private car would seldom be necessary, and factories would run on a fraction of their present voracious energy requirements.

Domestic efficiency
With a programme of draught-proofing, insulating walls, roofs, floors, and windows, and improved electronic controls for heating systems, the energy needed to keep a conventional building warm in winter could be cut in half. For new buildings, design and construction techniques exist which can virtually do away with the need for space heating or cooling. New, specially-coated glazing, for instance, can cut heat loss from a building sevenfold compared even with double glazing. The most efficient mass-produced fridges and freezers use 75% less electricity than previously-available models while compact fluorescent lightbulbs use 70-80% less electricity than standard filament bulbs for the same light output.

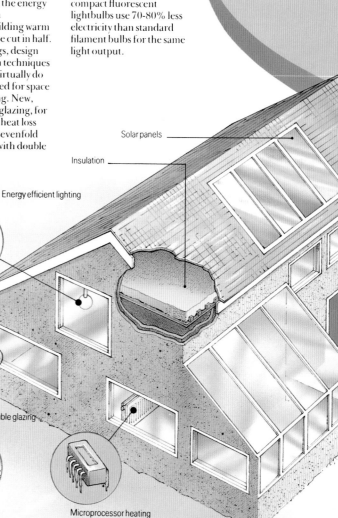

Domestic

Solar panels

Insulation

Energy efficient lighting

Double glazing

Microprocessor heating controls

Passive heating design

Industry

The top 500 transnational companies generate half the greenhouse emissions produced by world industry. Their conservation, energy efficiency, recycling, and waste management policies are fundamental to the health of the planet. For years, energy-intensive industries have seen efficiency as a way of reducing costs. Some have developed ways to recycle products and to use wastes as fuel. Significant energy savings can come, too, from using more efficient electric motors, new sensors and control devices, and advanced heat-recovery systems, such as Combined Heat and Power. Cleaner, more efficient process technologies can yield large energy savings, but require extra capital expenditure.

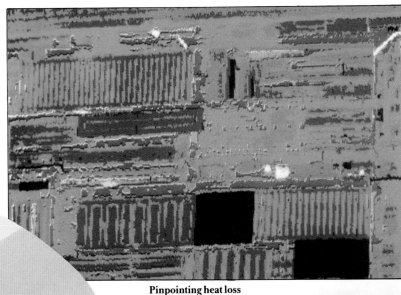

Industry

Transport

Pinpointing heat loss

The image, above, shows heat loss from an industrial complex. Aerial infra-red photography helps energy managers pinpoint where they can spend their conservation budgets to best effect. White, yellow, and red areas show points of greatest heat loss.

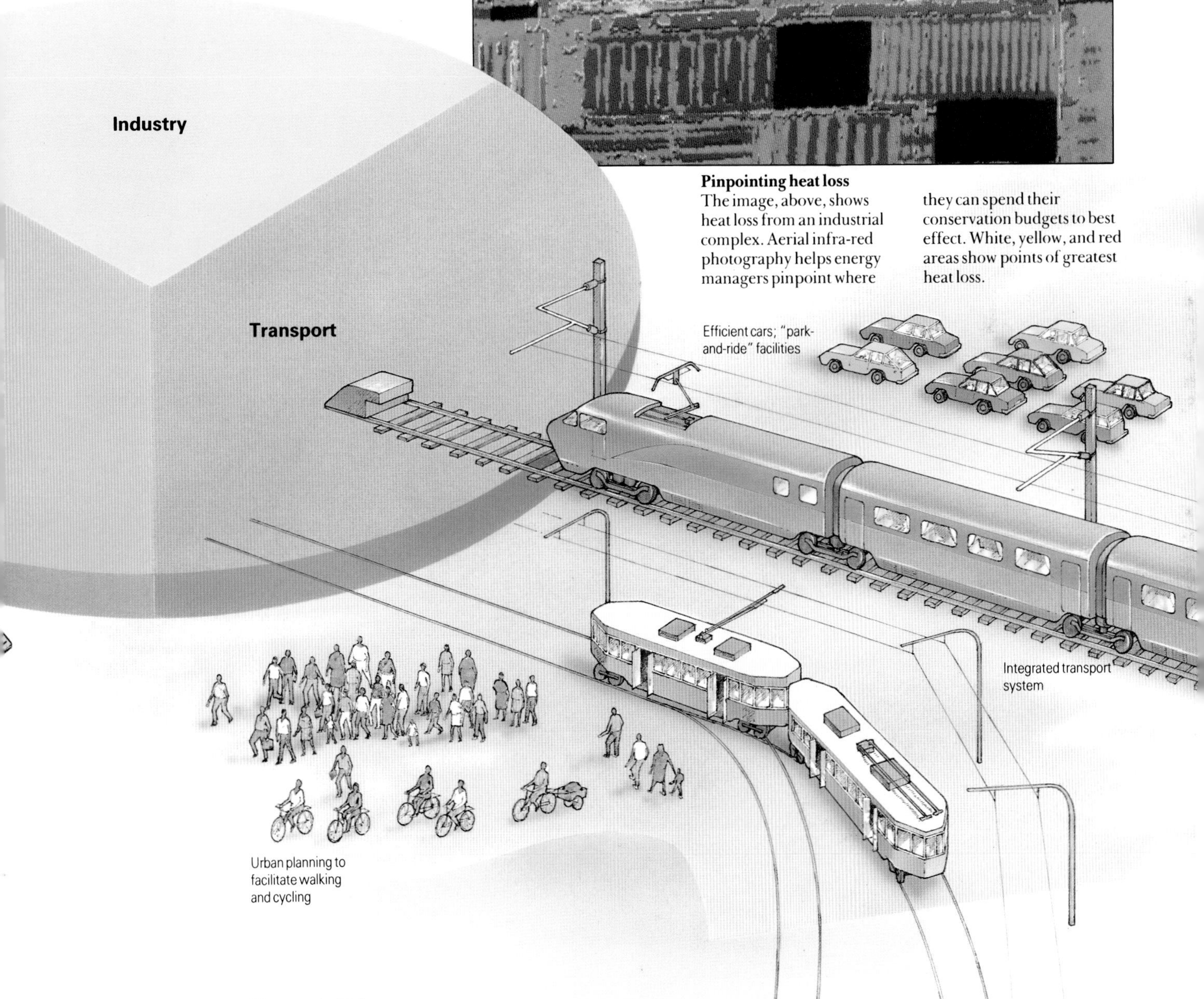

Efficient cars; "park-and-ride" facilities

Integrated transport system

Urban planning to facilitate walking and cycling

Transport efficiency

A rapidly-growing energy user, the transport sector is responsible for about 20% of global fuel use, and a third of total energy consumption in industrial countries. We can turn to high-efficiency cars, which make use of electronic tuning and streamlined bodywork. But mass transport systems – like buses, trams, and trains – are far more energy efficient. Integrated transport systems are demand-led, providing transport links tailored to need, such as "park-and-ride" facilities at rail and bus stations. Such services, together with land-use planning to reduce home to work distances and transport demand, are vital to a new energy path.

retain its current energy use at present levels with no reduction in productivity in the rich world, and still accommodate economic growth in the poorer countries.

Supplying energy to the South
Developing countries, with international aid, have recently begun to face up to the fuelwood crisis. But, if we are to meet the challenge, we must plant trees five times faster in the Third World as a whole, and in the worst-off areas, such as those of West Africa, between 15 and 50 times as fast.

Not that the task is as simple as raising money – although development agencies, led by the World Bank, are finally getting round to the job. Local

people need to be helped to confront challenge in ways *they* think are relevant. Too often the approach has been a "top down" affair, with urban bureaucrats trying to impose their plans on rural communities. Result: the planted trees are not cared for – or are poached. So at least as important as money is a "grass-roots" spirit, that mobilizes the initiative of the local people.

Fortunately, an expanded strategy along these lines is under way, known as "community forestry". It looks at village woodlots not only as sources of fuelwood, but as safeguards for topsoil, as windbreaks to protect crops, and as sources of food.

At the same time, we could do much through the simple expedient of improved cooking stoves that

Managing energy in the South

Managing energy policy in the developing countries of the South is, in many ways, much more complex than managing it in the industrialized North. Northern energy strategies are chiefly determined by price competition between four main fuels (electricity, oil, coal, and gas) supplied by large, often monopolistic, public or private suppliers. An extensive infrastructure provides each consumer with a range of options for meeting their energy needs. In the South, the biomass fuels, dung, wood, and charcoal, join the fuel mix in the poorer countries, playing a far larger role than fossil fuels or electricity. Often, there is no market at all in the primary fuel and where there is a market it is local or regional. The penetration of electricity is typically small and confined to major cities. Fossil fuels, at world market prices, are rarely cheap enough for anything but the most essential uses.

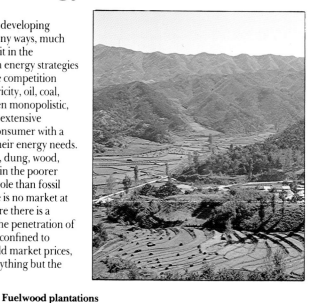

Reforestation in South Korea
Since wood is the primary fuel in the South, successful reforestation programmes are a key to future energy security. Both China and South Korea, operating in vastly different economic contexts, have mounted successful programmes, the South Korean programme in particular illustrating many of the lessons that must be learnt. Nearly 70% of Korea is now covered in trees, half of them young, and some 40,000 ha a year have been planted since 1976. Up until 1973, however, Korea's efforts at reforestation failed. Later, a new approach was tried, combining strong government support – it paid 65% of the costs – with intensive local participation. Village committees with locally elected leaders selected the best sites for planting and offered villagers a chance to share in the profits from the scheme. Grass-roots involvement is an essential condition for effective reforestation.

Fuelwood plantations
A plantation of fast-growing trees, such as *Leucaena*, can yield up to 50 tonnes of wood per hectare each year. They are especially useful for meeting urban fuelwood needs but are expensive to set up.

Village woodlots
Village woodlots are part of a new approach to reforestation known as "social" or "community" forestry. Trees planted on village or private land produce wood and other products for village, rather than commercial, use. Foresters educate and advise, but the villagers manage.

Fuel from sugarcane
One-quarter of the cars now sold in Brazil run on pure ethanol produced from the fermentation of sugarcane – the rest run on a 20% ethanol blend. Each tonne of cane produces 70 litres of ethanol. Finding an indigenous liquid fuel for transport is a pressing Southern energy problem.

generate more heat while burning less wood. Most stoves deliver less than 20 percent of potential heat, many only 5 percent (a developed-world stove is 70 percent efficient). An improved-design stove can cut heat losses by almost half, while costing a mere $10. But getting enough built at a low-enough cost, and getting them used by people with deep-rooted cooking traditions, is not proving easy.

Other forms of biomass are available besides wood. Most developing nations are in the tropics, where plants generally thrive. Sugarcane, cassava, and maize contain enough sugar or starch to generate ethanol through fermentation, while fast-growing water weeds and algae produce methane. Brazil, Zimbabwe, and the Philippines already obtain much

of their liquid fuel from biomass – and a "petroleum plantation" will never run dry.

Of course, the sun's energy can be tapped through other means, such as photovoltaic cells and solar pumps. Regrettably, however, these may well remain too costly for the poor. Alternative technologies range from established items such as hydropower and geothermal energy, to innovative devices such as high-tech windmills and tidal power. Some of these look promising, others less so. But there are plenty of options to help developing countries through their energy squeeze – provided the investment and the technology can be made available from the North. The recent rate of oil-field discovery has been nine times greater in the South than

Photovoltaics
In a recent experiment, each house in an Indonesian village was equipped with solar cells and energy efficient lights and appliances. The villagers will pay for these over ten years at the same interest rates that the power utility pays for power plants. Even though the utility had run a power line right past the village, it was found that it was cheaper not to hook up, but to use the stand-alone voltaics instead.

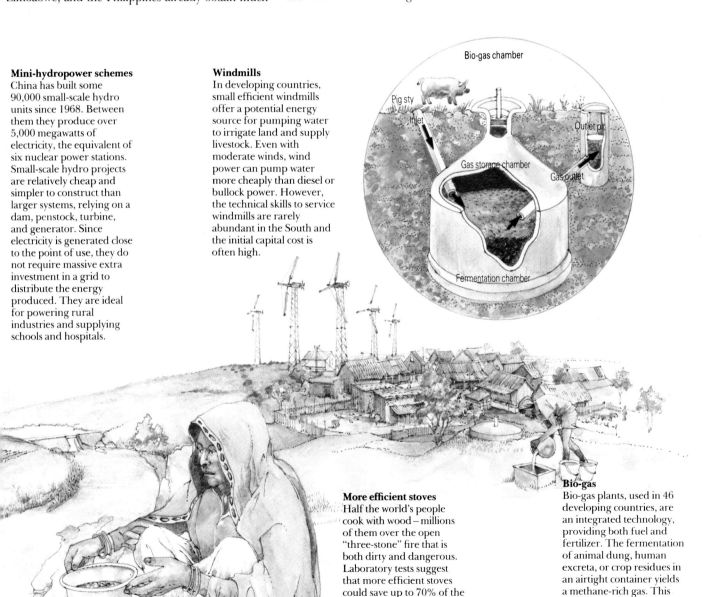

Mini-hydropower schemes
China has built some 90,000 small-scale hydro units since 1968. Between them they produce over 5,000 megawatts of electricity, the equivalent of six nuclear power stations. Small-scale hydro projects are relatively cheap and simpler to construct than larger systems, relying on a dam, penstock, turbine, and generator. Since electricity is generated close to the point of use, they do not require massive extra investment in a grid to distribute the energy produced. They are ideal for powering rural industries and supplying schools and hospitals.

Windmills
In developing countries, small efficient windmills offer a potential energy source for pumping water to irrigate land and supply livestock. Even with moderate winds, wind power can pump water more cheaply than diesel or bullock power. However, the technical skills to service windmills are rarely abundant in the South and the initial capital cost is often high.

More efficient stoves
Half the world's people cook with wood – millions of them over the open "three-stone" fire that is both dirty and dangerous. Laboratory tests suggest that more efficient stoves could save up to 70% of the wood used in an open fire. But too many programmes to provide efficient stoves fail. Bringing a stove from the laboratory into widespread use is a very complex process. To succeed, the needs and wishes of the intended users must be considered at every stage.

Bio-gas
Bio-gas plants, used in 46 developing countries, are an integrated technology, providing both fuel and fertilizer. The fermentation of animal dung, human excreta, or crop residues in an airtight container yields a methane-rich gas. This can be used to heat stoves, light lamps, run machinery, or produce electricity. The residue left by bio-gas production can be used as a fertilizer or as a component in animal feed. Sewage systems built to collect human wastes for bio-gas production also help to improve hygiene.

elsewhere. In an interdependent world, it makes economic and political sense for the North – which uses and wastes most energy – to ensure that the South has the resources needed to harness its energy resources on a sustainable basis.

Climate of agreement

Recognition of the global nature of problems such as climate change has resulted in a new kind of co-operation between governments in negotiating agreements. We are witnessing the birth of global "laws of the air" – but some bring more birth pains than others.

The treaty to protect the ozone layer is widely hailed as a landmark in environmental diplomacy. A first international agreement in 1985 created a mechanism for co-operation on research. After alarming scientific evidence had been gathered, the Montreal Protocol on Substances that Deplete the Ozone Layer was signed in 1987, committing signatory nations to a timetable of cuts in emissions of ozone-depleting CFCs and halons. In just over a year it had received enough ratifications to enter into force. As further worrying information came to light, the treaty was amended, and the CFC phase-out was accelerated to 1995.

International diplomacy is famed for moving at a glacial pace. Yet faced with a serious global problem, the world community moved quickly from a commitment-free convention to a binding treaty affecting powerful commercial interests and everyday products. What was the secret of the Montreal Protocol's success? Public concern, backed by the international scientific community, UNEP, and NGOs such as FoE and Greenpeace, formed a powerful consensus for action. The US already had the strictest national ozone protection legislation, and urged the more reluctant Europeans, Soviets, and Japanese to act. But perhaps the single most vital factor in the treaty's success was the creation of a fund to help developing countries make the transition to CFC substitutes. These countries argued that it was unfair of the North to expect the South to incur the cost of switching to substitutes to solve a problem it had no role in creating. Without their participation, growing CFC use in the developing world would have overwhelmed the treaty's reductions. So, at the last minute, a fund to help the developing countries was agreed and many of them, including China, subsequently ratified the convention.

Less successful was the 1992 Framework Convention on Climate Change, where intransigence on the part of the US (responsible for a quarter of the world's carbon dioxide output) prevented any binding targets to reduce carbon emissions by 2000, as many had hoped. This was despite calculations from scientists of the IPCC that a 60 percent cut in greenhouse gas emissions is an immediate necessity to stabilize their concentrations. Nevertheless the Convention does

The laws of the air

The atmosphere – like the oceans – recognizes no boundaries. Problems caused in one country or region spread to others. Issues such as acid rain, climate change, and ozone layer depletion can only be addressed by international action and agreement. New panels, protocols, and conventions regarding atmosphere and climate provide some hope and positive models for international diplomacy and agreement. The Convention on Long-Range Transboundary Air Pollution is well-positioned to deal with international air pollution, particularly since its members include all European and North American countries which account for 80% of global sulphur pollution. The Montreal Protocol on Substances that Deplete the Ozone Layer is proof that co-operation can succeed in heading off global environmental problems.

The 30% Club – 1985
The 1985 CLRTAP Protocol on Sulphur Emissions commits nations to reduce sulphur emissions from 1980 levels by 30% by 1993. Many nations have achieved this and are going further. Others, notably the US and UK, refuse to join the club.

NOx and VOC – 1988/91
A 1988 CLRTAP protocol calls for a freeze on nitrogen oxide (NOx) emissions at 1987 values by 1995, and the EC intends to cut emissions by 30% by 1998. Another 1991 CLRTAP protocol will also see emissions of smog-causing Volatile Organic Compounds (VOCs) from 21 nations cut by 30% by 1999.

Global air circulation
The Equator receives more of the sun's energy than the poles because the sun is directly overhead. The air at the Equator heats up, expands, and rises to high altitudes. This air cools and descends in the sub-tropics on either side of the Equator. Winds at low altitudes return air towards the Equator to replace the rising air, completing a circulation pattern. Further polewards, the pattern of circulation is more complex. Overall, warm air is transferred polewards, and is replaced by cold air moving towards the equator. Because the planet is spinning, the circulating air is dragged with it in great spirals. The air system carries water vapour and heat around the globe, but it also carries pollutants.

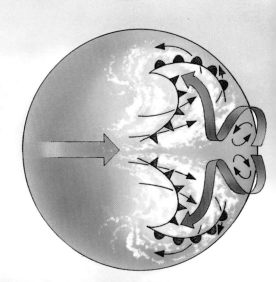

CLRTAP
The Convention on Long-Range Transboundary Air Pollution (CLRTAP) was the first international agreement on air pollution, signed by all European and North American countries.

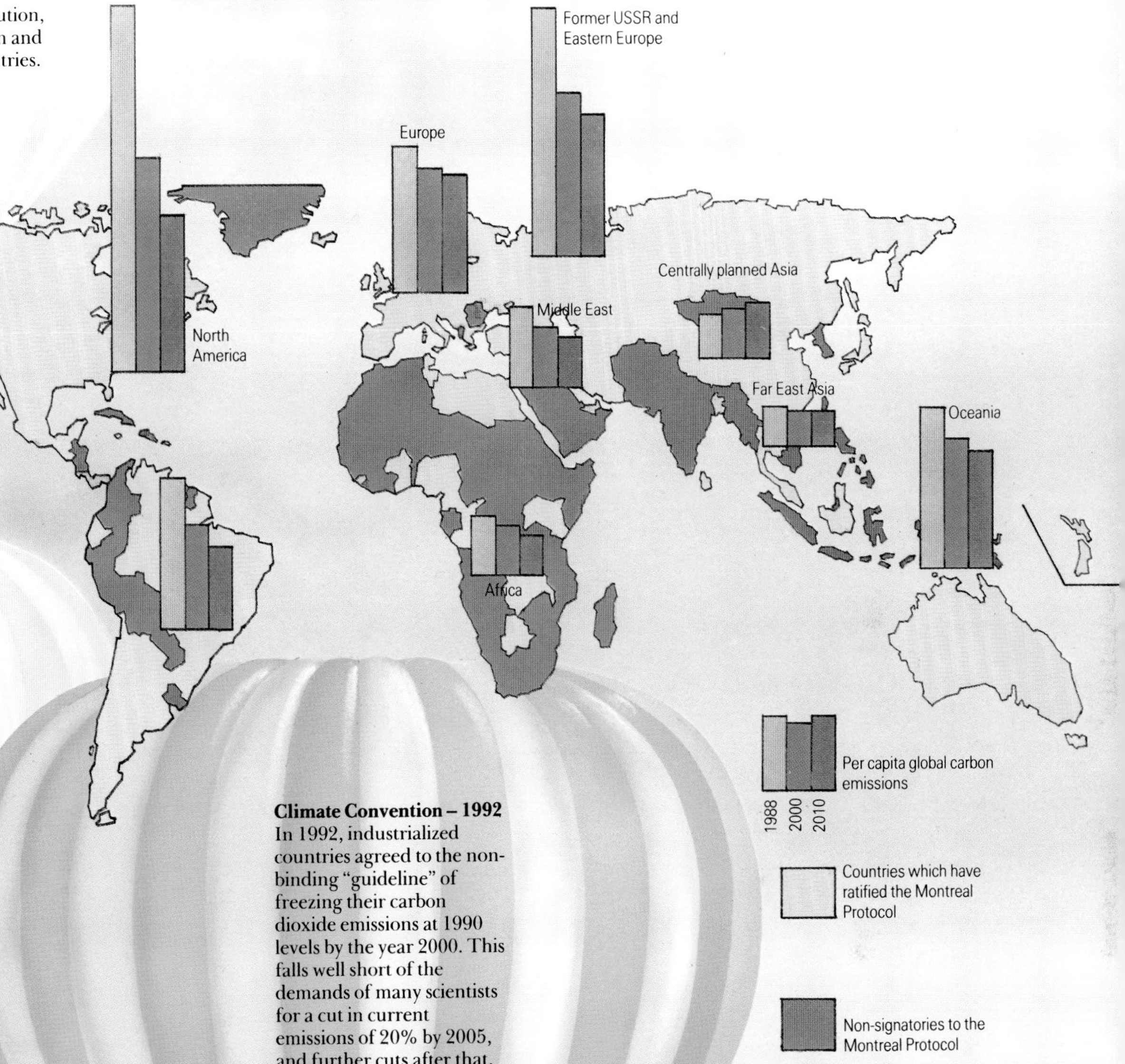

Former USSR and
Eastern Europe

Europe

Centrally planned Asia

North
America

Middle East

Far East Asia

Oceania

Africa

Montreal Protocol – 1987
Since its initial signing in 1987, the Montreal Protocol has been ratified by 75 countries, and has been revised as more data on the perilous state of the ozone layer came to light. 34 industrialised nations will cease production of the chemicals CFC-11 and CFC-12 by 1995, and by 2000 most of the rest will be eliminated.

Per capita global carbon
emissions

1988
2000
2010

Countries which have
ratified the Montreal
Protocol

Non-signatories to the
Montreal Protocol

Climate Convention – 1992
In 1992, industrialized countries agreed to the non-binding "guideline" of freezing their carbon dioxide emissions at 1990 levels by the year 2000. This falls well short of the demands of many scientists for a cut in current emissions of 20% by 2005, and further cuts after that. However, the agreement does require governments to submit detailed action plans of how carbon emissions will be curtailed.

Bilateral agreements
The international dimension of air pollution is well illustrated by Germany's decision to pay for sulphur dioxide scrubbers on Czech power plants contaminating German airspace, and Sweden similarly assisting Poland. The money so invested reaps cleaner air in these West European countries than by spending it on their own industries.

Reducing carbon emissions
What measures are necessary, and possible, to stabilize global carbon dioxide concentrations? The Worldwatch Institute has formulated a set of targets to narrow the disparity that exists among national per capita emission levels (see map, above). Nations with the highest emission rates would have to make the greatest cuts, while poorer nations in Africa and Asia would be able to increase theirs. Despite the increasing global population, these targets would cut carbon dioxide emissions by 12% by 2000. With CFC production under the Montreal Protocol phased out by then, the world could be on course to stabilizing carbon concentrations by mid next century.

establish a Scientific and Technical Committee to guide the UN on climate warming strategies. As the science becomes firmer, so targets and commitments will have to be set and enforced.

Our growing demand for water

Water is an infinitely renewable resource. By the end of the century, we shall need between two and three times as much as in 1980. Through careful management, everybody can enjoy enough and to spare – though if we carry on as we do now, another 15 years could find some 30 nations in deep trouble.

Fortunately, however, we have plenty of options. The thirstiest sector, irrigation, accounts for 70 percent of all water use, producing as much as 35 percent of our food from 17 percent of our croplands. Irrigation is rarely more than 30 percent efficient, even in the US. But in California and Israel, farmers make three times better use of scarce water, through "trickle-drip" irrigation. This technique applies small amounts of water direct to the parts of the plant that need it most. This results in much more food being grown with only half as much water, and with reduced risk of salinization. Only about 5,000 sq km of cropland now receive the benefits of this Blue Revolution, or less than half of one percent of all irrigated lands. Plainly there is much scope for the future.

More perplexing yet equally urgent, is the political dimension. Of 200 major rivers, almost three-quarters are shared by two countries, and the rest by three or more. Already there are disputes between India and Bangladesh over the Ganges, and Brazil and Argentina over La Plata, among at least a dozen such conflicts – and we can expect there will be many more instances, unless we can rise to the challenge in a co-operative spirit. A river basin represents a natural unit for management of a resource that recognizes no national boundaries. Fortunately we can draw on "blueprint experience" of success stories such as those of the Danube and Nile. In the case of the Nile, Sudan and Egypt work together to extract maximum benefit from a river that, despite its length, is small by its year-round volume. Despite some adverse repercussions of the Aswan Dam, Egypt sustains its agriculture and emergent industries through virtually 100 percent use of the Nile's river flow.

If irrigation agriculture now takes the largest share of water supplies, industry accounts for the fastest-growing share, while domestic demand is likewise soaring – an intersectoral source of conflict. The case of Israel is instructive, a country that already exploits 95 percent of its water resources, and that seeks to make more efficient use through recycling. A full one-fifth of water used by industry and households is recovered, mainly for irrigation. In 1962, the final amount of water used solely to generate $100 of manufactured goods was 20 cubic

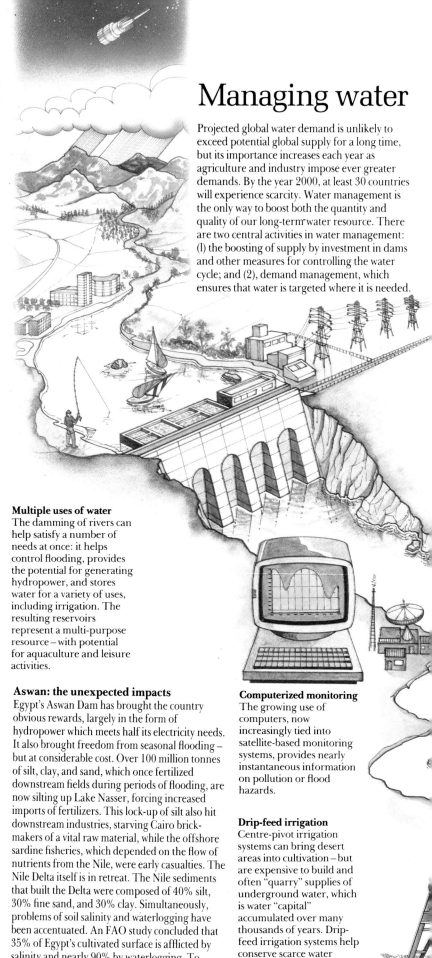

Managing water

Projected global water demand is unlikely to exceed potential global supply for a long time, but its importance increases each year as agriculture and industry impose ever greater demands. By the year 2000, at least 30 countries will experience scarcity. Water management is the only way to boost both the quantity and quality of our long-term water resource. There are two central activities in water management: (l) the boosting of supply by investment in dams and other measures for controlling the water cycle; and (2), demand management, which ensures that water is targeted where it is needed.

Multiple uses of water
The damming of rivers can help satisfy a number of needs at once: it helps control flooding, provides the potential for generating hydropower, and stores water for a variety of uses, including irrigation. The resulting reservoirs represent a multi-purpose resource – with potential for aquaculture and leisure activities.

Aswan: the unexpected impacts
Egypt's Aswan Dam has brought the country obvious rewards, largely in the form of hydropower which meets half its electricity needs. It also brought freedom from seasonal flooding – but at considerable cost. Over 100 million tonnes of silt, clay, and sand, which once fertilized downstream fields during periods of flooding, are now silting up Lake Nasser, forcing increased imports of fertilizers. This lock-up of silt also hit downstream industries, starving Cairo brick-makers of a vital raw material, while the offshore sardine fisheries, which depended on the flow of nutrients from the Nile, were early casualties. The Nile Delta itself is in retreat. The Nile sediments that built the Delta were composed of 40% silt, 30% fine sand, and 30% clay. Simultaneously, problems of soil salinity and waterlogging have been accentuated. An FAO study concluded that 35% of Egypt's cultivated surface is afflicted by salinity and nearly 90% by waterlogging. To crown all of this, the water-based parasitic disease schistosomiasis (p. 119) has exploded among people living around Lake Nasser and along the new irrigation canals. Thorough impact studies could have lessened many such problems.

Computerized monitoring
The growing use of computers, now increasingly tied into satellite-based monitoring systems, provides nearly instantaneous information on pollution or flood hazards.

Drip-feed irrigation
Centre-pivot irrigation systems can bring desert areas into cultivation – but are expensive to build and often "quarry" supplies of underground water, which is water "capital" accumulated over many thousands of years. Drip-feed irrigation systems help conserve scarce water resources, but have other benefits too. They remove the threat of parasitic disease spread by irrigation canals, and reduce soil salinization – which claims over 1 million ha a year.

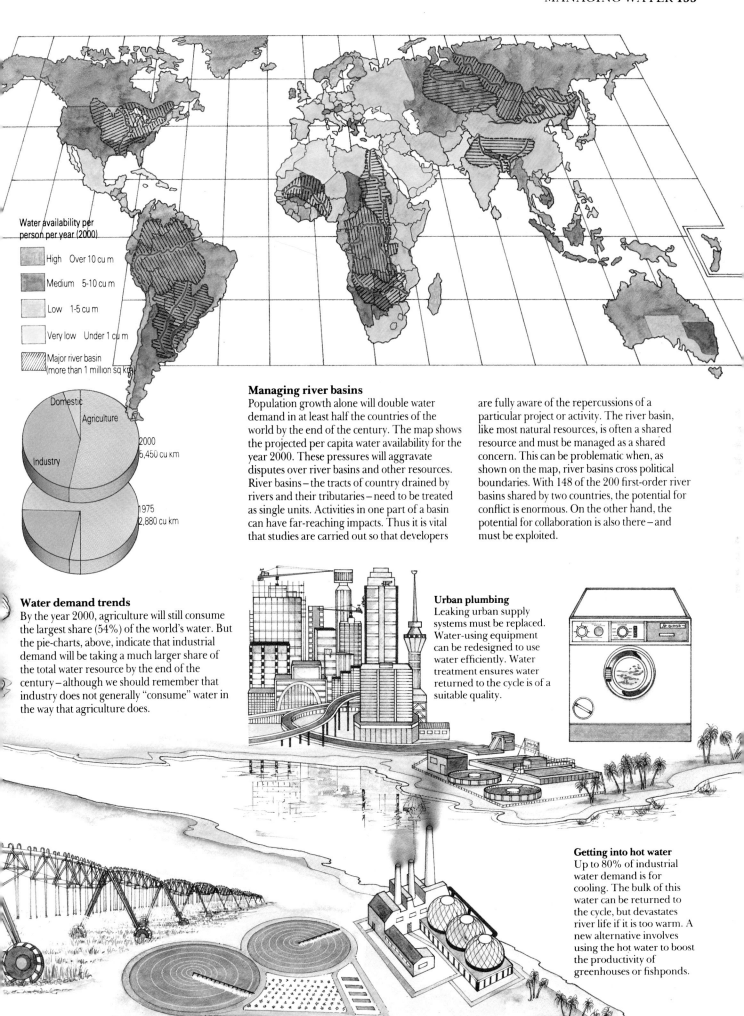

Water availability per person per year (2000)

High   Over 10 cu m

Medium   5-10 cu m

Low   1-5 cu m

Very low   Under 1 cu m

Major river basin (more than 1 million sq km)

Domestic

Agriculture

Industry

2000
5,450 cu km

1975
2,880 cu km

Water demand trends

By the year 2000, agriculture will still consume the largest share (54%) of the world's water. But the pie-charts, above, indicate that industrial demand will be taking a much larger share of the total water resource by the end of the century – although we should remember that industry does not generally "consume" water in the way that agriculture does.

Managing river basins

Population growth alone will double water demand in at least half the countries of the world by the end of the century. The map shows the projected per capita water availability for the year 2000. These pressures will aggravate disputes over river basins and other resources. River basins – the tracts of country drained by rivers and their tributaries – need to be treated as single units. Activities in one part of a basin can have far-reaching impacts. Thus it is vital that studies are carried out so that developers are fully aware of the repercussions of a particular project or activity. The river basin, like most natural resources, is often a shared resource and must be managed as a shared concern. This can be problematic when, as shown on the map, river basins cross political boundaries. With 148 of the 200 first-order river basins shared by two countries, the potential for conflict is enormous. On the other hand, the potential for collaboration is also there – and must be exploited.

Urban plumbing

Leaking urban supply systems must be replaced. Water-using equipment can be redesigned to use water efficiently. Water treatment ensures water returned to the cycle is of a suitable quality.

Getting into hot water

Up to 80% of industrial water demand is for cooling. The bulk of this water can be returned to the cycle, but devastates river life if it is too warm. A new alternative involves using the hot water to boost the productivity of greenhouses or fishponds.

metres, in 1975 only a little over one-third as much, the difference being made up by re-use within industry and by transfer to agriculture.

The UN Water Decade

The International Drinking Water Supply and Sanitation Decade, 1981-90, encapsulated an impulse that we should all take pride in – for the first time in history a concerted effort to supply some of the most basic facilities to more than one billion people in the developing world. It was an ambitious aim, not simply because of its grandiose scope but because few political leaders bothered with the issue. Yet the community of nations got together in 1978 to talk about taps and latrines, and to devise a plan to supply them on a grand scale.

The aim of the Decade
To bring about a real improvement in health through an integrated approach to sanitation and water management.

The UN has called the 1980s the "lost decade of development". The downturn in the world economy, the doubling of long-term debts, and population growth conspired to prevent the Decade from achieving its objective of universal access to water and sanitation. And yet 1300 million people received water supply, and 750 milllion received sanitation facilities, for the very first time.

One of the greatest impacts of the Decade was the higher priority given to water supply and sanitation issues at national and international levels. Many developing country governments learned that health programmes based on glossy new hospitals in cities only divert attention from real priorities such as clean water in rural areas – the zone where there is most

Clean water for all

The scope of the UN's Water and Sanitation Decade objective was massive. It was launched at the UN General Assembly with the aim of "providing water and sanitation for all" by 1990, but it was soon clear that full implementation would be a remote possibility. Rapid population growth, migrations of people from rural areas to cities, war, famine, and drought were added to the problems of debt and the global economic downturn which thwarted the Decade's objective. However, the Decade achieved remarkable results, especially in rural water supplies, where the number of people with access to facilities increased by 240%; the number with new sanitation facilities increasing by 150%. The map, right, shows the original targets of a number of developing countries with the actual 1990 percentage of population covered. The momentum of the Decade has gathered speed as the international community heads towards the goal of Health (and water and sanitation) for All by the year 2000. But for there to be any lasting impact, available funds must also be channelled into a variety of related areas (as indicated below).

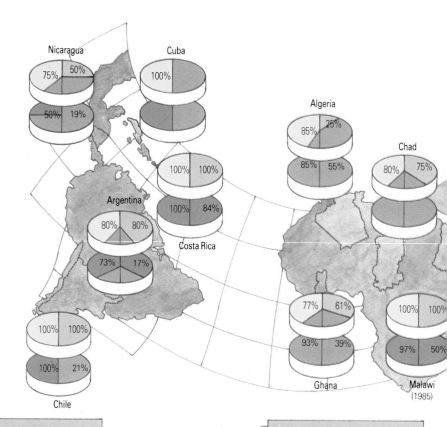

Funds
Secure and reliable funding is vital for the huge investments required by the Decade. This means a combination of local funds and those derived from central governments, overseas donors and banks. Between a third and a fifth of the investment will come from external sources.

Policy
The policy aims to: "foster complementary water supply and sanitation projects; focus on under-served populations; generate replicable and self-sustaining programmes; develop socially relevant and affordable systems; and link supply and sanitation with other improvements."

Education
The supply of clean water and sanitation must be linked to environmental health education programmes.

Grass roots
Special attention must be paid to training people at the village level. Trainees should be selected from within the communities.

Local funding
Funds to be made available to provide care and health education at a local level, avoiding over-centralized facilities.

need. Instead, they will get a better return on each "health dollar" if they invest it in preventative medicine, primarily in the form of plentiful supplies of clean water. Ironically, it was the lack of financial and human resources that eventually forced governments and external agencies to adopt new approaches to promoting safe water and sanitation. Many more facilities could be built with existing resources, and their use and maintenance improved, if the beneficiaries were involved at all stages of development and operation. The agencies began to be sensitive to the key roles that could be played by women, community leaders, and others.

The Decade also brought about some remarkable improvements in drilling, water treatment, and sanitation technologies. As a result, the new technological solutions are cheaper and more efficient than those installed when the Decade began.

With the Decade now over, the international community still faces the enormous challenge of how the pumps, pipes, and pits are to be dug, drilled, and laid in order to reach a new target: "Health for All" by the year 2000. The total cost of attaining universal water and sanitation coverage is estimated at $36 billion per year. However, just one-third of this sum per annum would enable 80 percent of those currently unserved to be reached.

The Decade may have failed in its numerical objective of water and sanitation for all but it was not only about numbers, it was about people too. It succeeded in focusing attention on the user communities

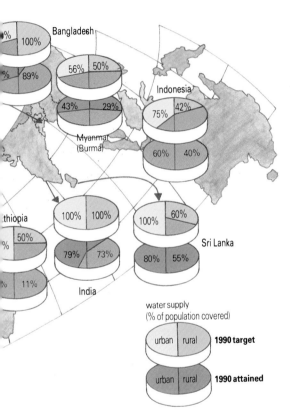

Bangladesh 100%
89% 56% 50%
43% 29%
Indonesia
75% 42%
Myanmar (Burma)
60% 40%
thiopia 100% 100%
50% 100% 60%
79% 73% Sri Lanka
11% 80% 55%
India

water supply
(% of population covered)

| urban | rural | **1990 target** |
| urban | rural | **1990 attained** |

Better health
Clean water supplies, improved sanitation, and relevant hygiene education produce a significant improvement in health.

Child Mortality
Clean water could save many of the 4.6 million children who die each year of diarrhoea.

Stable population
Reduction in child mortality leads to more effective central planning and resource allocation, and a fall in birth rate.

Malawi – getting it right
Malawi, one of the poorest communities, is the only Third World country likely to achieve the objective of clean water for its entire population. Their target date is the year 2000. Two major factors have contributed to Malawi's success: village-based self-help schemes and the choice of appropriate technologies. For a large water project, a committee is formed involving local villagers and government officials, who provide technical support. The community which is to benefit from the project provides the labour and local organization. Prior to 1977, the government favoured borewells. These, however, were expensive to sink and required constant maintenance. Where sufficient ground water was available, gravity-piped water schemes were found to be a more suitable alternative. Malawi is still in the process of looking for the perfect pump and so far more than 200 designs have been tested.

as active participants in the development process, rather than their being merely passive recipients as before.

Towards a conserver society

While reviewing energy, water, and other elements, we have seen how we need to make better use of our resources. We have enough and to spare, provided we are less profligate in our ways. Fortunately, there is massive scope for us to improve. The challenge lies not so much with "technical fixes", but with our approach to our world around us. Hitherto we have engaged in something of a Wild West economy, supposing that there are always pastures new beyond the horizon. Now we know there are no new horizons to explore and exploit: our planet is a closed eco-system, and we are running up against the boundaries of our biosphere. So a more appropriate image is Earth as a spaceship, where most materials have to be recycled. For us, "moving on" will be a case of leaving behind the throwaway society, and advancing to a conserver society.

To qualify as citizens of a conserver society, we must shift entrenched attitudes and thinking. We need to recognize that there is rarely such a thing as "waste": rather there are materials that sometimes end up in the wrong place.

The transition has already begun. The European steel industry re-uses scrap metal with energy savings as high as 50%; in the case of copper, 90%; and of aluminium, 95%. Recycling a glass container saves only 8%; but in parts of the US, a citizen buying a bottle of soda or beer now pays a deposit against return of the empty bottle. If all drinks containers in the US were to be re-used, the annual savings would amount to 0.5 million tonnes of glass – plus almost 50 million barrels of oil used in production processes. In Japan, OPEC spurred an increase in recycling of raw materials from 16% to 48% in just five years. Major new businesses are emerging to exploit waste chemicals and oil. The thrifty Chinese claim they re-use 2.5 million tonnes of scrap iron each year, and about three million tonnes of waste paper. Forward-looking manufacturers are designing products from tea-pots to cars with final disassembly and recycling in mind.

All this recycling also helps with the problem of waste disposal. An average American generates about one tonne of waste each year. Three-quarters of this garbage is buried in thousands of landfills, with half the remainder incinerated.

In the main, the conserver society depends on the commitment of individuals, and of industry and other commercial interests. But they can be inspired by government incentives and penalties. Furthermore, governments can foster the anti-waste campaign by initiatives to eliminate "concealed waste", even by helping to make planned obsolescence obsolete. Increasing product lifetime through better design, repair, and re-use is more effective than recycling, since it doesn't require processing recycled materials.

Waste into wealth

Every living organism uses energy to process raw materials and, in doing so, it often produces some form of waste. In natural systems, such wastes are soon exploited by other organisms, in a perpetual cycle of re-use. Human communities process materials on a grand scale, using enormous quantities of energy and other resources in doing so. We also produce mountains of waste, the bulk of which passes through the system just once. The case studies, far right, illustrate the current situation for some materials which are recycled, at least in part. In many Third World countries, fortunately, a high proportion of many waste materials is re-used. But there is a darker side to the picture: on the outskirts of Cairo, for example, tens of thousands of people sort through the city's waste in conditions of indescribable squalor, while in India scrap collectors are killed or maimed during their hunt for spent cartridge cases on army-firing ranges. Some developing countries are trying to work out ways in which they can improve the working conditions of their recyclers. As for the developed countries, although the "throwaway society" still flourishes, new recycling industries are emerging, generating new employment and boosting energy efficiency.

Sorting waste for re-use
The key to material recovery and recyling is separation, whether performed by "rubbish sifters" (*zabbaleen*) or by the latest automatic sorting equipment. Japan, which now recycles about 10% of its municipal waste, provides many examples. Ueda City pays an entrepreneur a portion of the money saved when s/he recycles wastes which would otherwise have been landfilled. Residents sort the material into combustible and non-combustible fractions. Glass is removed by hand, ferrous metal by magnet.

Jobs from junk
Compared with incineration and landfilling, recycling is cheaper and creates more jobs, due to its much lower capital requirements. Already, recycling is a more important employer than metal mining in the US. It is estimated that in the US at least 30,000 people are involved in recycling aluminium alone – twice the employment in the primary aluminium production industry. The recycling industry offers the potential for many more jobs in the future.

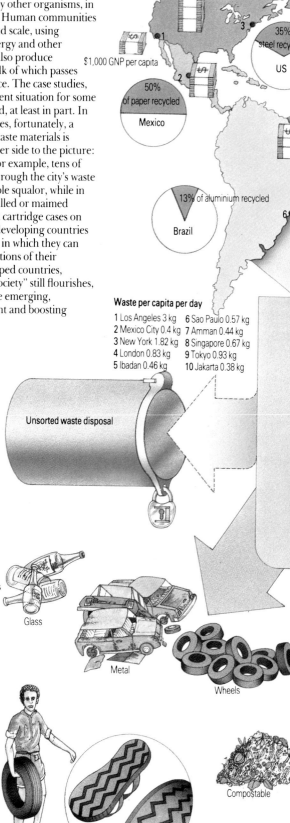

$1,000 GNP per capita

35% steel recyc
US

50% of paper recycled
Mexico

13% of aluminium recycled
Brazil

Waste per capita per day

1 Los Angeles 3 kg		6 Sao Paulo 0.57 kg
2 Mexico City 0.4 kg		7 Amman 0.44 kg
3 New York 1.82 kg		8 Singapore 0.67 kg
4 London 0.83 kg		9 Tokyo 0.93 kg
5 Ibadan 0.46 kg		10 Jakarta 0.38 kg

Unsorted waste disposal

Glass

Metal

Wheels

Compostable

Recovered materials

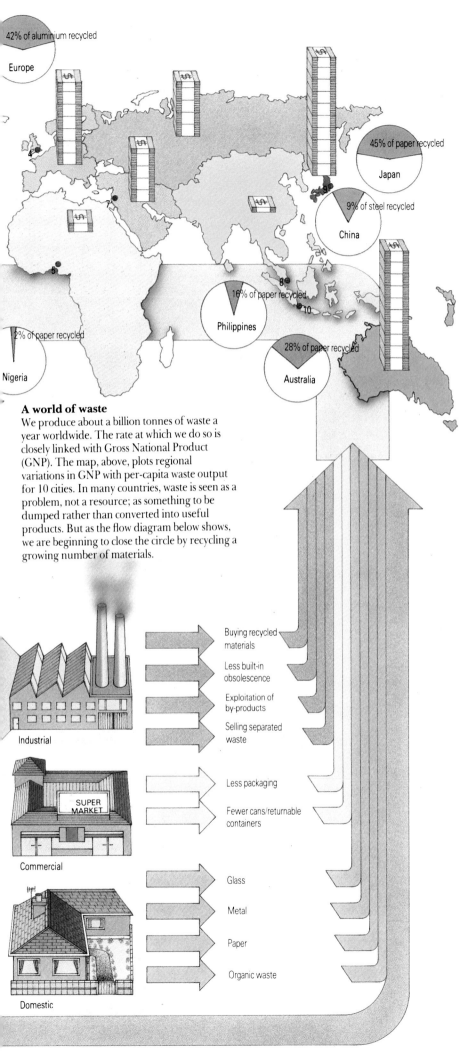

42% of aluminium recycled

Europe

45% of paper recycled

Japan

9% of steel recycled

China

16% of paper recycled

Philippines

2% of paper recycled

Nigeria

28% of paper recycled

Australia

A world of waste

We produce about a billion tonnes of waste a year worldwide. The rate at which we do so is closely linked with Gross National Product (GNP). The map, above, plots regional variations in GNP with per-capita waste output for 10 cities. In many countries, waste is seen as a problem, not a resource; as something to be dumped rather than converted into useful products. But as the flow diagram below shows, we are beginning to close the circle by recycling a growing number of materials.

Industrial

Buying recycled materials

Less built-in obsolescence

Exploitation of by-products

Selling separated waste

Commercial

SUPER MARKET

Less packaging

Fewer cans/returnable containers

Domestic

Glass

Metal

Paper

Organic waste

Recycling iron

The world steel industry uses scrap for about 45% of its iron requirements, with some countries reaching 60-75%. The world recession has hit the scrap industry hard, but new technologies are increasing the demand for scrap. The most buoyant sector of the US steel industry, for example, uses the "minimill", which is based on electric-arc furnaces and the use of scrap. The US dominates the world ferrous scrap market.

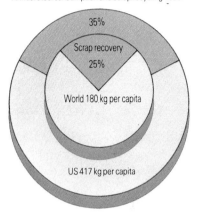

Annual steel consumption and scrap recycling 1988

35%

Scrap recovery
25%

World 180 kg per capita

US 417 kg per capita

Recycling paper

Only 25% of the world's paper is currently recycled, although there is no good technical or economic reason why this recycling rate should not be doubled by the year 2000. Recycling half of the world's paper consumption would meet almost 75% of new paper demand – and would release 8 million ha of forest from paper production. Fibre-rich countries, such as Canada and Sweden, are not in the front-rank of paper recyclers. Recycling rates are much higher in such fibre-poor countries as Japan (45%), Mexico (50%), and the Netherlands (43%).

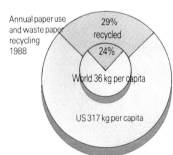

Annual paper use and waste paper recycling 1988

29% recycled

24%

World 36 kg per capita

US 317 kg per capita

Recycling aluminium

Throw away an aluminium soft-drink container and you throw away the energy equivalent of half a can of gasoline. It is estimated that 80% of all aluminium used could be recycled, but less than 30% of world production comes from scrap. In 1980, aluminium scrap moving across national boundaries totalled 800,000 tonnes (representing over 5% of world production), a trade valued at $600 million. Over half the aluminium cans used in the US are now recycled, and even more dramatic results will be possible as deposits are charged on cans.

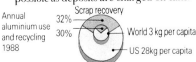

Annual aluminium use and recycling 1988

Scrap recovery
32%

30%

World 3 kg per capita

US 28kg per capita

EVOLUTION

Introduced by Paul Ehrlich

Professor of Population Studies, Stanford University

The human population has reached almost five billion people by, in essence, "burning its capital"–destroying and dispersing a one-time bonanza of fossil fuels, minerals, deep soils, water, and biological diversity. It is the loss of biological diversity that may prove the most serious; certainly it is the most irreversible. *Homo sapiens* depends on the genetic variety present in the many millions of species and billions of their populations for a huge range of services, many of them absolutely essential to the support of civilization.

We are causing some loss of genetic diversity directly, by wiping out or severely over-harvesting certain species and populations, from rhinoceroses, elephants, spotted cats, and whales to orchids and cacti. But the major danger comes indirectly, from habitat destruction. Humanity is paving over, ploughing under, chopping down, damming, poisoning, and otherwise ruining habitats at a truly horrifying rate. It is primarily this that threatens the mountain gorilla, the California condor, and a host of lesser known organisms. The destruction of tropical forests alone could easily reduce the organic diversity of our planet by 50 percent in the next few decades.

Attempts are being made to arrest the decay of Earth's organic diversity: the Convention on International Trade in Endangered Species has delayed the extermination of some prominent species and, at least on paper, a world system of "Biosphere Reserves" is being organized. But our efforts may be too little and too late. It is one thing to draw lines on maps around areas of tropical forest and declare them protected, quite another to find the will and resources to preserve them in perpetuity.

Even today, too much attention is focused on the preservation of *species*, whereas the loss of genetic diversity *within* species is probably a more immediate and vital issue. The biochemical make-up of plants, for instance, is notoriously variable geographically–and plant chemicals have provided a great many of society's medicines and industrial products. By the time species are recognized as endangered, their utility to human beings is often greatly compromised: populations are too small to be significant in ecosystems, gene pools are too depleted to permit successful domestication, and useful characteristics are lost.

What *should* be done is clear. In developed nations, disturbance of any more land should be forbidden, and creation of exotic monocultures, be they golf courses, wheat fields, or tree plantations, restrained everywhere. Priority should be given to making the already huge areas occupied by humanity more hospitable to other organisms, and above all, to arresting the steady flux of poisons into the air and water, especially acid rain. Programmes to protect and re-establish relatively natural ecosystems are doomed to failure while poisons remain unchecked.

The problem is vastly more serious in the less developed nations, where raw population pressure and poverty make the continued destruction of natural systems seem inevitable. Poor nations, like poor people, must look always to their immediate needs. It is only by finding ways to help them provide for those needs that the rich nations can support conservation in the poor nations, and thus (in the long run) save themselves–for human survival is intertwined with the survival of genetic diversity, and thus with the fate of every natural system.

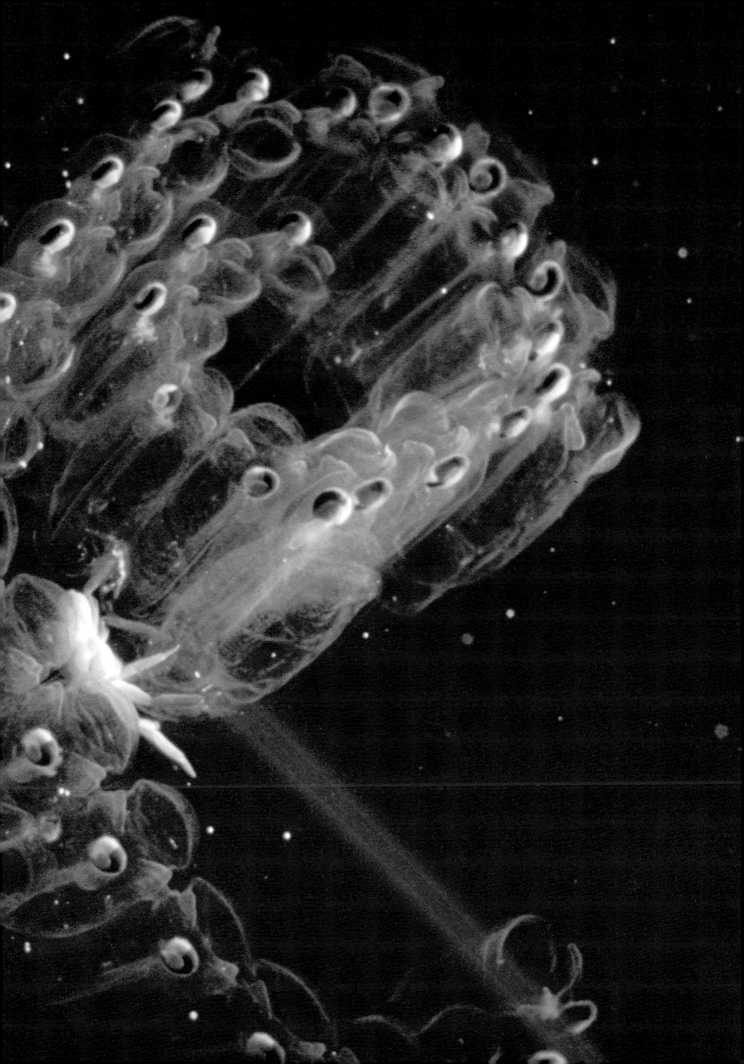

THE EVOLUTIONARY POTENTIAL

The first flickerings of life emerged 3.6 billion years ago, shortly after the planet coalesced from a swirling mass of gases into solid state. From a primeval impulse of DNA, the "building block" of life, there has evolved a steadily expanding stream of life-forms, swelling to a flood as species have become ever-more numerous and diverse. The creative flow has been far from constant, however. It got off to a slow start, and as recently as the late Permian era 225 million years ago, there were no more than 350,000 species or so – mostly marine creatures. But thereafter the spread of life on to land led to an outburst of creativity that has ultimately produced many millions of species.

Not that the prehistoric parade has merely grown larger. Certain categories have emerged to dominate from time to time, then have been relegated to the sidelines. Our modern world is sometimes considered as primarily a world of mammals. But mammals, which constitute only one in a thousand of today's community, are relative newcomers. They did not reach prominence until the end of a reptilian era that lasted for 160 million years.

The evolutionary process has culminated in today's array of life, a vast resource of material on which natural selection can work to generate still more abundant and complex manifestations of life. Our present life pool represents but a small part of the potential panoply of life that will steadily develop if *Homo sapiens*, a newly dominant species, allows the process to persist with the creative capacity that has been at work virtually since the beginning of life.

Peering into the dim past, then, we can discern a procession of ever-changing life-forms, new ones appearing as old ones fade from the scene. The average duration of a species has been only about five million years, which means that of the half billion or so species that have ever lived, at least 98 per cent have disappeared.

As we gaze on the life pool of today, let us reflect that it does not amount to a mere conglomeration of species that go their separate ways. They depend on each other for multiple services. Plants, for example, supply more than food for plant eaters. They also help to maintain the mix of gases in our atmosphere – oxygen, nitrogen, carbon dioxide, and so forth – that support all forms of life. Even the lowliest manifestations of life, bacteria, serve to recycle nutrients that keep plants going. In short, the life pool is much more than the sum of its parts.

The life pool

All life is one. This is not a cliche, but a biological reality. Over 3.6 billion years, a teeming variety of life-forms has evolved, yet in every living cell there are common features of nucleic acids that encode inheritance, and adenosine triphosphate that provides energy. In every living organism, hormones and similar compounds carry vital chemical messages. In turn, organisms are linked together in intricate ecosystems. What happens to one can affect all: our present biosphere has evolved only after photosynthesis in early algae and plants began to release essential oxygen into the atmosphere. Evolution produces vast diversity, yet it links all living forms in a single process.

We, as members of the species *Homo sapiens*, are the first life-forms to be able to modify the evolutionary process, yet we do so largely in ignorance and by accident. We take small account of the intricate food chains that maintain both species and ecological harmony, or of the rich store of genetic information that species represent. Indeed, scientists have classified perhaps no more than 5% of the planet's organisms (see below).

The biosystem is a library of survival strategies; it makes up an entire literature, a language, and a tradition, every unique part of which may bear upon any other. Built up over billions of years of selection and extinction, it is not to be tampered with lightly.

Undiscovered species
5-30 million
(mostly invertebrates)

Known species
c.1.7 million

A census of species
Science has identified 1.5 million species of animals and 300,000 of plants. But millions of others remain unclassified – even mammal species, such as bats. Each year too, 20 new species of reptiles are found. Exploration is adding continuously to the numbers classified.

Invertebrates
1,400,000
(insects 1,000,000)

Vertebrates 47,000

Vascular plants
250,000

Amphib
Fish
Lampreys
Sea squirts
Echinoderms

Insects

Centipedes
Millipedes

Crustaceans

Spiders:
scorpions

Segmented worms

Squid: octopus

Bivalves: molluscs

Single-shell
molluscs

Brachiopods

Flat worms

Jellyfish: corals

Sponges

Single-celle
animals

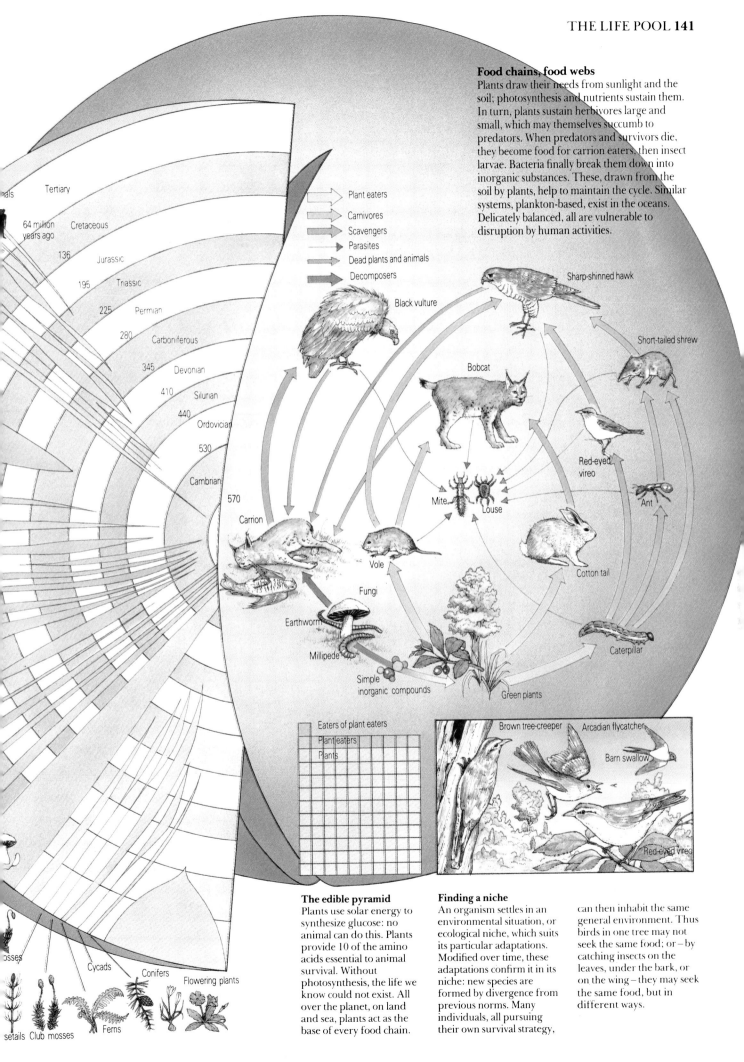

Food chains, food webs

Plants draw their needs from sunlight and the soil; photosynthesis and nutrients sustain them. In turn, plants sustain herbivores large and small, which may themselves succumb to predators. When predators and survivors die, they become food for carrion eaters, then insect larvae. Bacteria finally break them down into inorganic substances. These, drawn from the soil by plants, help to maintain the cycle. Similar systems, plankton-based, exist in the oceans. Delicately balanced, all are vulnerable to disruption by human activities.

Plant eaters
Carnivores
Scavengers
Parasites
Dead plants and animals
Decomposers

Black vulture
Sharp-shinned hawk
Short-tailed shrew
Bobcat
Red-eyed vireo
Ant
Mite
Louse
Carrion
Vole
Cotton tail
Fungi
Earthworm
Millipede
Caterpillar
Simple inorganic compounds
Green plants

Tertiary
64 million years ago
Cretaceous
136
Jurassic
195
Triassic
225
Permian
280
Carboniferous
345
Devonian
410
Silurian
440
Ordovician
530
Cambrian
570

osses
Cycads
Conifers
Flowering plants
setails Club mosses
Ferns

Eaters of plant eaters
Plant eaters
Plants

Brown tree-creeper
Arcadian flycatcher
Barn swallow
Red-eyed vireo

The edible pyramid

Plants use solar energy to synthesize glucose: no animal can do this. Plants provide 10 of the amino acids essential to animal survival. Without photosynthesis, the life we know could not exist. All over the planet, on land and sea, plants act as the base of every food chain.

Finding a niche

An organism settles in an environmental situation, or ecological niche, which suits its particular adaptations. Modified over time, these adaptations confirm it in its niche: new species are formed by divergence from previous norms. Many individuals, all pursuing their own survival strategy, can then inhabit the same general environment. Thus birds in one tree may not seek the same food; or – by catching insects on the leaves, under the bark, or on the wing – they may seek the same food, but in different ways.

It is a single unifying phenomenon of our planet Earth – the only planet known to feature the genetic code of life, DNA.

Co-operative evolution

The life pool is the product of billions of years of evolution. We humans think of ourselves as at the pinnacle of this process – more advanced, more important, and, above all, more powerful than lower forms of life – especially microbial life, which seems to us hardly to be alive at all.

But recent science has overturned these views to give a startling and humbling picture of our true place in nature. In the words of US scientist Lynn Margulis: "Far from leaving microorganisms behind on an evolutionary ladder, we are both surrounded by them and composed of them". Microbes are everywhere: in soil; in every living structure; the foundation and stuff of all ecosystems. They are the rulers of our cells and bodies.

The familiar division of life-forms into plants and animals is far less significant than the fundamental division of microbial life into two types, dauntingly named the prokaryotes (bacteria) and the eukaryotes (modern complex cells). The evolution of microbial life stretches back over 3.7 aeons to the origins of life on Earth. For the first 2 aeons, the only life-forms were bacteria. Yet these lowly creatures invented all the chemistry of life: photosynthesis to tap the sun's energy; fermentation to release the energy of dead matter; and oxygen breathing. And in so doing they transformed the Earth, atmosphere, and oceans. Microbes have created the world we know.

Science is also making remarkable discoveries about evolution. To the classic mechanisms of random genetic mutation and survival of the fittest, must now be added gene swapping, the habit of bacteria of borrowing bits of genetic material floating in the medium around them. By this means bacteria have conducted a "fast track" evolutionary process for 3.5 billion years.

Moreover, the Darwinian view of competition as the single driving force of evolution – *nature red in tooth and claw* – is now seen as naive. Co-operation to mutual advantage has proven an equally successful evolutionary strategy – and one, again, pioneered by microbes. Long ago, bacterial cells once foreign and hostile to each other began entering into partnership, sharing functions and genes through a process termed endosymbiosis. And so arose the modern eukaryotic cells and hence the evolution of plant and animal life, as the eukaryotes, with their complex cell structure and ability to specialize, formed new associations in multi-celled communities and organisms. We, *Homo sapiens*, are the product of microbial interaction and co-operation. And this process of symbiotic evolution continues even today.

Most challenging of all is the new view of the evolution of the Earth itself. Far from seeing its changing geology and evolving biota as two separate

The web of Gaia

Far from being "lords of creation", we humans are a part of nature, beings woven from the web of Gaia. Our bodies are microbial collectives, communities of co-operative cells, and the library of life's history is written in us. As Charles Darwin said, "Each living creature must be looked at as a microcosmos, a little universe". We are accustomed to think of the geological history of the Earth, and the history of life, as separate strands. But we are learning more and more that this is not so. Rather, the evolution of life and the environment have progressed hand in hand, each changing the other. Cells are the fundamental units of life, and they need a narrow range of conditions – temperature, moisture, chemistry – to survive. By exchanging energy and materials with the environment, cells not only sustain their internal state, but alter neighbouring conditions. Over the aeons, the whole environment is thus shaped and sustained by cell activity. The gases of the air, the chemistry of soil and water, the very rocks beneath our feet, are a product of this long interaction. The evolution of cellular life is mapped, right, from the beginnings 3.7 billion years ago, through the long age of bacteria, to the symbiotic origins of modern cells, and thence to multi-cellular plant and animal life and the biosphere of Gaia we know today.

Bacteria's long reign
The history of Earth's environment is written in the bacteria, who shaped it alone for the first 2 billion years. Ancient archebacteria reflect the conditions in the early Earth: there are salt-loving types; forms that thrive in sulphur springs or hot deep-sea vents; and fermenters that live without oxygen, breaking down dead matter in sediments. The younger "true" bacteria reveal the dramas that followed: blue-green algae with their great invention, photosynthesis, and its by-product, oxygen, that profoundly changed the world; and oxygen-breathing aerobic bacteria, the first consumers.

Consumers

Plants

1 billion years ago

0.7

Endosymbiosis

Co-operation and community have been central to life's evolutionary strategy. All the larger organisms are communities of cells. It now seems that the cells themselves are a product of ancient experiments in communal living. Evidence is increasingly accepted that certain key organelles in eukaryotic cells are the descendants of once free-living bacteria, that long ago took up residence inside other bacteria, known as urkaryotes, in a symbiotic relationship. According to this *endosymbiosis* theory, the first such resident was an aerobic bacterium, ancestor of the mitochondrion present in all modern cells. Later, this host/mitochondrion partnership was joined by another resident, probably a photosynthesizing cyanobacterium – ancestor of the chloroplast. From this new partnership all plants evolved. This process of symbiotic evolution continues even today.

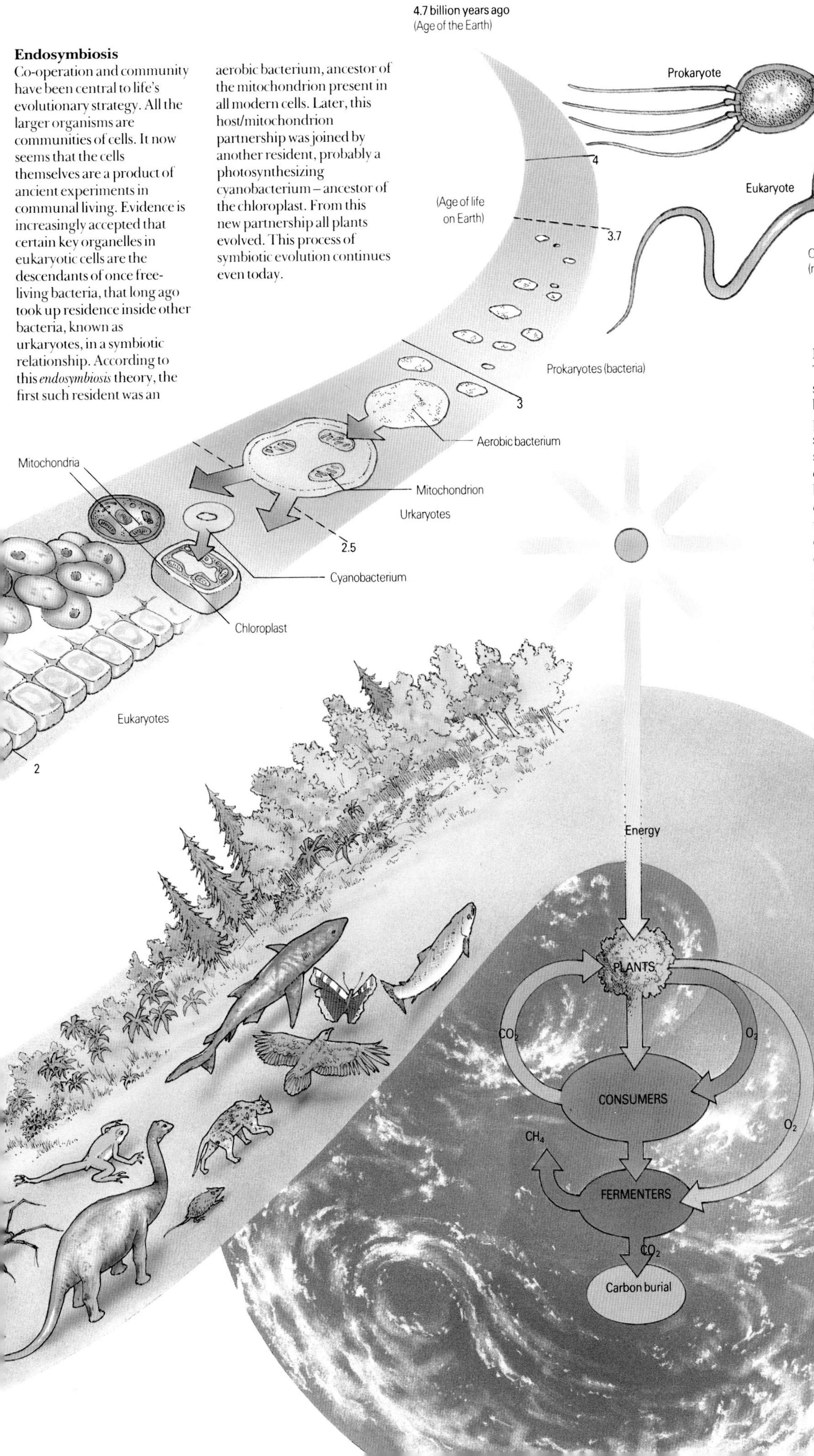

4.7 billion years ago
(Age of the Earth)

Prokaryote

Eukaryote

Nucleus

Organelle
(mitochondrion)

(Age of life
on Earth)

4

3.7

Prokaryotes (bacteria)

3

Aerobic bacterium

Mitochondria

Mitochondrion

Urkaryotes

2.5

Cyanobacterium

Chloroplast

Eukaryotes

2

Prokaryotes and eukaryotes

The unit of life is the cell. The simplest organisms are the bacteria. They are prokaryotes with a cell structure consisting of a single, walled cell filled with cytoplasm, no membrane-bound nucleus, and no organelles (separate bodies within the cell). A single chromosome is coiled at the centre. By contrast, the vast majority of cells that make up modern plants and animals are eukaryotes. They have a complex cell structure with a separate, enclosed nucleus containing many chromosomes wound round a protein core, and a full complement of organelles with distinct functions, such as mitochondria (all cells) and chloroplasts (plants).

Systems of exchange

The chemistry of life and the environment is dominated by the three basic strategies invented by bacteria: photosynthesis, aerobic respiration, and fermentation. Plants and blue-green algae can use the sun's energy directly in photosynthesis. They convert carbon dioxide and water to oxygen and organic matter. Animals and microbial consumers derive energy from respiration. They use up most of the oxygen and organic matter, and release carbon dioxide (CO_2). Fermenters gain energy by digesting dead matter in the soil, releasing CO_2 and methane (CH_4) which reacts with any remaining oxygen to produce CO_2. Some carbon is buried deep, effectively providing a "sink" for CO_2, which leaks in from volcanoes. These three systems of exchange are largely in balance. So profound and pervasive is their influence, that they can almost be seen as the metabolism of one, giant living system, Gaia.

Energy

PLANTS

CO_2

O_2

CONSUMERS

O_2

CH_4

FERMENTERS

CO_2

Carbon burial

strands, science now begins to describe them as forming one integral process, which over the aeons has woven the seamless web of Gaia.

The natural habitat

Our blue planet, the only planet known to support any spark of life, supports it in extraordinary abundance and variety. From the tallest mountain peak to the deepest ocean trench, from rainless desert to dripping rainforest, there is scarcely a corner of the Earth's surface which does not reveal some variation on life's theme. Indeed, the diversity of our evolutionary resource defies imagination.

Clearly, this evolutionary resource is anything but uniform. Life's most basic impulses send it racing down endlessly branching paths. Wherever we go, we encounter environmental variety. Each woodland, each stretch of open savannah, each wetland or montane ecosystem, has its own distinctive communities of plants and animals. But life has evolved clear strategies to meet such environmental challenges – and we see the results in recognizable associations such as the "spruce-moose" biome of North America's boreal forests.

Moreover, just as environments change, so life strategies are themselves in constant flux, developing ever-new variations on the vital theme. The key factors which help shape the patterns of life, and particularly vegetation patterns, include temperature and rainfall, together with two closely connected factors – latitude and altitude.

Once we know how hot or wet an environment is, or how cold and dry, and provided we know its location, we can make a fair guess about what forms of plant life are likely to predominate. Of course, other factors, such as soil type and topography, can also have an important influence. But, as a general rule, we know that at the Equator we shall find year-round warmth and moisture, and we shall expect to encounter evergreen rainforest.

As we move towards the tropics, rainfall becomes more seasonal, and we expect to see deciduous trees able to counter periodic drought. As the average rainfall decreases, we encounter increasingly open woodland, giving way to bush and then grassy plains. Finally, after moving through various forms of scrub, we find continental deserts. Other kinds of desert are found in unlikely places. The coastal deserts of Chile and Namibia, for example, are formed where cold offshore currents cool the air and prevent it rising high enough to produce rain.

Moving out from the tropics towards the poles, we find these vegetation patterns echoed by different forest formations and different types of grassland. The conditions are colder and winter becomes a force to be reckoned with. We encounter communities which are less fecund and varied than their tropical counterparts. Here, life strategies branch out from distinctive adaptations in response to seasonal change: caribou migrate to less harsh areas, birds head for distant horizons, while bears

Life strategies

Combinations of warmth and rainfall generate diverse life strategies. To portray these life strategies, biologists use an idealized or super-continent (right). It balloons to the north, reflecting the great land masses of Eurasia and North America. It tapers towards the Equator, and extends like a teardrop into the southern hemisphere, to represent the shape of southern Africa and South America.

Compare this super-continent with a real-world map. We can see that if Central America were to bulge out along the Tropic of Cancer, instead of almost dwindling away, we would have another vast desert corresponding to the Sahara; and if Africa bulged out along the Equator, we could expect a rainforest of Amazonian proportions.

Life strategies are broadly classified into large domains, known as biomes. These biomes reflect basic bio-communities which have evolved mainly in response to local variations in climate and topography.

Freshwater ecosystems

From river-source to sea, many kinds of life strategies are invoked here. When the river is narrow and its current fast, plants cannot survive and most fish cannot thrive. When the river bed is less steep, plants root in the mud which lies in sheltered spots near the river banks. The current is still fast, but fish like trout or minnow can breed here. Further downstream, where the banks slope more gently, the water is turbid, and the bed is full of sand and gravel, bream flourish. Finally, in the more salty and muddy estuarine region, flounder and smelt predominate, while salmon pass through on their spawning and feeding migrations.

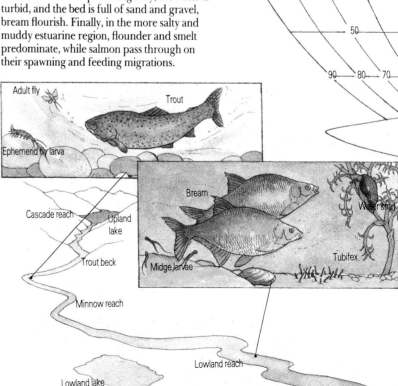

Boreal forest

The boreal forest, accounting for 11% of the super-continent, fringes the northern sub-polar regions. Moose browse on spruce and birch, falling prey in turn to the timber wolf.

Temperate forest

Growing as tall as their boreal counterparts, the trees of the temperate forest cover 9% of the super-continent. They comprise a mixture of deciduous and coniferous trees, and shelter a bewildering array of insects, birds, and animals.

Savannah

Herbivores in their millions, such as the wildebeest and zebra, are found on open savannah, which covers another 11% of the super-continent – their predators include the lion, cheetah, and hyena.

Desert

Deserts cover almost one-quarter of the land surface (semi-desert 13%, hot desert 8%, and cold and coastal deserts, 2%). The cacti, roadrunner, and chuckwalla lizard, above, are found in the hot deserts of North America.

Tropical forest

About one-seventh of the super-continent is covered by tropical forests. Rainforests hold an exceptionally high number of species; tropical deciduous forests do not have as many but are still highly diverse.

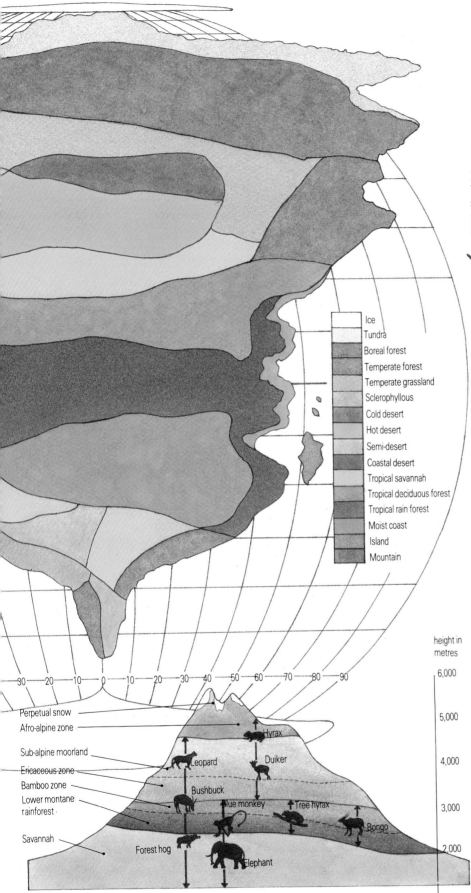

Ice
Tundra
Boreal forest
Temperate forest
Temperate grassland
Sclerophyllous
Cold desert
Hot desert
Semi-desert
Coastal desert
Tropical savannah
Tropical deciduous forest
Tropical rain forest
Moist coast
Island
Mountain

height in metres

6,000

5,000

4,000

3,000

2,000

Perpetual snow
Afro-alpine zone
Sub-alpine moorland
Ericaceous zone
Bamboo zone
Lower montane rainforest
Savannah

Hyrax
Leopard
Duiker
Bushbuck
Blue monkey
Tree hyrax
Bongo
Forest hog
Elephant

Montane ecosystem

In the montane ecosystems of Africa, there is a complex succession of vegetation zones between the savannah, at the lowest level, and the permanent snowfields which cap such mountains as Mount Kenya, Mount Kilimanjaro, and the Mountains of the Moon. Moving up through the haunts of the elephant, forest hog, blue monkey, and bushbuck, we find far fewer animals as the snow comes closer.

and squirrels go underground. In some areas of the world we can move through a bewildering succession of life strategies within a surprisingly short distance, for example, as we climb through a montane ecosystem.

Shaping our food base

Until 10,000 years ago, our ancestors had little effect on the plants and animals around them. Then came a quantum leap with more profound implications than any during the previous one million years, both for humankind and for a growing spectrum of once wild species. Often accidentally, our ancestors found ways to domesticate a few animals and plants, entering into an evolutionary partnership with selected species. No other living creature has accomplished this much control over the evolutionary process, and it has been an advance to alter the face of the Earth.

The first essential step occurred when our ancestors concentrated on those forms of wild grasses – cereals – that promised to yield most food. They planted varieties with larger grains, with shorter growing seasons, and with other attributes that would serve their needs. They began to apply selection pressures that hitherto had remained the domain of "natural" evolution. Before long, they had developed types of wheat, for example, whose seed heads did not "shatter" and drop their seeds to the ground when ripe: these early breeders preferred to keep the seeds for their own use. This meant that without human help, wheat plants could no longer propagate themselves.

A similar pattern was repeated with the animals, such as sheep and cattle. Selected for their docility among other traits, domesticated strains must soon have lost their capacity to survive on their own. Not only tamed by humans, but weighed down with flesh or milk, today's cattle breeds would be hard put to survive in the wild.

As one domestication followed another, our ancestors assembled a group of species that supplied them with a growing range of benefits. From these early successes in domestication, there evolved the phenomenon of full-scale agriculture. Today, a considerable proportion of the Earth's surface is devoted to staple crops or to huge herds of livestock. As a result of this "globalization" of basic food sources, a growing proportion of the world's population feeds from a common bowl.

Curiously enough, our ancestors contented themselves with just a handful of domesticates. Although hunter-gatherers had exploited hundreds of plants and dozens of animals, these early agriculturalists confined their attentions mainly to a total of less than 50 species. These basic food sources still meet our needs today. Thirty crops supply 95 percent of our nutrition, and a mere eight, led by wheat, maize, and rice, provide three-quarters of our diets. Our meat and milk come from a still smaller range of species. Thus we practise an agriculture that is, in

Partners in evolution

Homo sapiens has derived great benefit from a special partnership with Earth's plant and animal species. Intensive farming of a minority of species has released a large proportion of humankind for work other than food-growing, and thousand-fold increases in population have become possible.

Purposeful crossbreeding to produce improved varieties is a modern development. Previously, agriculturalists would select the more productive species and leave evolution to work its own advances by natural crossbreeding. As a result, the earliest domestications occurred in the areas of greatest diversity shown on the world map, right, where the probability of "crosses" leading to improved varieties was greatest. The areas are not rigidly defined; some species were being domesticated at the same time though in different places.

North Am
Turkey

Meso America
Maize
Tomato
Cassava
Sweet potato
Turkey

Areas of origin:
domestic plants and animals

Maize: dispersal

Travelling partners
Our most valued and adaptable crops have travelled with us in early migrations, invasions, explorations, and to colonial settlements. Some have spread to the far corners of the Earth. Others, such as coffee, now dominate lands far from their point of origin. The map shows the peregrinations of corn and wheat.

Animal farm

As with crop plants, humankind has evolved a special partnership with animals: a small number of highly efficient animals provides all our needs. Dairy cows can yield up to 4,800 litres per lactation with current farming methods. They are fed high-energy concentrates by computer-controlled feeders. The price of this efficient production is the near-total dependence of such animals on their keepers.

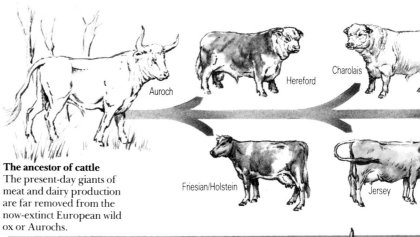

Auroch

Hereford

Charolais

Friesian/Holstein

Jersey

The ancestor of cattle
The present-day giants of meat and dairy production are far removed from the now-extinct European wild ox or Aurochs.

The corn story

Originally allowed to grow in fields of primitive corn because it enhanced the yield, the wild Mexican grass Teosinte became the ancestor of modern corn. The diagrams, right, which are based on modern breeding experiments, reveal how the hard outer casing of the Teosinte "spike" is gradually made softer. Eventually, we arrive at the giant modern variety (far right), drawn to scale – its kernel fruit totally exposed. Without human intervention to remove and plant its kernels, modern corn would become extinct in a few generations: the seedlings would be so densely clustered that they would compete among themselves for water, soil, and nutrients, and fail to reach reproductive size.

Teosinte

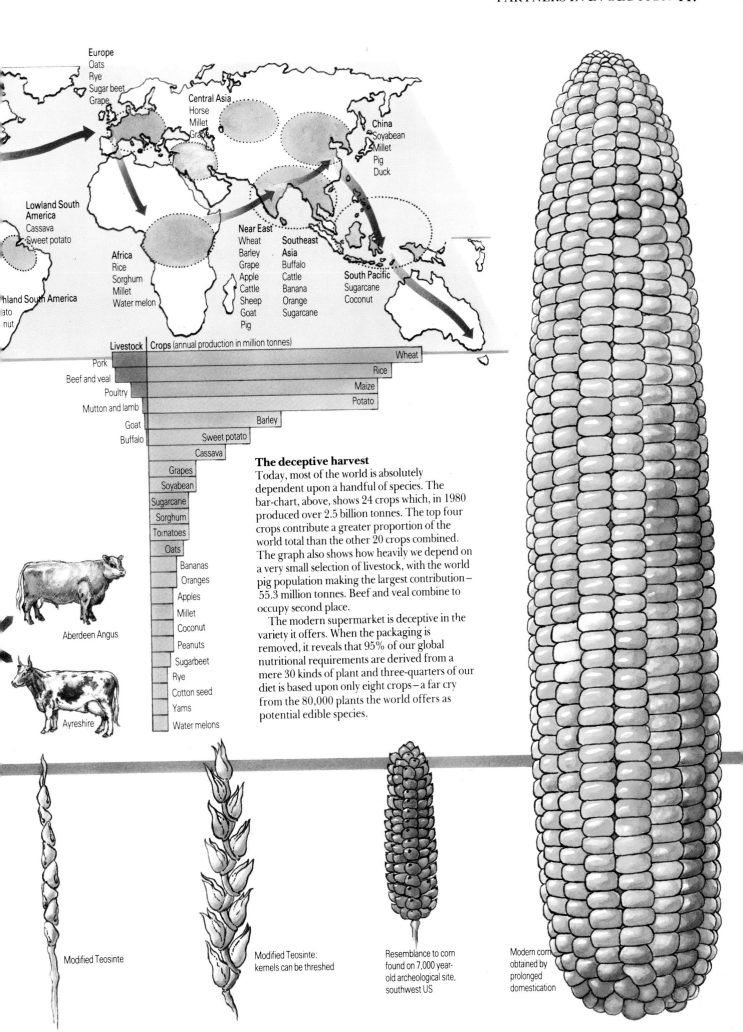

Europe
Oats
Rye
Sugar beet
Grape

Central Asia
Horse
Millet
Grape

China
Soyabean
Millet
Pig
Duck

Lowland South
America
Cassava
Sweet potato

Near East
Wheat
Barley
Grape
Apple
Cattle
Sheep
Goat
Pig

Southeast
Asia
Buffalo
Cattle
Banana
Orange
Sugarcane

South Pacific
Sugarcane
Coconut

Africa
Rice
Sorghum
Millet
Water melon

hland South America
ato
nut

Livestock | Crops (annual production in million tonnes)

Pork
Beef and veal
Poultry
Mutton and lamb
Goat
Buffalo

Wheat
Rice
Maize
Potato
Barley
Sweet potato
Cassava
Grapes
Soyabean
Sugarcane
Sorghum
Tomatoes
Oats
Bananas
Oranges
Apples
Millet
Coconut
Peanuts
Sugarbeet
Rye
Cotton seed
Yams
Water melons

Aberdeen Angus

Ayreshire

The deceptive harvest

Today, most of the world is absolutely
dependent upon a handful of species. The
bar-chart, above, shows 24 crops which, in 1980
produced over 2.5 billion tonnes. The top four
crops contribute a greater proportion of the
world total than the other 20 crops combined.
The graph also shows how heavily we depend on
a very small selection of livestock, with the world
pig population making the largest contribution –
55.3 million tonnes. Beef and veal combine to
occupy second place.

The modern supermarket is deceptive in the
variety it offers. When the packaging is
removed, it reveals that 95% of our global
nutritional requirements are derived from a
mere 30 kinds of plant and three-quarters of our
diet is based upon only eight crops – a far cry
from the 80,000 plants the world offers as
potential edible species.

Modified Teosinte

Modified Teosinte:
kernels can be threshed

Resemblance to corn
found on 7,000 year-
old archeological site,
southwest US

Modern corn
obtained by
prolonged
domestication

terms of its array of sources, not much more than a Neolithic agriculture. What prospects could lie ahead of us if we were to develop more of these evolutionary partnerships, realizing more of the potential of our genetic resources?

Tapping diversity

Diversity is the hallmark of life on Earth, whether we compare different species, races, and populations within a species, or different individuals within a population. Just as the innumerable species that make up the life stream are bewilderingly diverse to the eye, so even outwardly identical organisms prove to be astonishingly diverse if we concentrate our gaze on the level of the gene.

One individual organism may have thousands of genes, each gene influencing some inherited trait such as height, weight, rate of growth, or resistance to disease. Even if the individuals within a species are counted in millions or billions, such statistics shrink in scale when compared with the total number of potential genetic permutations those individuals could produce. And it is this diversity, passed on by genes, which holds the key to a species' ability to adapt so readily in response to environmental pressures. Naturally, only a tiny fraction of this genetic potential is ever expressed, since most organisms die before they can reproduce. Even so, this genetic potential is one of our world's most valuable resources.

Jojoba
Long considered a desert weed, jojoba produces a wax which retails to Japan for $3,000 a barrel as a substitute for sperm whale oil. As jojoba plantations come "on stream", the price should fall steeply.

The genetic resource

Our genetic resources represent an extraordinarily well-stocked library. Each species can be seen as a single book on one of an unknown number of shelves, each page a slice of its gene pool.

So far, we have made surprisingly little use of this library, concentrating on a few volumes of immediate interest. Yet even these few volumes have provided us with uncounted benefits, some of which are described here.

Tragically, this vast, valuable, and irreplaceable gene library is being vandalized. Complete volumes, indeed entire shelves, are being lost in bouts of habitat destruction, and key sections of the library are now in danger of being gutted. We are losing genetic information and materials whose value we can only guess at.

We must conserve and exploit this unique genetic resource. Once its value is fully recognized, its beneficiaries will be more likely to secure its future.

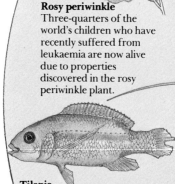

Rosy periwinkle
Three-quarters of the world's children who have recently suffered from leukaemia are now alive due to properties discovered in the rosy periwinkle plant.

Tilapia
Tilapia fish may soon usurp carp as the choice of the world's fish farms. The East African tilapia fish, with 164 species in Lake Malawi alone, converts food to flesh faster than most other fish and is increasingly used in aquaculture projects.

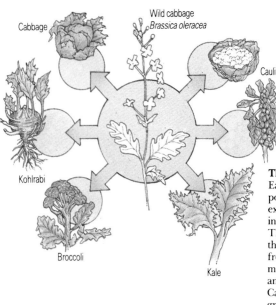

Cabbage
Wild cabbage
Brassica oleracea
Cauliflower
Brussels sprout
Kohlrabi
Broccoli
Kale

The gene pool
Each species has a gene pool from which an extraordinary array of individuals can be selected. The examples here show the variety that is enjoyed from the gene pools of wild mustard, *Brassica oleracea*, and the dog family, Canidae. If all but pedigree greyhounds disappeared from the Earth, we could say that "the dog" had been conserved – but imagine the loss of genetic variety.

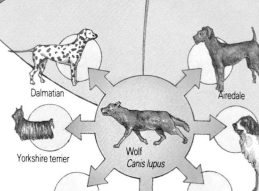

Dalmatian
Airedale
Yorkshire terrier
Wolf
Canis lupus
St Bernard
Dachsund
Basset
Greyhound

Guayule
Guayule, another wild shrub, grows in areas with scant rainfall and produces a natural rubber. The shrub's genetic diversity is extreme, thus an exceptionally important resource.

Molluscs
Molluscs are first-rate pollution monitors. Other marine organisms may prove useful for viral research

Every day, and almost always unconsciously, we use products whose very availability and usefulness reflect that genetic wealth. Our staple food crops achieve marvels of productivity every year, not simply because we have developed new fertilizers and pesticides, but because of genetic improvements achieved by the world's crop breeders.

Similar advances, with similar roots, are achieved every day in medicine. Every time we buy a drug or other pharmaceutical, there is almost a 50:50 chance that we can thank the genetic resources of a wild species for its efficacy. The value of medicinal products derived from such sources now approaches $40 billion every year. For example, if you had leukaemia as a child in 1960, or if you were suffering from Hodgkin's disease or a number of other cancers, you had a one-in-five chance of long-term survival. Today, thanks to two drugs developed from the rosy periwinkle, a tropical forest plant, you would have a four-in-five chance.

Similarly, industry processes and packages this genetic wealth. Every time we polish the furniture, pull on our jogging shoes, or hit a golf ball, walk across a carpet or ride high in a jet, we can thank genetic resources that in one way or another contribute to such products.

But while today's benefits are astonishing enough, they represent the tip of an iceberg. So far, scientists have taken only a preliminary look at some 10 percent of the 250,000 plant species, a considerable

Sharks
Many species of shark are proving valuable for research into liver ailments and certain cancers

Armadillos
The only animal known to contract leprosy, the armadillo helps to prepare a vaccine for all sufferers.

Manatees
The Florida manatee has slow-clotting blood. This characteristic has led to new insights in the area of haemophilia research.

Value of the wild
Every hour of every day, a growing number of plants and animals unwittingly underpin and improve the quality of our lives. A single gene from Ethiopia protects California's barley crop, worth $150 million a year, against yellow dwarf disease. Aspirin, probably the world's most widely used drug, has been developed from a chemical blueprint supplied by willow bark.

There are many other examples. The widely used contraceptive pill stems from the discovery of diosgenin in wild Mexican yams. Yet we are destroying this wealth as fast as it is being discovered. The pollution of the Great Lakes of Africa, for example, which contain as many as 400 tilapia fish species, could destroy this wild gene pool and reduce our chances of breeding superior fish. Similarly, despite their value to haemophilia research, the ungainly Florida manatee has been reduced in number to a mere 850. Perhaps worst of all, is the certain knowledge that we are destroying species even before their full value has been estimated.

Genetic engineering
By unlocking the potential of our genetic resources, genetic engineers could have an impact greater than that of atomic scientists. The new gene-manipulation techniques, now becoming available, could accelerate breeding programmes – and achieve genetic combinations which would scarcely be possible in nature.

For the moment, as the examples, right, show, we benefit from accelerated plant-breeding programmes. But the use of tissue culture and gene-transfer techniques promises to launch a new era. If, in the future, we could transfer the nitrogen-fixing ability of the bean to wheat, for example, we might be able to dispense with costly fertilizers. The agricultural biotechnology industry could be worth $100 billion a year by the late 1990s – but genetic engineers could ultimately be left with much less raw genetic material to work with if the world's wild gene pools are depleted.

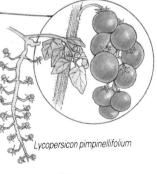

Lycopersicon pimpinellifolium

Helianthus petiolaris

Manihot glasiovii

Tomato
Without the resistance to Fusarium wilt provided by the wild Peruvian tomato, many of today's most successful commercial tomato varieties would not have remained viable.

Sunflower
The symbol of a healthy diet, the sunflower now ranks as one of the world's most important oil crops. Sunflower breeders have drawn heavily on the gene pool of the wild sunflower, *Helianthus petiolaris*.

Cassava
The yield of this vital food crop has been boosted experimentally by as much as 18 times since disease-resistance traits were transferred from its wild cousin, *Manihot glasiovii*.

number of which have already proved to be of enormous economic importance. And we have scarcely begun to investigate the potential of the animal kingdom.

Librarians of the rainforest

How are we to unlock the secrets of the wild? How can we best penetrate the ancient "library" that is the genetic resource?

The people of Surinam say: "Na boesi, ingi sabe ala sani" – "In the jungle, the Indian knows everything". It is here, amongst the populations of traditional and vanishing cultures, that one important answer lies. Traditional cultures are the "librarians" of our "gene library" and each holds a stack of index cards carefully drawn up after thousands of years of patient trial and error.

Keys to the wild

As we become increasingly remote from natural environments, and from the resources they contain, so we become ever-more dependent on the knowledge and skills of those live in close harmony with their immediate environment. Many of our foods and pharmaceuticals were first investigated because of their use by indigenous peoples. Instead of bulldozing aside these "human keys" to the wild, we should see them as a vital, intrinsic element of the ecosystems we are trying to conserve.

Even with the extraordinarily sophisticated equipment now available for extracting and analysing animal, microbial, and plant materials, our basic problem is simply knowing where to start. There are many examples of the ways in which traditional societies have helped focus our search for new medicines, new foods, and other products. By respecting and conserving human diversity, we will ensure that we can tap the hidden wealth in conserved areas.

There are other ways to "unlock the wild". We can try random screening of plants, for example, as done in the US. But this ultimately entails thousands of screening tests, much time, and a great deal of money. A better option is to search through the literature of traditional cultures. Or we can carry out studies of unpublished ethnobotanical data, which, combined with spot tests, often lead us to species worthy of further research.

But best of all is for ethnobotanists to concentrate on fieldwork. One study of Amazonian rainforests by two US scientists has led to the identification of more than 1,000 plants which have potential economic benefits. The project was dedicated to examining traditional uses of forest plants by native dwellers of the rainforests. The research revealed that at least six plants were used as contraceptives by

Heart-stopping poison from Guyana
Indians in a Guyana rainforest have used a special fish poison for centuries which, rather than leaving the fish prey to other fish eaters such as piranhas, drastically changes their behaviour – causing them almost to fly out of the water. Dr Conrad Gorinsky has pointed to the possibilities of this drug in heart surgery, since it is able to stop the heart without killing the organ. The heart can be re-started after surgery. The plant species responsible for this poison was discovered in 1774 by Christophe Fusee Aublet, but was wrongly described as having the characteristics of a very close but inactive relative. This type of confusion is not uncommon. To the Indians of Guyana, Aublet's *Clibadium sylvestre* (Aubl.) is quite distinct and can be identified by its smell and the different bracts on its leaves.

the forest's inhabitants. Other discoveries, to name but a few, included a cure for fungal skin infection, a high-protein coconut fruit that could be used for feed and fertilizer, and a seed that could be used to make soap.

Yet forest peoples, probably the only humans who successfully manage forests on a sustainable basis, are in a state of crisis. These "librarians" look set for dismissal, their library threatened with closure. Experience shows that these cultures, evolved and developed over thousands of years, can disappear at frightening speed. In Amazonia alone, more than 90 different tribes are thought to have died out this century. Many more forest peoples will cease to exist as cultural entities by the end of the century.

One way out of this tragedy is to put the world of the "gene librarians" on a sound economic footing. Already several African and Asian nations have begun to encourage the development of traditional medicine as an important component of their public health care programmes – the World Health Organization estimates that 80 percent of people in the developing world rely on traditional medicine. Indigenous medicines are relatively cheap, and readily accepted by the local populace. Developing countries often cannot afford to spend millions of dollars on importing medicines. Some countries, such as Thailand and Nepal, have developed their medicinal plants into a means of earning foreign exchange, each exporting medicines worth over $1 million a year. Even in Germany, a few forests are managed to encourage the growth of purple foxglove, the source of digitalis.

Mayapple

Medicine men and toothache-trees
Local experts, like the medicine man shown above, are being consulted in their hundreds by WWF and IUCN in an attempt to collect information on the uses of plants by traditional societies. In Tanzania, one scientist has discovered a tree used by local people to cure toothache – not only is it a completely new species, but its genus has not been recorded in Africa. In Amazonia, an ethnobotanic team has catalogued more than 1,000 plants used by South American rainforest Indians which have economic potential as food, medicines, or industrial raw materials. Tubocurarine, for example, is used as a muscle relaxant. Derived from the pareira plant, it is found wild in the tropical forests of southern Brazil, Peru, Colombia, and Panama.

Indians, mayapple, and cancer
North American Indians have provided one key to an important wild resource – the mayapple. The vanishing tribes of North American Indians, centuries ago, unlocked the secrets of this remarkable plant species. Used by Penobscot Indians to treat warts and by Cherokees to treat deafness and kill parasitic worms, the secret properties of the mayapple were then passed on to the colonists. Today they are under the microscopic eye of modern chemists. A new drug, VePesid, a semi-synthetic derivative of podophyllotoxin – a natural extract of mayapple – is now being used to treat testicular cancer. It has already had remarkable results and, in some instances, has attained up to 47% recovery rate for treated patients. The mayapple story is not yet over. Following the advice of North American Indians, it has also been successfully used to repel potato beetles. Moreover, the mayapple shows some interesting activity towards important viruses such as Herpes 1, Herpes 2, Influenza A, and measles.

EVOLUTION IN CRISIS

On the timescale of evolution, our species scarcely registers. But in terms of its impact on the resources of the planet, and on the genetic heritage which is the living foundation of the future, our brief history is all too significant. It is a story of destruction.

Habitat destruction

So far as we know, ours is the only green planet in the universe. Yet we, the latest inhabitants, are depleting this rich heritage at a rate that will grossly impoverish the Earth's green cover in a twinkling of geological time.

We have already seen (pp 38-43) that we are expanding deserts and eliminating the remaining tropical forests at a critical rate. We look set to destroy much of our inheritance of coral reefs, mangroves, estuaries, and wetlands. We shall plough up enormous areas of grassland. Still other areas will be paved over, chopped down, dug up, drained, and poisoned – or will be "developed" until their natural potential is homogenized out of existence. All this we shall do in the name of human welfare.

Unless we achieve a remarkable turn-round in the stewardship of our heritage, the next century will witness a speeding up of the destructive process as human numbers continue to surge and human appetites for raw materials grow keener. Yet who can blame the subsistence farmers who, in order to feed their starving families, knowingly degrade the natural resource base of tomorrow's livelihood? Perhaps more culpable are the super-affluent citizens of the developed nations who, by demanding ever-greater flows of natural resources from all over the world, are just as destructive.

Our great-grandchildren may well look out on a planet that has suffered far greater depletion than during the course of a major glaciation. Were they, by then, to have learned how to live in ecological accord with the Earth's life-support systems, they will surely find that the damage of 150 years of human activity will take millennia to restore.

We are finding that some tropical forests, once removed, do not readily re-establish. When soil cover disappears, together with critical stocks of nutrients, the forest's comeback is pre-empted. Desertification is also irreversible, except at massive cost. Moreover, certain sectors of the biosphere, notably tropical forests, coral reefs, and wetland ecosystems, serve, by virtue of their biotic richness and their ecological complexity, as "powerhouses"

The irreplaceable heritage

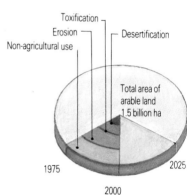

Living organisms are the heirs of 3.6 billion years of evolution. Just as the complexities of their inheritance almost defy calculation, so the consequences of our disruption of their networks of biological interdependence cannot readily be analysed or predicted. Yet humanity often seems to regard this heritage with indifference at best. Mismanagement now threatens to destroy entire sectors of the biosphere, inflicting grievous injury on our very life-support systems.

Various pressures bring this about; poverty in one part of the world is matched by excessive consumerism elsewhere. But the outcome tends to be the same – wholesale over-exploitation of humanity's environment. Entire ecosystems are undermined: over-grazing and brush clearance constantly extend the deserts; coastal wetlands, drained for agriculture, spill toxic chemicals instead of nutrients into the sea, while industrial wastes and sewage aggravate their impact; each minute, around 30 ha of tropical forest are destroyed, diminishing the habitat of thousands of species; in Europe, intensive cultivation eliminates woodlands and hedgerows, together with their myriad organisms.

The balances and linkages that are the very process of life on Earth thus come under threat. What if they finally unravel?

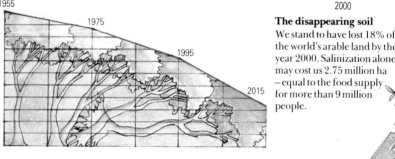

The disappearing soil
We stand to have lost 18% of the world's arable land by the year 2000. Salinization alone may cost us 2.75 million ha – equal to the food supply for more than 9 million people.

The shrinking forest
One-quarter of Brazil's forests have gone; more than four-fifths of Thailand's forests have been eliminated. The graph shows the decline in area of closed tropical forest (see also pp 38-9).

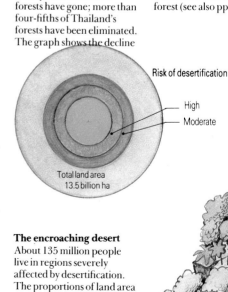

Risk of desertification
— High
— Moderate

Total land area
13.5 billion ha

The encroaching desert
About 135 million people live in regions severely affected by desertification. The proportions of land area at risk are shown above (see also pp 42-3).

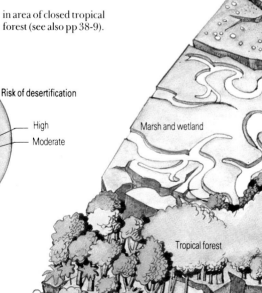

Marsh and wetland

Tropical forest

Loss of UK habitats since 1949	%
Lowland meadows	95
Lowland heath	50-60
Ancient lowland woodlands	30-50
Lowland wetlands (fens)	50
Hedgerows	25
Upland heaths and grasslands	30

Britain's vanishing countryside

Every year, 4,000 small farms in Britain are taken over by larger units. The environmental results are clear. Every year 3,200 km of hedgerow and almost 75,000 ha of heathland disappear, with all the plants and animals they support. In just over 40 years, 224,000 km of hedgerows have been torn out, and half the long-established deciduous woodlands destroyed. Over 80% of the flower-rich lowland meadows have been ploughed up or built over. Already more than 300 plant species are officially listed as endangered.

Threatened Cape flora

The Cape's floristic kingdom covering 1.8 million ha is one of the 6 most significant concentrations of flora on Earth. In this region occur 68% of the 2,373 South African plants under severe threat.

Acid rain in Scandinavia

In a lake with a pH of 5.5 (lower figures on the pH scale indicate greater acidity), most fish perish; with a pH of 4, the original lake ecosystem can die. Some 14,000 of Sweden's lakes are significantly acidified (see pp 116-7).

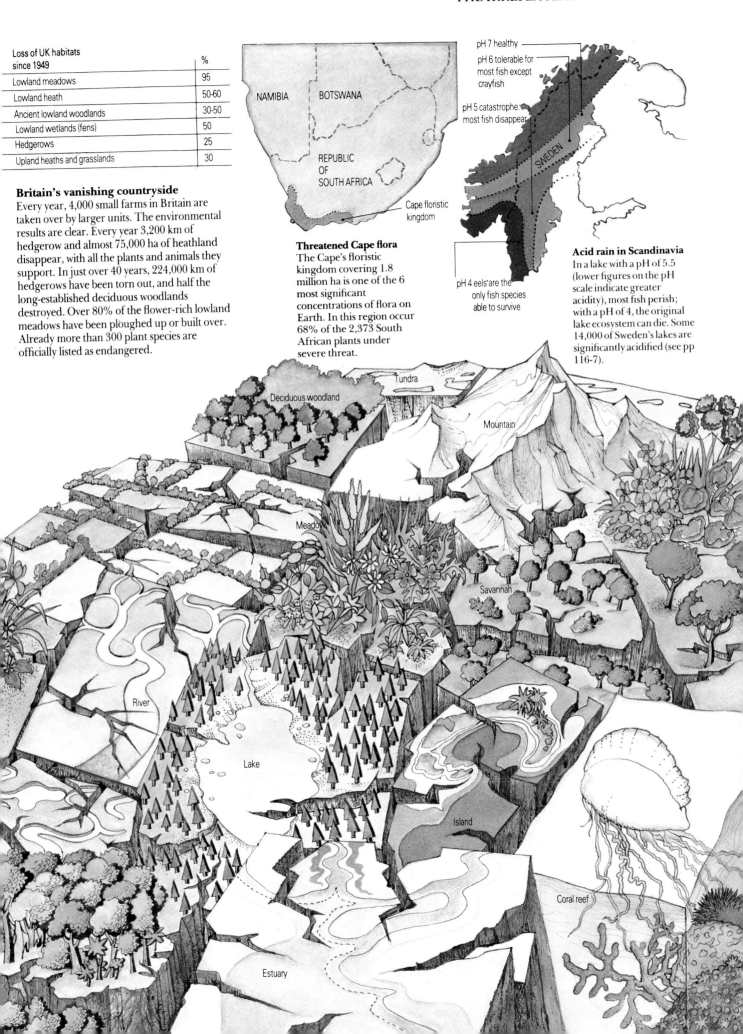

NAMIBIA
BOTSWANA
REPUBLIC OF SOUTH AFRICA
Cape floristic kingdom

pH 7 healthy
pH 6 tolerable for most fish except crayfish
pH 5 catastrophe: most fish disappear
SWEDEN
pH 4 eels are the only fish species able to survive

Deciduous woodland
Tundra
Mountain
Meadow
Savannah
River
Lake
Island
Coral reef
Estuary

of evolution. By destroying these salient sectors of the biosphere, we are impoverishing the future course of evolution.

Loss of species

Well over 90 percent of all species that have ever lived have disappeared. They have become extinct through natural processes, often being superseded by "better" or "fitter" species. After humankind first appeared on the scene and learnt to hunt animals for food, commerce, and sport, often disrupting natural environments in the process, the extinction rate started to soar, until eventually it reached about one species a year in the early 20th century.

Today, when natural environments are being degraded and destroyed on every side, the extinction rate has surely reached between 50 and 100 species a day. Within the next few decades we stand to witness the extinction of millions of species – perhaps as many as 50 per cent of the total number.

In its scale and compressed time span, this process of extinction will represent a greater biological débâcle than anything experienced since life began. It will massively exceed the "great dying" of the dinosaurs and their kin, together with associated organisms, 65 million years ago, when a sizeable share of Earth's species disappeared.

We know that there are about 45,000 species of vertebrates on Earth, and about 250,000 species of higher plants, with somewhere around 100,000 species of lower plants. The great bulk of the 5-30 million species currently thought to exist are invertebrates, of which insects comprise almost 80 percent. And it is among the invertebrates, particularly among the insects, that the major extinctions are now occurring.

When conservationists used to say that we were losing one species a year, they were referring almost exclusively to mammals and birds, which in total account for only about 13,000 species. Botanists have compiled a reasonably detailed account of the plants currently threatened with extinction. They have come up with a total of 25,000 species. Further, they believe that, as a rough rule of thumb, there are between 20 and 40 animal species for every one plant species, dependent on those plants for their survival. So for every plant that disappears, many more animal species may eventually disappear.

We know that of the plant species, about one in ten can be categorized as threatened, and the same sort of proportion holds for mammals and birds. It also apparently holds for other vertebrates, such as fish, reptiles, and amphibians. If we extrapolate this proportion to the invertebrates, we have good reason to suspect that somewhere between 500,000 and one million species are under threat right now.

Ironically, the reduced stock of species which survives this wave of extinctions is likely to include an unusually high number of opportunistic species, able to move into ecological niches vacated by

The destruction of diversity

We are in the early phase of what looks likely to be an unprecedented era of extinctions. Extinction has always been a way of life on Earth, but the present wave of extinctions natural rate. Indeed, the rate of species loss is so great that it threatens to disrupt evolution itself.

If large-scale habitat disruption and destruction continue to accelerate, we run a real risk that the diminished stock of species will not represent an adequate resource base on which natural selection can work to rebuild the rich panoply of life. So far as we can discern from the fossil record, the "bounce back" period could well extend over several million years. The process of species formation will clearly continue, even accelerate in places, but it will be outrun by extinction. We should be worried about the loss of diversity for its own sake and because it threatens existing and potential future resources. But the implications of the headlong destruction of species for the future course of evolution are more worrying still. "Death is one thing", as Drs Soule and Wilcox neatly put it; "an end to birth is something else".

The flame of life
From its first flickerings, an estimated 3.6 billion years ago, the flame of life has burned ever brighter, dying down during the major periods of extinction, but leaping back to even greater heights thereafter. Today the flame of life is threatened by intense human pressures: habitat destruction is gathering pace, while the superpowers are developing weapons which threaten to extinguish all higher life-forms. Even if we manage to avoid nuclear Armageddon, we risk snuffing out an extraordinary number of life-forms.

1700

Millions of years ago					
Cambrian 570	Ordovician 530	Silurian 440	Devonian 410	Carboniferous 345	Permian 280

Habitat destruction

Competition

Over-exploitation

Annual rate of species loss

1

6

400

Unravelling the tapestry of life

Like a child tugging threads from some
enormous piece of tapestry, we continue to tear
at the web of life with little if any knowledge of
the possible impact. Perhaps those who dug
limestone out of caves near Kuala Lumpur and
reclaimed land from swamps up to 40 km away
realized that they were destroying the roosting
sites and feeding grounds of a bat, *Eonycteris
spelaea*. But there was one thing they did not
know: this single bat species is responsible for
pollinating one of Southeast Asia's most highly
prized fruit crops, derived from the durian tree.
The annual durian crop, worth some $120
million, is now at risk.

c. 10,000

Today's dodos

Everybody has heard of the
dodo, extinct some 200
years since, but few are
aware of the whole range of
life now under threat – or
of what this implies. When
the dodo disappeared, at
least one tree species,
reliant on the dodo to help
its seed germinate, slid
towards extinction.

The graph of destruction

Current estimates suggest that we are losing
between 50 and 100 species a day from the 5-30
million species thought to exist. By the time
human populations reach some sort of ecological
equilibrium with this one-Earth habitat, at least a
quarter of all species could have disappeared. The
loss could even be higher, possibly one half.
Habitat destruction is now the most important
cause of species loss. If present trends continue,
particularly the loss of tropical rainforests, we can
expect far higher annual rates of species
extinctions (see graph, right).

c. 50,000

Jurassic	Cretaceous	Tertiary	The future
195	136	64	

recently extinct species or to thrive on our rubbish. Previous opportunists have included the house-fly, the rabbit, the rat, and "weedy" plants. We run a risk of creating a pest-dominated world.

A crisis of uniformity

Each mammal, insect, or plant pushed into the abyss of extinction takes with it genes which could have proved very valuable. As species disappear, so do gene pools – and so does the prospect of exploiting those genes to improve our future welfare. Losses through extinction are irreversible, although our ignorance about the economic potential of so many of our genetic resources means we are often totally unaware of what we have lost.

Worse, the erosion of genetic diversity is greater than the statistics on species loss alone would suggest. If a species with one million individuals is reduced to only 10,000 (which may still be enough to ensure the species' survival), it may have lost 90 percent of its races, populations, and other genetic sub-units, with a concomitant loss of half of its genetic diversity. This "concealed" erosion of genetic diversity is generally overlooked, although ultimately it could represent as grave a threat as the loss of the species itself.

More immediately, we are losing wild and semi-domesticated plants whose genes are essential if we are to maintain our major crop plants. Of the estimated 80,000 edible plants, only about 150 have been cultivated on a large scale – and less than 20 provide 90 percent of our food. The more we bend these 20 plants to our needs, the more vulnerable they become – and the more they need infusions of genes from outside their own shrinking gene pools.

There has been a parallel decline in the number of strains of many farm animals. In pursuit of immediate profits, the livestock industry tends to concentrate on a limited number of breeds, with the result that we are heading towards emergency levels of "homogenized breeding". In Europe and the Mediterranean basin, 118 cattle varieties are under threat – and only 30 breeds are holding their own.

The case of the Cornish chicken illustrates the error of allowing older strains to die out. This fast-growing subspecies of the domestic fowl was superseded by new breeds which laid more eggs or tasted better. Then, as the Cornish chicken faded away, breeders decided they needed an infusion of its genes to boost the growth rate of the very chickens which had usurped its place on the farm.

Sometimes, too, a plant or animal can act as a signpost to hidden genetic riches. So it has been with the rosy periwinkle, a Madagascar forest plant which has given us two potent anti-cancer drugs. So rare are the critical alkaloidal chemicals in most forms of the plant, that pharmacologists once had to process 500 tonnes of plant material to extract one kilogram of drug. Now a West Indian variant has been found which contains 10 times more of the alkaloid, enhancing the production process.

Traditional varieties
Small, but well supported, the sack of traditional maize hangs from a thick, multi-stranded rope, representing many plant varieties. Genetic diversity defends this plant population against pest and pathogen.

New varieties
Today's maize dangles from a dangerously thin thread. The bigger the sack, the more readily it could fall: the US harvest almost suffered this fate in 1970 when a maize fungus disease threatened over half of all major maize lands. At the eleventh hour, a "technical fix" was achieved by drafting in a more resistant strain from Mexico, the ancestral home of maize.

Genetic erosion

"The products of agro-technology are displacing the source upon which the technology is based. It is analagous to taking stones from the foundation to repair the roof."

PROFESSOR GARRISON WILKES
UNIVERSITY OF MASSACHUSETTS

Habitat destruction is not the only threat to the process of evolution. Genetic erosion is depleting the gene base of many existing crop plants and farm animals. Productive diversity is replaced by dangerous homogeneity, and future avenues of agricultural development are cut off.

The scale of the potential loss is indicated by the recent history of *Zea diploperennis*, a rare perennial maize discovered in 1978. Found in a few hectares of farmland in the Sierra de Manantlan, Mexico, this hitherto unknown variety was down to some 2,000 plants – and the elimination of its habitat continues. Yet its genes could open up the prospect of perennial maize production and increased resistance to at least four of the seven most important maize diseases, all of which could lead to billion dollar savings.

There are many more examples. Take the Rio Palenque Research Station in Ecuador. In area it amounts to as little as 170 ha, yet it supports 1,025 plant species – the highest recorded concentration of plant diversity on Earth. Regrettably, this last patch of wet forest of coastal Ecuador is being undermined by local people who enter the forest to cut wood for fuel and construction.

1840's Irish potato blight, 2 million die

1860's Vine diseases crippled Europe wine industry.

1870-90 Coffee rust robbed Ceylon of valuable export.

1942 Rice crop destroyed, millions of Bengalis died.

1946 US oat crop devastated by fungu epidemic.

1950's Wheat stem rust devastated U

1970 Maize fungus threatened 80% of corn hectarage.

Counting human cost
The potential impacts of monoculture collapse have become ever greater, as increasing numbers of people come to depend on a shrinking number of crop varieties. If the world maize crop failed tomorrow, it would do a great deal more than cut our supplies of food and feed. It would also impinge on many other products to which maize makes a contribution, such as aspirin, penicillin, tyres, plastics – even the "finish" on these very pages.

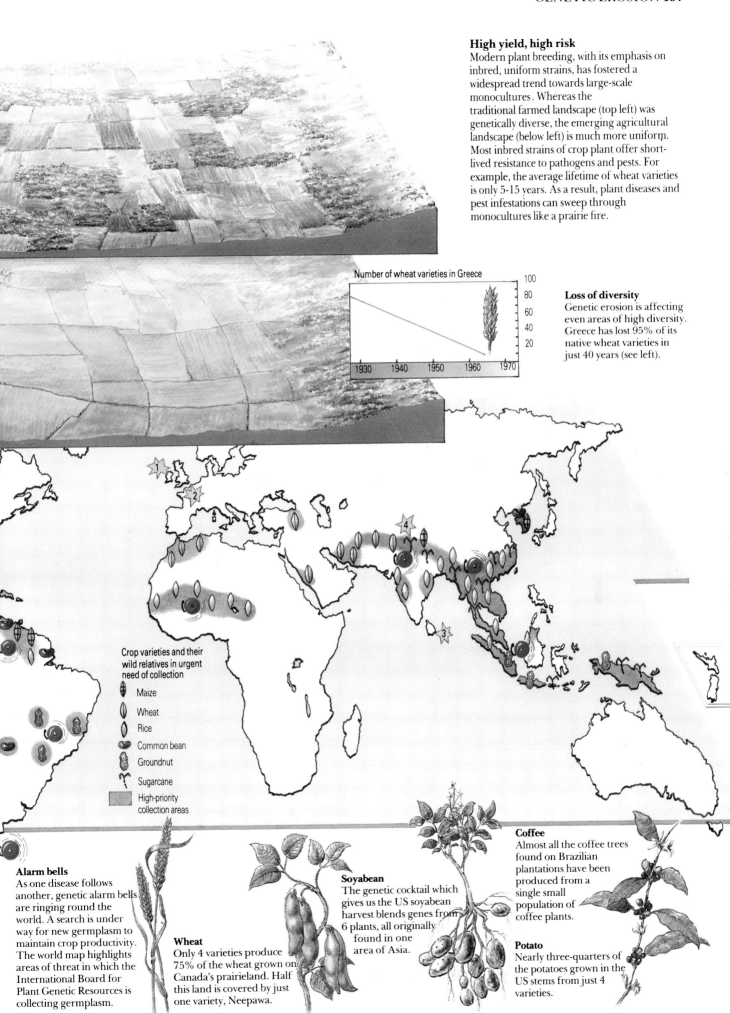

High yield, high risk

Modern plant breeding, with its emphasis on inbred, uniform strains, has fostered a widespread trend towards large-scale monocultures. Whereas the traditional farmed landscape (top left) was genetically diverse, the emerging agricultural landscape (below left) is much more uniform. Most inbred strains of crop plant offer short-lived resistance to pathogens and pests. For example, the average lifetime of wheat varieties is only 5-15 years. As a result, plant diseases and pest infestations can sweep through monocultures like a prairie fire.

Number of wheat varieties in Greece

100
80
60
40
20

1930   1940   1950   1960   1970

Loss of diversity

Genetic erosion is affecting even areas of high diversity. Greece has lost 95% of its native wheat varieties in just 40 years (see left).

Crop varieties and their wild relatives in urgent need of collection

- Maize
- Wheat
- Rice
- Common bean
- Groundnut
- Sugarcane
- High-priority collection areas

Alarm bells

As one disease follows another, genetic alarm bells are ringing round the world. A search is under way for new germplasm to maintain crop productivity. The world map highlights areas of threat in which the International Board for Plant Genetic Resources is collecting germplasm.

Wheat

Only 4 varieties produce 75% of the wheat grown on Canada's prairieland. Half this land is covered by just one variety, Neepawa.

Soyabean

The genetic cocktail which gives us the US soyabean harvest blends genes from 6 plants, all originally found in one area of Asia.

Coffee

Almost all the coffee trees found on Brazilian plantations have been produced from a single small population of coffee plants.

Potato

Nearly three-quarters of the potatoes grown in the US stems from just 4 varieties.

Towards a lonely planet

Human beings, uniquely intelligent, have achieved a self-awareness that both gives them power over and separates them from the life around them. The natural world is seen as theirs to exploit and despoil. The biological and mineral storehouses of the Earth, once thought inexhaustible, are ruthlessly plundered. In the process, the web of interdependence from which humanity itself is derived begins to fall apart. Yet in this rage to prosper lie the seeds of deep despair. As the human species creates around it a constantly growing desolation, it is in danger of finding itself isolated on a desecrated planet. Its only companions will be the cowed species it has domesticated and the rodent survivors, wily and vicious, that have resisted its assault. Racing along our current path of development, we ignore at our peril the warnings sounded by those whose eyes are not blinkered by short-term priorities.

"Man has long lost his ability to foresee and forestall. He will end by destroying the earth."
Albert Schweitzer

"Extinction does not simply mean the loss of one volume from the library of nature. It means the loss of a loose-leaf book whose individual pages, were the species to survive, would remain available in perpetuity for selective transfer and improvement of other species."
Professor Thomas Eisner
Cornell University

"For as long as Man continues to be the ruthless destroyer of lower living beings, he will never know health or peace. For as long as men massacre animals, they will kill each other. Indeed, he who sows the seeds of murder and pain cannot reap joy and love."
Pythagoras

"I owe an allegiance to the planet that has made me possible, and to all the life on that planet, whether friendly or not. I also owe an allegiance to the 3½ billion years of life that made it possible for me to be here, and all the rest of you too. We have a responsibility to the largest population of all, the hundred of billions of people who have not yet been born, who have a right to be, who deserve a world at least as beautiful as ours, whose genes are now in our custody and no one else's."
David R. Brower, Chairman, Friends of the Earth

"Unwittingly for the most part, but right around the world, we are eliminating the panoply of life. We elbow species off the planet, we deny room to entire communities of nature, we domesticate the Earth. With growing energy and ingenuity, we surpass ourselves time and again in our efforts to exert dominion over fowl of the air and fish of the sea.

We do all this in the name of human advancement. Yet instead of making better use of lands we have already to our use, we proclaim our need to expand into every last corner of the Earth. Our response to natural environments has changed little for thousands of years. We dig them up, we chop them down, we burn them, we drain them, we pave them over, we poison them in order to mould them to our image. We homogenize the globe.

Eventually we may achieve our aim, by eliminating every "competitor" for living space on the crowded Earth. When the last creature has been accounted for, we shall have made ourselves masters of all creation. We shall look around, and we shall see nothing but each other. Alone at last."
Norman Myers

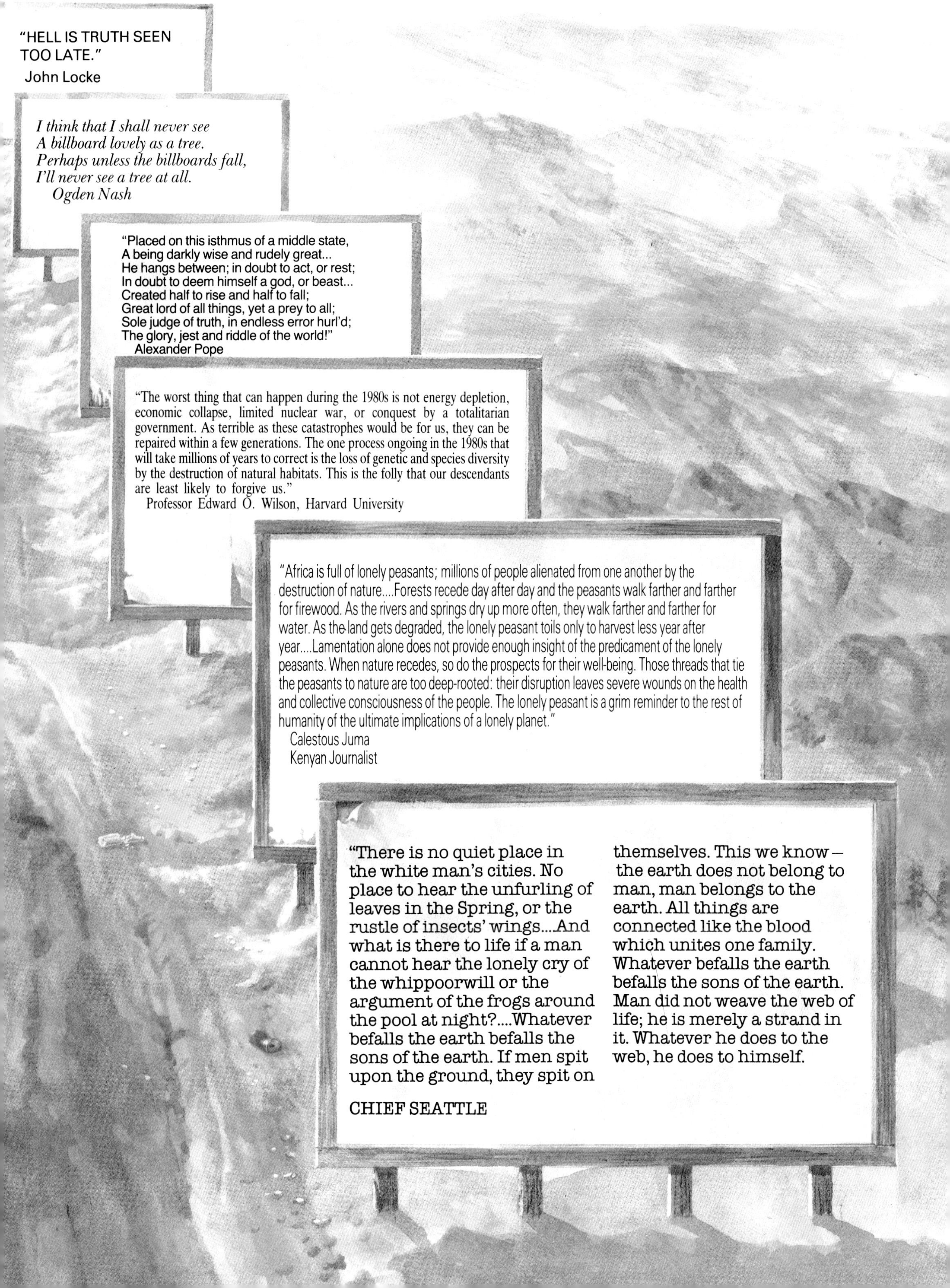

"HELL IS TRUTH SEEN
TOO LATE."
John Locke

I think that I shall never see
A billboard lovely as a tree.
Perhaps unless the billboards fall,
I'll never see a tree at all.
    Ogden Nash

"Placed on this isthmus of a middle state,
A being darkly wise and rudely great...
He hangs between; in doubt to act, or rest;
In doubt to deem himself a god, or beast...
Created half to rise and half to fall;
Great lord of all things, yet a prey to all;
Sole judge of truth, in endless error hurl'd;
The glory, jest and riddle of the world!"
    Alexander Pope

"The worst thing that can happen during the 1980s is not energy depletion, economic collapse, limited nuclear war, or conquest by a totalitarian government. As terrible as these catastrophes would be for us, they can be repaired within a few generations. The one process ongoing in the 1980s that will take millions of years to correct is the loss of genetic and species diversity by the destruction of natural habitats. This is the folly that our descendants are least likely to forgive us."
  Professor Edward O. Wilson, Harvard University

"Africa is full of lonely peasants; millions of people alienated from one another by the destruction of nature....Forests recede day after day and the peasants walk farther and farther for firewood. As the rivers and springs dry up more often, they walk farther and farther for water. As the land gets degraded, the lonely peasant toils only to harvest less year after year....Lamentation alone does not provide enough insight of the predicament of the lonely peasants. When nature recedes, so do the prospects for their well-being. Those threads that tie the peasants to nature are too deep-rooted: their disruption leaves severe wounds on the health and collective consciousness of the people. The lonely peasant is a grim reminder to the rest of humanity of the ultimate implications of a lonely planet."
  Calestous Juma
  Kenyan Journalist

"There is no quiet place in the white man's cities. No place to hear the unfurling of leaves in the Spring, or the rustle of insects' wings....And what is there to life if a man cannot hear the lonely cry of the whippoorwill or the argument of the frogs around the pool at night?....Whatever befalls the earth befalls the sons of the earth. If men spit upon the ground, they spit on themselves. This we know — the earth does not belong to man, man belongs to the earth. All things are connected like the blood which unites one family. Whatever befalls the earth befalls the sons of the earth. Man did not weave the web of life; he is merely a strand in it. Whatever he does to the web, he does to himself.

CHIEF SEATTLE

EVOLUTION IN MANAGEMENT

Our ability to disrupt the planet has been well tested. Now, we are challenged by a unique opportunity to turn that same massive ability to the task of large-scale managment of Earth's living resources, on a sustainable basis, combining two imperatives – development and conservation.

Protecting our heritage

To safeguard the world's wildlife and wildlands, we have established a growing number of parks and reserves. Some, such as Tanzania's Serengeti and Australia's Great Barrier Reef, have been set up to protect wildlife and its habitats. Others, such as Yosemite in the US and Nepal's Mount Everest Park, seek to protect spectacular scenery.

The protected-areas movement emerged just over 100 years ago, with the establishment of the Yellowstone National Park in the US. But it has only really taken off during the last 25 years, as country after country has recognized the need to safeguard pristine nature before it is too late. We still urgently require more protected areas: our present network is less than one-third of our overall needs.

Conserving the wild

Only a very small percentage of the world's land surface has been set aside to protect wild species and their genetic resources. Worse, the parks and reserves which we do have are far from representative of major types of ecosystems. Almost half the total conserved area can be found in North America, in the boreal forest and semi-frozen areas of Greenland and Arctic Canada. Of nearly 200 biogeographical provinces in the world, one in eight is not represented by a single park or reserve – and a similar number are represented by only one or two protected areas. Among the most poorly protected biogeographical provinces are tropical moist forests, grasslands, and Mediterranean-type zones. We are a long way from achieving the extensive, strategically sited network of protected areas which we need. Some of the most important gaps are highlighted here.

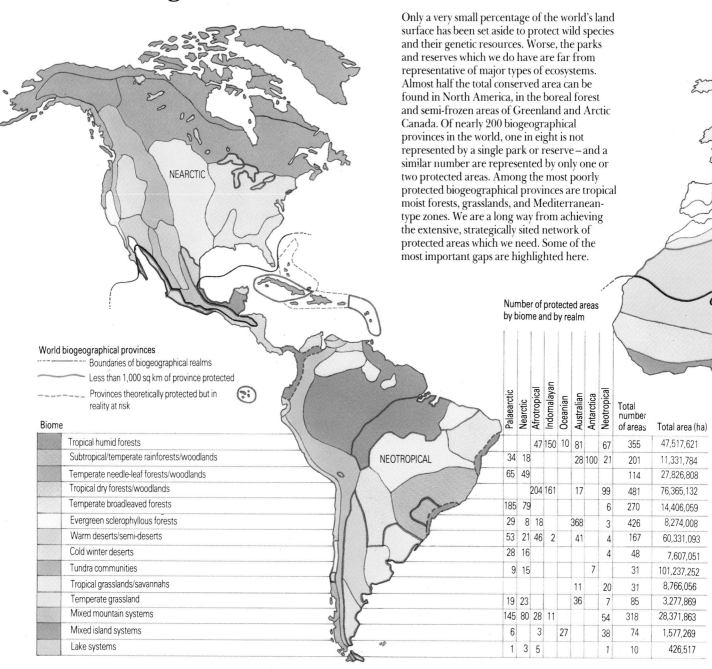

World biogeographical provinces
- ---------- Boundaries of biogeographical realms
- ∿∿∿∿ Less than 1,000 sq km of province protected
- - - - - Provinces theoretically protected but in reality at risk

Number of protected areas by biome and by realm

Biome	Palaearctic	Nearctic	Afrotropical	Indomalayan	Oceanian	Australian	Antarctica	Neotropical	Total number of areas	Total area (ha)
Tropical humid forests			47	150	10	81		67	355	47,517,621
Subtropical/temperate rainforests/woodlands	34	18				28	100	21	201	11,331,784
Temperate needle-leaf forests/woodlands	65	49							114	27,826,808
Tropical dry forests/woodlands			204	161		17		99	481	76,365,132
Temperate broadleaved forests	185	79						6	270	14,406,059
Evergreen sclerophyllous forests	29	8	18			368		3	426	8,274,008
Warm deserts/semi-deserts	53	21	46	2		41		4	167	60,331,093
Cold winter deserts	28	16						4	48	7,607,051
Tundra communities	9	15					7		31	101,237,252
Tropical grasslands/savannahs						11		20	31	8,766,056
Temperate grassland	19	23				36		7	85	3,277,869
Mixed mountain systems	145	80	28	11				54	318	28,371,863
Mixed island systems	6	3			27			38	74	1,577,269
Lake systems	1	3	5					1	10	426,517

NEARCTIC

NEOTROPICAL

Meanwhile, just as natural landscapes are being modified, so the protected-areas movement is adapting to the growing pressures. The traditional, purist approach has been to establish reserves from which all forms of human exploitation are banned. Increasingly, however, conservationists recognize that it will become ever-more difficult to declare large tracts of land "off limits" to human use and development. We know what we want to protect parks *from*: now we must devote more thought to what we are protecting them *for*. If existing parks are to survive, let alone be joined by new protected areas, they must be seen to be meeting the real needs of people – not just the esoteric interests of nature enthusiasts. This is all the more urgent in the Third World, where there is the greatest need to

achieve a comprehensive parks system – and where there is the greatest pressure on existing protected areas from land-hungry farmers.

Fortunately, it is possible to demonstrate that parks do indeed serve the cause of development. In northern Sulawesi, for example, a rainforest park has been set up which will mean that Indonesia will forfeit revenues from uncut timber. But the government is backing the initiative on the grounds that it will protect a rainfall-catchment zone which is critical for several million rice-paddy farmers in the valley bottomlands below. In several African parks, where dams have been built to supply water for the wildlife, local people raise fish for market. Elsewhere, dry-season grazing for livestock may be permitted, or some subsistence hunting tolerated.

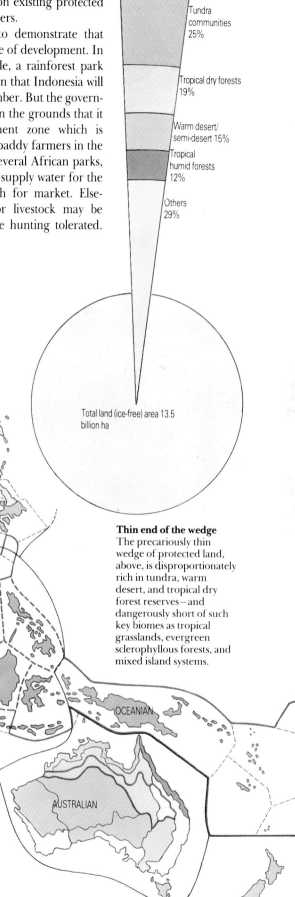

Total protected area 5%

Tundra communities 25%

Tropical dry forests 19%

Warm desert/ semi-desert 15%

Tropical humid forests 12%

Others 29%

Total land (ice-free) area 13.5 billion ha

Thin end of the wedge
The precariously thin wedge of protected land, above, is disproportionately rich in tundra, warm desert, and tropical dry forest reserves – and dangerously short of such key biomes as tropical grasslands, evergreen sclerophyllous forests, and mixed island systems.

PALAEARCTIC

INDOMALAYAN

AFROTROPICAL

OCEANIAN

AUSTRALIAN

ANTARCTICA

Tool-kit for conservation
Biogeographers, by providing maps like the one above, help conservationists to identify priorities and achieve a representative spread in the ecosystems protected.

In all the instances cited, local people do not clamour for the park to be abolished. Rather, they support its survival.

Conserving the unknown

The best way to preserve threatened gene pools is to protect the relevant habitats, by designating them as national parks or game reserves. A second option involves protecting the gene pools off-site, whether in botanical gardens, zoos, or gene banks. This second approach offers many apparent advantages, but it also suffers from some critical drawbacks.

Almost all our present gene banks are off-site, including seed and germplasm storage facilities, clonal plantations, seed orchards, and rare-breeds farms. Such facilities will play an increasingly important role, but their contribution will be limited by a number of key factors. First, there is the sheer size of the task. The diversity of the wild genetic resource is so great that it seems highly unlikely that off-site gene banks could ever save more than a fraction of the total resource.

More importantly, the seeds of some crop species and their wild relatives cannot be dried prior to storage, because they are killed in the process. Gene banks, for example, cannot preserve many seed plants. This applies especially to most vegetatively propogated plants, such as potato, cassava, members of the orchid family, and tree species such as apples and pears that do not breed true from seeds. Many tropical plant species, such as the cacao tree, can be conserved outside the wild only with the greatest difficulty. Some wild species also prove much more difficult to maintain and regenerate than their domesticated relatives, including a number of peanut and sunflower species. Other species, such as barley, beans, and maize, sustain genetic damage during long-term storage.

In addition, the entire stock of a gene bank can be destroyed by a prolonged power cut or by human negligence. Worst of all, a plant's evolution is effectively frozen while it enjoys the refrigerated safety of the gene bank. Outside, evolution goes on – and the protected plant may emerge to find that its wild pathogens and pests have evolved new forms of attack to which it is now unduly vulnerable.

So off-site gene banks can only provide part of the answer to genetic erosion. Ultimately, the only viable long-term approach must involve safeguarding gene pools in the wild. But even today, very few on-site conservation areas coincide with the most valuable concentrations of genetic resources – the so-called "Vavilov centres". Even if we can get such "gene parks" set up in the right places, other solutions are needed to protect the traditional crop varieties which are still cultivated by small-scale farmers throughout the developing world.

One idea which has been proposed to boost on-site conservation of such primitive strains is a tax on corporations selling genetically improved seed. Global sales of the seed industry now exceed $25

Preserving the genetic resource

There are three main options for the preservation of genetic diversity. We can protect it *on site*, by conserving the entire ecosystem in which it naturally occurs. If this is not possible, we can store part of the organism, such as its seed or semen, *off site*, in some form of gene bank. Or we can keep whole organisms *off site* – in an aquarium, botanical garden, culture ollection, plantation, or zoo. Most progress has been made with off-site conservation of crop genetic resources, through an international network of "base collections" covering more than 20 of the world's most important crops. But many species do not take kindly to life in botanical gardens, zoos, or the refrigerated world of the gene bank. So, while all three options have a vital role to play, the first must be our priority.

Gene banks
The seeds of many plant species, especially those with small, dry seeds, can often be stored in a dormant state for long periods at a humidity level of 5% and at a temperature of -20°C. A small gene bank can protect many thousands of species.

Botanical gardens
The world's botanical gardens play a vital role, but they have a problem of storage capacity. The Royal Botanic Garden at Kew, England, receives around 2,000 accessions of seeds each year, yet its total holding of about 25,000 species has scarcely changed over 40 years. As new material arrives, older specimens are discarded.

Zoos
Zoos have recorded a string of successes with their captive breeding of some threatened species (see left). But there is a rule of thumb that a vertebrate stock of less than 50 individuals is liable to carry the seeds of its own destruction, with harmful genes accumulating rapidly in inbred populations. And many species refuse to breed at all in captivity.

Rare-breeds centres
At least 20 breeds of British farm animal have become extinct during the 20th century. To counter this trend, rare-breeds survival centres have been set up, serving as living museums. They provide an opportunity for the general public to underwrite genetic conservation, by paying to see breeds on which their own future may well depend.

On-site protection

The focus of on-site protection programmes tends to be on species of popular appeal, and on species under threat; on unique ecosystems; on ecosystems which are representative of a particular type of habitat; or on some combination of all these approaches. The key objective is to protect as much genetic diversity as possible.

On-site gene banks score over off-site gene banks for a number of reasons, not least because the evolution of species in on-site reserves can continue uninterrupted, providing the breeder with a dynamic reservoir of genes that confer resistance to pests or pathogens.

On-site reserves also serve as living laboratories, allowing the breeder to study a species' ecology. This can throw up valuable information which might otherwise be overlooked; several crucial characteristics of wild tomatoes have surfaced in this way, including their tolerance of saline soils, high temperatures, and humidities, and their resistance to insects and disease.

Land races and wild genes

Among the most threatened gene pools are those of "land races", a term covering primitive or traditional plant cultivars and animal breeds. With most wild species, extinction tends to be a fairly gradual process, but land races, developed by local farmers, often enjoy only very limited distribution and can be lost in a single episode of habitat destruction.

Wild gene pools are now increasingly recognized as vital resources for future plant and animal breeding. Variety, in genetic terms at least, is more than the spice of life – it is the key to future survival.

Preserving diversity

The theories of Nikolay Vavilov are at the heart of many conservation programmes. He suggested that the centre of origin of a cultivated plant is to be found in the region where its wild cousins show maximum adaptiveness. The "Vavilov Centres" are prime targets for conservation. The map shows these centres in modified form.

The gene library

At present, international seed companies can borrow and benefit from the gene libraries of the world (many in developing countries) at virtually no charge. As the wild resource becomes more valuable, a strong case develops for making the end-user pay a "lending fee" for "books borrowed". Putting a value on the wild resource in this way will give developing countries a clear economic incentive to preserve the wild habitats.

billion a year, so that even a one percent tax on sales would generate $250 million a year. The revenues could be used to subsidize peasant farmers, enabling them to maintain some of their traditional varieties alongside higher-yield varieties.

Legislating for life

There are three obvious ways to put wildlife conservation on a legal footing. The first is to establish laws for individual species, such as the vicuna, or for a small group of species, such as marine mammals. The second is to establish a treaty for a region, as in the case of the Convention on Conservation of European Wildlife and Natural Habitats. The third is to establish a treaty of worldwide scope, such as the Convention on Wetlands of International Importance. Of these three, the first is easiest to establish, while the third tends to have the most significant impact.

When more than one nation is involved, negotiations become complex and enforcement can be a problem. Since many species occur in several countries at once (the cheetah, for example, is found in 20-odd African countries) or migrate from one country to another, it is often essential that several nations make common cause together.

The most recent breakthrough is CITES – the Convention on International Trade in Endangered Species. This treaty has been adopted by over 110 nations. It tackles the widespread, illicit trade in threatened species, whether they be tropical fish bound for the world's aquaria or spotted-cat skins destined for wealthy shoulders. The legitimate trade is worth billions of dollars a year, and while the illicit trade has been cut back by CITES, it remains big business. Back-sliding nations such as Japan still import huge quantities of tortoiseshell, crocodile skins, musk, and other wildlife products – despite the fact that they have signed CITES.

Meanwhile, wildlife continues to disappear faster than ever. The problem is not so much the poacher or other law-breaker, with rifle or trap in hand: instead, it is the settler, the subsistence farmer, persons without any malign intent towards wildlife. But they use tools such as the axe or plough, which are ultimately far more destructive. Clearly, we need a global treaty which builds on the undoubted successes of CITES, a treaty which protects critical habitats. A Convention, perhaps, on the Protection of Endangered Species – to be known as COPES?

The idea would be that each nation should accept responsibility for all species within its borders. In return, a nation would be able to apply for support from the community of nations to enable it to do a better job. Many developing nations harbour enormous concentrations of species – Panama, for example, may contain as many species as the entire US – but they lack the financial and scientific resources to protect them properly.

Wherever species may be found, they are part of everyone's heritage – so we should all share the cost

Laws and conventions

International law, based on treaties and conventions, was not used to protect wildlife until less than a century ago. The first wildlife treaties were largely concerned with economically important species – and with eliminating species viewed as pests. Two years before the first European wildlife treaty was signed, in 1902, a Convention was concluded which aimed to protect African game for trophy hunters and ivory traders – and which called for the destruction of such "noxious" pests as crocodiles, lions, leopards, hyenas, poisonous snakes, and birds of prey. Since then, there has been a growing spate of treaties and conventions, the most important of which is almost certainly CITES – the Convention on International Trade in Endangered Species. Enforcement is a continuing weakness, with countries like Japan signing CITES, and then proceeding to turn a blind eye to illicit dealing in wildlife products.

The Biodiversity Convention
Signed by almost all countries, but not the US, at the 1992 Earth Summit in Rio, the Biodiversity Convention is essentially a deal under which Western countries gain access to biological resources, such as chemicals and genes from wild plants, in return for transferring money and technical assistance for conservation and sustainable use of wild areas in the South.

☐ Coverage of UNEP Regional Seas Programme

CITES
Signed by over 110 countries since 1973, CITES prohibits international commercial trade in the rarest 600 or so species of animals and plants, and requires licences from the country of origin for exports of about another 200 groups. CITES has clamped down on the trade in many endangered species – try, for example, to buy a tiger-skin coat in London, Paris, or New York. And the parties to CITES voted to ban the trade in ivory from 1990 – poaching had reduced Africa's elephant population by half in the previous decade. But wildlife is big business and illegal trade continues. A single orchid or Amazonian parrot can fetch $5,000. A fur coat made from South American ocelots can sell for $40,000 in Germany. Rhino horns are worth their weight in gold. The world map, right, shows nations that are legally bound to enforce CITES and some examples of illegal trade which continue to undermine the impact of the Convention.

1 Walrus A demand for walrus ivory threatens this species.
2 American ginseng Exported to Far East for its alleged medicinal properties.
3 Hyacinth macaw Smuggled into Bolivia for export.
4 Nile crocodile Trade in wild populations now prohibited.
5 Whipsnake Skins go to the US, despite an export ban.
6 Pangolin Skin is traded illegally to supply the US boot industry.
7 Clown fish Large numbers are exported from the Philippines by the aquarium trade.

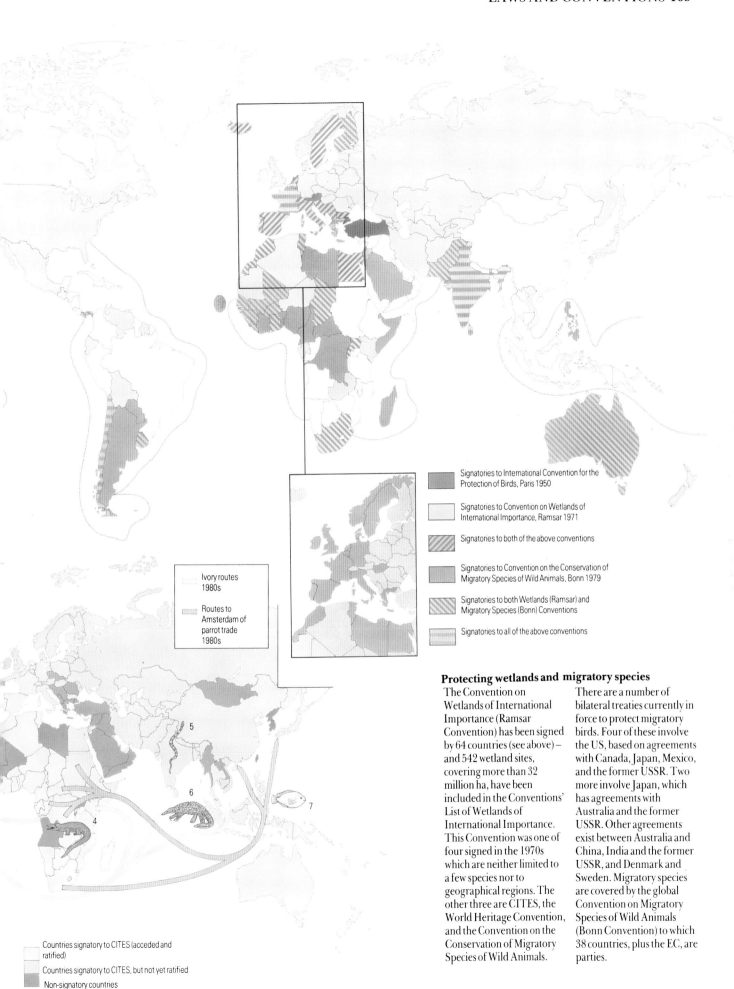

Signatories to International Convention for the Protection of Birds, Paris 1950

Signatories to Convention on Wetlands of International Importance, Ramsar 1971

Signatories to both of the above conventions

Signatories to Convention on the Conservation of Migratory Species of Wild Animals, Bonn 1979

Signatories to both Wetlands (Ramsar) and Migratory Species (Bonn) Conventions

Signatories to all of the above conventions

Ivory routes
1980s

Routes to
Amsterdam of
parrot trade
1980s

Countries signatory to CITES (acceded and ratified)

Countries signatory to CITES, but not yet ratified

Non-signatory countries

Protecting wetlands and migratory species

The Convention on Wetlands of International Importance (Ramsar Convention) has been signed by 64 countries (see above) – and 542 wetland sites, covering more than 32 million ha, have been included in the Conventions' List of Wetlands of International Importance. This Convention was one of four signed in the 1970s which are neither limited to a few species nor to geographical regions. The other three are CITES, the World Heritage Convention, and the Convention on the Conservation of Migratory Species of Wild Animals.

There are a number of bilateral treaties currently in force to protect migratory birds. Four of these involve the US, based on agreements with Canada, Japan, Mexico, and the former USSR. Two more involve Japan, which has agreements with Australia and the former USSR. Other agreements exist between Australia and China, India and the former USSR, and Denmark and Sweden. Migratory species are covered by the global Convention on Migratory Species of Wild Animals (Bonn Convention) to which 38 countries, plus the EC, are parties.

of their protection. To date there is no adequate mechanism allowing the richer nations to help others on a suitable scale.

Strategies for a sustainable future

We have a fairly good idea of how we should set about protecting our heritage of wildlife and genetic resources. The key will be to establish many more parks and reserves – perhaps covering as much as 10 percent of some regions of the world. But what, meanwhile, happens to the other 90 percent? If we allow these areas to be "developed out of existence", we shall undermine a sizeable proportion of our renewable resource base. And the protected areas will almost certainly be swamped by tides of humanity seeking new resources.

So we must expand our vision beyond the traditional concerns of conservation. We must pay increasing attention to the entire biosphere and to the many life-support systems which we have tended to take for granted, even as we make ever-growing demands upon them. In short, we need a strategy for a new conservation that embraces all life on Earth.

Happily, we now have some excellent ideas, summed up in the term *eco-development*. This term refers to development which takes into account the ultimate health of our planetary ecosystem. The basics of this new approach have been spelled out in two pioneering assessments and manifestos, the World Conservation Strategy (WCS) of 1980, and its successor of 1991 – Caring for the Earth.

Conservation and development, the WCS suggested, are two sides of the same coin: conservation cannot succeed without sustainable development, and development cannot be sustained without conservation. Brave words – but how can we convert them into action? Since 1980, the WCS has been tried and tested as the basis for national conservation policies in more than 50 countries. Simultaneously, international agencies, such as the UN and the World Bank, have become more aware of conservation needs and opportunities. The UN set up the World Commission on Environment and Development, and its "Brundtland Report" of 1987 contributed to the recognition of the imperative of sustainable development.

In 1991, IUCN, UNEP, and WWF published Caring for the Earth; a new conservation strategy structured around various principles upon which sustainable living can be built. It calls for a new commitment from governments, organizations, and individuals to the ethic of sustainability and lists the conservation of the Earth's vitality and diversity as a core principle for sustainable living. By the end of the century, it advocates, all countries should have adopted comprehensive strategies to safeguard their biological diversity, and a system of protected areas covering at least 10 percent of each of its main ecological regions.

National Conservation Strategies are under way in developing countries such as Nepal and Zambia. Developed countries too are carrying out similar exercises. We have started in the right direction.

Towards a new conservation

When it was launched in 1980 by IUCN, UNEP, and WWF, the World Conservation Strategy was a revolutionary document, presenting a single, integrated approach to global problems. It rested on three important propositions. First, species and populations must be helped to retain their capacity for self-renewal. Second, the basic life-support systems of the planet – climate, water cycle, and soils – must be conserved if life is to continue. Third, genetic diversity is a major key to the future – it too must be preserved.

The world applauded the Strategy when it first appeared, and it formed the basis for national conservation policies in more than 50 countries.

Yet in the decade since 1980 the complexity of the problems the world faces became clearer, and the need to act even more pressing. In 1991, the same organizations launched a new conservation strategy – Caring for the Earth, extending and emphasizing their original message, setting targets, and suggesting methods for the implementation of the strategy.

Although it is a strategy addressed to the whole world community, the world leaders who attended the June 1992 "Earth Summit" were an important audience. The contribution of national strategies to sustainability is considerable, as the case of Nepal, right, demonstrates.

National Conservation Strategies in action
Conservation strategies are being prepared to suit the needs of different countries. In New Zealand and Uganda, priority is given to monitoring fish stocks and regulating levels of exploitation. In Zambia, the NCS looks to minimize the adverse effects of mining, while in Nepal, fuelwood and soil erosion are high on the list. The NCS for Nepal aims to put environmental planning on a permanent footing. At government level, this involves creating new committees and commissions (see right).

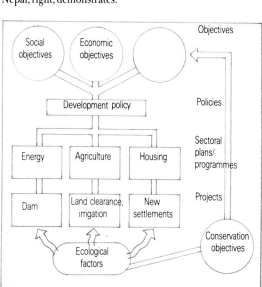

Integrating conservation
Most attempts to assess the environmental impacts of development have focused on specific projects, whereas many critical decisions are taken higher up the decision chain, left. Conservation objectives must be integrated with other main objectives in formulating national policies, before they crystallize into projects and programmes. When ecological factors are considered only at the bottom of the chain, their influence is limited, at best. If integrated at the top level of decision-making, they can have a highly positive influence.

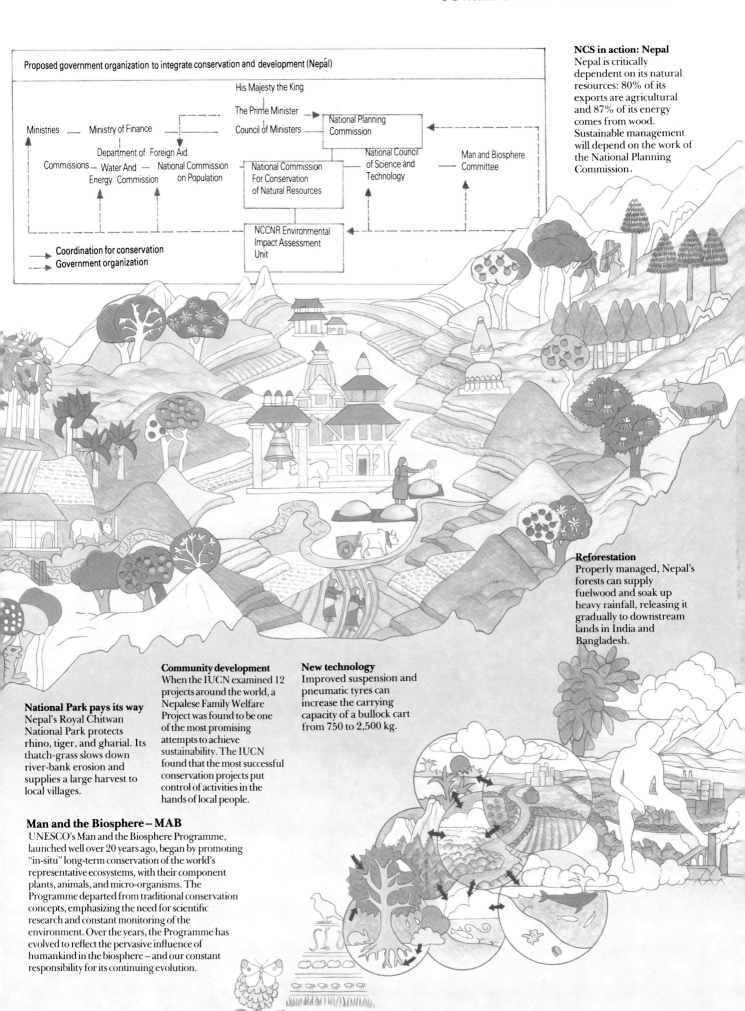

Proposed government organization to integrate conservation and development (Nepál)

His Majesty the King

The Prime Minister

Council of Ministers

National Planning Commission

Ministries — Ministry of Finance

Department of Foreign Aid

Commissions — Water And Energy Commission

National Commission on Population

National Commission For Conservation of Natural Resources

National Council of Science and Technology

Man and Biosphere Committee

NCCNR Environmental Impact Assessment Unit

→ Coordination for conservation

--→ Government organization

NCS in action: Nepal
Nepal is critically dependent on its natural resources: 80% of its exports are agricultural and 87% of its energy comes from wood. Sustainable management will depend on the work of the National Planning Commission.

Reforestation
Properly managed, Nepal's forests can supply fuelwood and soak up heavy rainfall, releasing it gradually to downstream lands in India and Bangladesh.

Community development
When the IUCN examined 12 projects around the world, a Nepalese Family Welfare Project was found to be one of the most promising attempts to achieve sustainability. The IUCN found that the most successful conservation projects put control of activities in the hands of local people.

New technology
Improved suspension and pneumatic tyres can increase the carrying capacity of a bullock cart from 750 to 2,500 kg.

National Park pays its way
Nepal's Royal Chitwan National Park protects rhino, tiger, and gharial. Its thatch-grass slows down river-bank erosion and supplies a large harvest to local villages.

Man and the Biosphere – MAB
UNESCO's Man and the Biosphere Programme, launched well over 20 years ago, began by promoting "in-situ" long-term conservation of the world's representative ecosystems, with their component plants, animals, and micro-organisms. The Programme departed from traditional conservation concepts, emphasizing the need for scientific research and constant monitoring of the environment. Over the years, the Programme has evolved to reflect the pervasive influence of humankind in the biosphere – and our constant responsibility for its continuing evolution.

HUMANKIND

Introduced by Dr Nafis Sadik

Executive Director of the UNFPA (United Nations Population Fund)

The end of the Cold War has brought renewed optimism about prospects for solving the world's critical problems through international partnership. But, as the following pages show, the challenges before us have not diminished. For too many people, the hope for a life of greater personal security remains distant. While some developing countries have made good progress in meeting their citizens' basic needs, others have come no closer to the goal of eliminating poverty. In fact, the number of people living in absolute poverty continues to increase every year, and economic security remains a critical, elusive need for a large part of the global population. Moreover, although the status and circumstances of women have improved in some countries, inequality continues to define gender relations in most parts of the world.

During the 1980s, we also became more aware of the importance of the environment and of the need to strike a balance between economic development and the preservation of natural resources and habitat. Environmental degradation and the threat of climate change are formidable problems, for which we have to find sustainable solutions. These will need to include drastically modified consumption patterns in the industrialized countries of the North, as well as measures to eliminate poverty and to slow population growth in the South.

Finding answers to these problems will require more human inventiveness and creativity than ever before. It will also take much greater political commitment, to ensure the success of remedial policies and programmes. This political determination has to come forward at the national and international level. Indeed, solutions to an increasing number of contemporary problems hinge on the ideal of global collaboration and the reality of global interdependence.

So far, collaboration has not been the principal mode by which we address such issues. More often than not, political, economic, or cultural barriers between countries have hindered a co-operative approach to international problem-solving. But the time has surely come when our ability to rise to higher levels of global partnership is being tested. The unrivalled degree to which humankind's problems have become global in character equally provides an unrivalled test of our determination to solve our problems together, now and in the future.

There are signs that we may be prepared to commit ourselves to higher levels of international collaboration. Many political barriers have come down over the past decade, thus presenting us with a good opportunity to improve global partnership. One example is the Earth Summit. Although its achievements were not perfect by any measure, the Earth Summit in Rio de Janeiro in 1992 showed to a degree not seen before that a global consensus exists on many issues, on which it is highly worthwhile and desirable to build.

THE HUMAN POTENTIAL

The Chinese proclaim that, "of all things, people are the most precious". And so they are. When one person is joined by a second, their joint capacity is not simply doubled: they can inspire one another, laugh together, and love each other. When all the peoples of the planet are considered together, they represent a capacity for labour, knowledge, creativity, conscious understanding, and happiness that cannot be measured. But, however the Chinese may delight in people as the finest manifestations of life's forces on Earth, they certainly do not believe that more must be better.

In fact, no other community matches the Chinese in their efforts to limit their numbers – on the grounds that more Chinese would result in poorer Chinese. The quality of human life is quite distinct from its quantity. There are five million Danes, for example, and over 810 million Indians – yet it would be grotesque to suppose that Indians are 160 times better off than Danes.

The size of a country's or region's population can, of course, increase its political influence – but this relationship is far from automatic (the control of wealth is far more significant). In 1900, 17 percent of the world's total population lived in Europe, whereas by the year 2000, only about 8 percent will do so. The developing world, which today is home to 75 percent of the global population, could account for 90 percent within 50 years. The weight of numbers may well effect a shift in focus towards the South.

In the South, the population is predominantly young, putting pressure on child health and education services. In the industrialized North, the size of the over-60's sector has social and economic implications which worry many governments. As longevity increases, nation after nation will be affected by the "grey wave", and we will have to develop new ways to care for the world's senior citizens. They represent a valuable reservoir of skills and experience.

In the light of social and economic advance, and technological change, the potential of the young people should be even greater. The percentage of young people in education has never been larger. The level of education, training, and creative enthusiasm in a population can powerfully influence the extent to which the potential of the human resource is released and harnessed. Some developing countries, notably China, have achieved near-miracles in improving the education, health, and life expectancy of their citizens, giving a

People potential

The greatest natural resource on the planet is the human race itself. And, in a world in need of ever-increasing care and protection, the full potential of every individual has never been more in demand. Releasing such potential is rarely easy, rarely comfortable, but the scale of the challenge we now face makes it essential. There is no shortage of urgent tasks, all of which are well within our capacity – providing we can mobilize the necessary political will, and physical energies, amplified through appropriate technologies. But lack of basic needs such as food, water, fuel, and shelter, plus social neglect, and often sheer prejudice, continue to obscure many elements of the human resource: one of the most blatant examples has been the failure to capitalize on the abilities of women; another is the neglect of the young and unemployed. We have the human resources we need, if only we are prepared to give them a chance.

Biceps and brains
The power and ability of people is not amenable to measurement. In theory, the 3.3 billion people aged 15-64 could till the world's cultivable land in under 3 days. But such muscle power, roughly equal to 2.5 billion kilowatts of energy per hour, pales beside creative and inventive power – the great art of the world, for instance, or the creation of artificial intelligence.

China
China has made real progress in developing people potential. In 1990 the total population was 1.1 billion, and China's GNP is relatively low, at $330 per person. Yet adult literacy is now 73%, a vast improvement on the pre-1949 picture. A comprehensive health service has boosted life expectancy to 70 years, and the increasingly recognized private sector is boosting employment. The darker side is that personal liberty has been curtailed, particularly in the aftermath of the 1990 Tiananmen Square massacre.

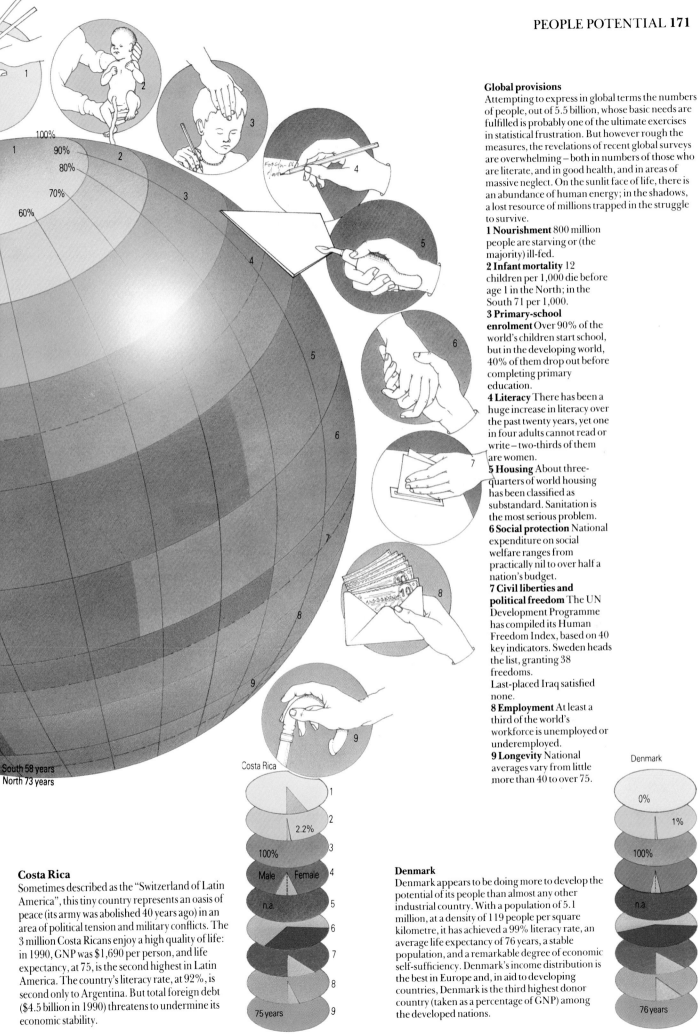

Global provisions

Attempting to express in global terms the numbers of people, out of 5.5 billion, whose basic needs are fulfilled is probably one of the ultimate exercises in statistical frustration. But however rough the measures, the revelations of recent global surveys are overwhelming – both in numbers of those who are literate, and in good health, and in areas of massive neglect. On the sunlit face of life, there is an abundance of human energy; in the shadows, a lost resource of millions trapped in the struggle to survive.

1 Nourishment 800 million people are starving or (the majority) ill-fed.

2 Infant mortality 12 children per 1,000 die before age 1 in the North; in the South 71 per 1,000.

3 Primary-school enrolment Over 90% of the world's children start school, but in the developing world, 40% of them drop out before completing primary education.

4 Literacy There has been a huge increase in literacy over the past twenty years, yet one in four adults cannot read or write – two-thirds of them are women.

5 Housing About three-quarters of world housing has been classified as substandard. Sanitation is the most serious problem.

6 Social protection National expenditure on social welfare ranges from practically nil to over half a nation's budget.

7 Civil liberties and political freedom The UN Development Programme has compiled its Human Freedom Index, based on 40 key indicators. Sweden heads the list, granting 38 freedoms. Last-placed Iraq satisfied none.

8 Employment At least a third of the world's workforce is unemployed or underemployed.

9 Longevity National averages vary from little more than 40 to over 75.

South 58 years
North 73 years

Costa Rica
2.2%
100%
Male | Female
n.a.
75 years

Denmark
0%
1%
100%
n.a.
76 years

Costa Rica

Sometimes described as the "Switzerland of Latin America", this tiny country represents an oasis of peace (its army was abolished 40 years ago) in an area of political tension and military conflicts. The 3 million Costa Ricans enjoy a high quality of life: in 1990, GNP was $1,690 per person, and life expectancy, at 75, is the second highest in Latin America. The country's literacy rate, at 92%, is second only to Argentina. But total foreign debt ($4.5 billion in 1990) threatens to undermine its economic stability.

Denmark

Denmark appears to be doing more to develop the potential of its people than almost any other industrial country. With a population of 5.1 million, at a density of 119 people per square kilometre, it has achieved a 99% literacy rate, an average life expectancy of 76 years, a stable population, and a remarkable degree of economic self-sufficiency. Denmark's income distribution is the best in Europe and, in aid to developing countries, Denmark is the third highest donor country (taken as a percentage of GNP) among the developed nations.

powerful boost to their capacity to contribute to their societies. The real challenge is to enable all people, old and young, to realize more of their potential. Freeing our human resources from the difficulties that prevent their participation represents our best hope of making the transition in good order and in good time.

The working potential

In early days, "work" and "activity" meant much the same thing – learning or teaching, preparing food or hunting for it, growing up or growing old, were all of equal importance. But in "advanced" societies, there has long been a dominant popular equation linking work to jobs, jobs to money, and

money to human worth. This is fast becoming questionable in the North – and was never entirely relevant in the South. Two key factors are contributing to the shift of values: the growth of the workforce, and the changing nature of work.

We can already enumerate the workforce of the year 2000, because its numbers are now among us: there is only a 15-year lag before the average newborn child becomes a new worker. The greatest increases will be in the South, adding some 840 million extra potential workers to the 1985 figure of 2 billion. The implication is that about 50 million new jobs will need to be created each year.

As for the changing nature of work, the industrialized nations have already undergone a massive

The world at work

Human skill and energy represent a resource which is potentially renewable in perpetuity. But tremendous shifts are now taking place in the way these skills and energy are used. Some 300 years ago it took over 90% of the world's labour force working on farms to feed a much smaller population, whereas today the proportion of population working on the land in many developed countries has fallen to below 10%, in some cases considerably less. Major changes in the structure and technological base of industries, fuelled partly by the information revolution and partly by market changes (pp 200-201), have created labour surpluses in many regions, coupled with shortages of certain skills.

Labour force estimates
The histograms, below left and right, show male and female workforces, in millions, for seven regions of the world. This is the economically active population – that is, all those above a specific age (15 in most countries) who have worked for profit or who have sought work.

Developed countries
These show an increasing proportion of women in the workforce. A small overall growth is projected.

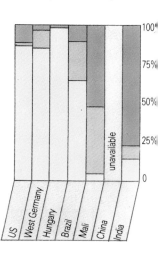

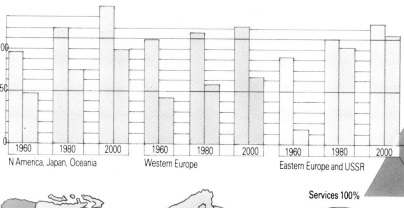

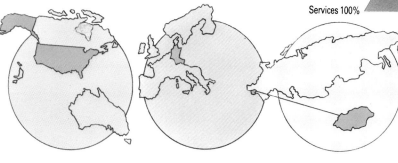

United States
The shift from traditional land-based employment is nowhere clearer than in the US. Agricultural work has shrunk dramatically (see triangle). Waged work has reached 90%. White-collar jobs are increasing, blue-collar ones falling.

Germany
Europe's strongest industrial nation has a high proportion of waged work, of which 12.8% counts as "professional and technical" work. Like the US, the recent trends in employment are moving towards the services sector.

Hungary
Hungary enjoys a high percentage of the population in the labour force. The workforce has shown a marked shift from agriculture towards industry. Some 45% of the workforce is female.

Wage labour trends
The International Labour Office (ILO) work categories reveal massive self-employment in the South, and dominant waged work in the North. (The categories, however, are not always readily comparable between nations.)

Waged
Self-employed
Other

transformation. There are freely entered contracts and increasing employee protection. The welfare state cushions the workforce to some extent from the impact of recession and unemployment. In the OECD region, services such as finance and business, wholesale and retail trades, and government employment now offer more jobs than traditional industries.

The premium set on waged employment by citizens of developed countries stems from the legitimacy it bestows in society. A job not only provides money, but psychological rewards. Nearly 80 percent of the workforce relies on waged work.

In traditional rural societies, by contrast, whole families invest their labour in a great deal of unpaid work, which escapes definition by conventional statistics. Of the measured workforce, agriculture still claims 70 percent in the least-developed countries, and 59 percent in all developing countries. Formal employment is supported by a massive informal sector, giving as much as 60 percent of employment in the biggest cities.

Work patterns are likely to change radically in the future. The key characteristic of many new technologies is that they amplify the productivity of workers, whether they work predominantly with their biceps or brains. Some believe that the result in developed economies will be widespread deskilling of workers, others that the future lies with retraining, job-sharing, and self-employment. For the

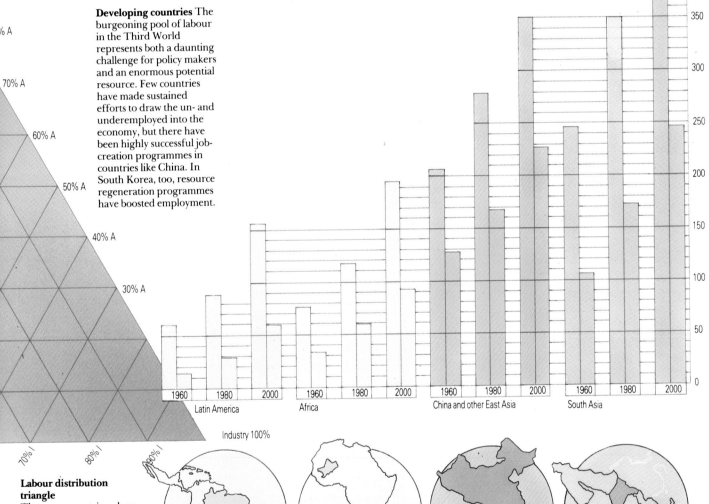

Developing countries The burgeoning pool of labour in the Third World represents both a daunting challenge for policy makers and an enormous potential resource. Few countries have made sustained efforts to draw the un- and underemployed into the economy, but there have been highly successful job-creation programmes in countries like China. In South Korea, too, resource regeneration programmes have boosted employment.

Labour distribution triangle
The seven countries whose labour distribution trends (1960-80) are plotted in the diagram above mostly mirror world trends in the three main areas of work. Arrows pointing down show agricultural work waning, with employment in industry and services increasing. Many Southern countries still have large agricultural workforces. Major shifts, as in Brazil and Hungary, reflect government economic programmes. Chinese data are available only for 1980.

Brazil
Brazil's economic transition has ridden on the back of rapid industrial growth. Surveys show a parallel rise in those with professional skills. There is also a thriving black-market sector (in part due to loss of agricultural work).

Mali
Mali's low unemployment and huge self-employment match a high farm-ownership ratio. The workforce focuses on bare subsistence: upgrading the farming economy could release resources for other developments.

China
Of the huge Chinese population, some 1.4 billion in 1990, at least 70% are peasants, owning their land communally, but encouraged to show private initiative. As China continues to industrialize, the peasant force will shrink.

India
With the world's second largest population, India ranks as the ninth largest economy. The workforce is predominantly self-employed. Population growth could swell the labour force to 385 million by 2000.

present, problems in the South are different. Here the need is to ensure that people have not only jobs, but adequate income. It is easy to forget, however, that many newly independent countries have already come very far, demonstrating that populations previously restricted to unskilled labour can produce, within a few decades, the professional and other skilled workers they need.

Knowledge versus wisdom

The human race can be justly proud of its science and learning. In the few centuries since Galileo deduced our position in the heavens, the ingenuity, imagination, and perseverance of our species have led to the detailed exploration of our vast solar system. As one quantum leap in understanding has followed another, new technologies have emerged in their wake, enabling us to deal far more effectively with the tasks we undertake. Like the proverbial lilypads multiplying on a pond, knowledge breeds knowledge at an exponential rate, carrying science forward even faster.

Recent times have witnessed an unprecedented spread of knowledge throughout the human population. The advent of machine printing sparked the information revolution, creating a base for mass education. Press, radio, and television have successively reached ever larger audiences. Today, almost one-fifth of the world's 5.5 billion people is participating in some form of education. Annual book production now totals about 850,000 titles: one person reading a book a day would take more than 2,300 years to read one year's supply.

Formal educational systems and school enrolment are, fortunately, growing faster than the school-age population: the number of developing world children in primary and secondary education almost tripled between 1960 and 1980. There remains, however, a huge challenge in developing countries, where many children do not have a school to attend.

But learning and knowledge is not necessarily wisdom. In acquiring new knowledge, traditional communities have been overwhelmed, losing some of the perceptions which previously sustained them. Some of our new skills run counter to all conventional wisdom: several nations now have the power to vaporize atomically large regions of the planet.

While science is pre-eminent, it is increasingly remote from the humanities. Indeed, Western civilization has almost relinquished the holistic approach to learning, understanding, and acting. The "systems theories", which attempt a unified overview, also tend towards specialization. Yet the keys to re-integration are beginning to emerge. Many now seek guidance from traditional societies on how to re-establish a balance between humankind and its environment. Our species is still in the adolescent phase, learning more facts than it knows what to do with. But the global challenges we face are beginning to provoke new levels of understanding, of wisdom, which are vitally needed if we are to manage our inheritance properly.

Homo sapiens

Every species other than *Homo sapiens* adapts its form and behaviour to the pressures of its environment. *Homo sapiens*, on the contrary, has achieved the remarkable feat of being able to adapt the environment instead, overcoming many natural limitations, through the development of technologies and cultures. While this has brought myriad benefits, the spiralling growth of knowledge has been possible only through a high degree of specialization (top right). We need to develop a new holism, appropriate for our advanced societies in the Gaian ecosystem.

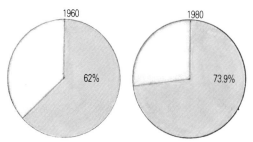

1960 62% 1980 73.9%

Primary-school enrolment

The global school
The '60s and early '70s were the great boom years in education: by 1980, 70% of the developing countries' primary-age children were in school, and 32% of secondary-age children. In the North, 94% of primary-age children were in school and 84% of secondary-age children.

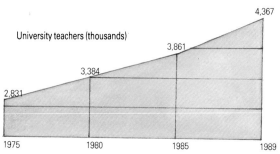

University teachers (thousands)

2,831 — 3,384 — 3,861 — 4,367

1975   1980   1985   1989

Adult education explosion
The late 1960s saw the global total of tertiary-level teachers surge from 1.36 to 2.06 million – a growth rate of 8.6% a year. This fell to 2.9% in the '80s. From a lower base, the developing country total grew by 5% a year, compared to 1.4% in the North. Today, the world total tops 4.4 million.

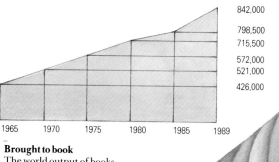

842,000
798,500
715,500
572,000
521,000
426,000

1965   1970   1975   1980   1985   1989

Brought to book
The world output of books has expanded rapidly to reach 842,000 titles in 1989, with two-thirds of this production in the North.

The information explosion
The output of new knowledge is growing side by side with economic and social development, if not with wisdom. Recent growth is being funded by research and development budgets now topping $150 billion, a major proportion going on military research and development. New technologies are being developed in order to store, index, and analyse the explosion of data.

Life sciences

Physical sciences

Techology

Medicine

Social sciences

History

Philosophy and religion

Arts

Literature

Gods

Occult Medicine
Social awareness

History Stories
Folklore ritual

Arts
and crafts

Knowledge of
the natural world

Number of scientific journals

1,000,000

100,000

10,000

1,000

100

10

Exponential growth in scientific journals
To plot this growth on a linear scale would require several lengths of paper. The scale is logarithmic to accommodate the rate of increase: since the 18th century these journals have been growing by a factor of 10 every 50 years.

1700   1800   1900   2000

Western and traditional knowledge
The explosive proliferation of knowledge in the North has produced millions of highly trained people, ill-informed of what is going on outside the confines of their own discipline. In a highly complex, rapidly changing world, this fragmentation of knowledge has emerged as a major obstacle to understanding. A great deal of effort has gone into interdisciplinary projects designed to re-integrate knowledge, and to develop multi-stranded solutions for multi-faceted problems. Building design, for example, increasingly involves team discussion between architects, other experts, and users. The new synthesizers of knowledge in the North have come to recognize the wisdom and experience embodied in more unified, holistic cultures.

CRISIS: THE INABILITY TO PARTICIPATE

Statistics on those excluded from the benefits of global society can be numbing. During the past year, effects of malnutrition have claimed tens of millions of lives; poverty, disease, worklessness, and refugee problems have marred hundreds of millions more, the numbers growing every day. The tragedy of this waste of human potential is that so much of it could be avoided, with fairly minor adjustments to our budgets, although large ones to our "priorities". What cannot be avoided, however, is the disastrous consequences for humanity of our runaway numbers, in some regions outstripping efforts to increase provision.

Population growth

We have witnessed huge increases in global food output since the 1950s, with per capita grain production peaking in 1984. Since then, however, per capita output has not kept pace with population growth, and fell in 1988 due in part to a drought afflicting the main grain-producing regions. Tragically, this scenario is likely to be repeated. Every year the world has to find the food to feed almost another 100 million mouths. The existing inequities of supply and the geographic mismatch between people and resources are seriously exacerbated by current population growth rates – generally highest in the poorer countries.

A key factor in the growth of human numbers is a population's age profile. A greater proportion of young people gives a greater potential for future growth in numbers. Some 45 percent of Africans, 36 percent of Latin Americans, and 33 percent of Asians are under 15 years old – a series of potential population time bombs.

Seemingly small annual growth rates can add up to potentially devastating increases in population. A 3 percent annual growth rate leads to a doubling of population every 23 years – even a 2 percent rise implies a doubling every 35 years. Demographers have plotted an "S" curve for Homo sapiens showing Africa at its highest annual growth point of 3 percent, with other regions slowing down.

The human species is unique in having developed artificial means to limit reproduction. Yet many nations fail to use this capability. Often people are forced into desperate measures: the global number of abortions, for example, is estimated to be at least 150,000 a day. All too often, a combination of poverty, very high child mortality, and lack of social

The numbers game

In a world with a finite capacity to support life, our seemingly infinite capacity for reproduction remains a central problem. In the last 100 years, world population increased from about 1.5 billion to 5.5 billion. UN estimates project a likely world population of 8.5 billion by the year 2025, and 10 billion by 2050. The long-range estimates indicate a population of 11.6 billion by the middle of the 22nd century.

How alarming are these figures? Assuming traditional farming methods are improved in the South, only the Middle East would be incapable of feeding its population in the year 2000. If all available land is used, together with an increase in fertilizer use, and surpluses are shipped across frontiers, most regions could support more than their current populations.

But a more realistic breakdown of the demand-and-supply picture, country by country, indicates a major crisis ahead. By the year 2000, 1.7 billion people are projected to be living in countries that cannot support their existing populations. Thirty of these countries will be in Africa, 14 in Central America, 6 in Asia, and 15 in the Middle East. If governments do not face the issue of curtailing population growth, their struggle against famine, disease, and high infant mortality will be even harder.

The population switchback

Many populations of wild animals explode and collapse regularly. Stability requires a finely tuned balance between numbers born and numbers dying (see below). The large "S" curve diagram, right, shows the current positions of human population in several regions. Although an obvious route to stability is to reduce the birth rate, this does not achieve an immediate balance. A nation with a "youthful profile" to its population clearly contains many potential parents. China's two-child family policy could not prevent its present population from growing to 1.8 billion prior to zero growth. So China is trying to re-establish the one-child family plan in an attempt to keep its eventual total to around 1.2 billion.

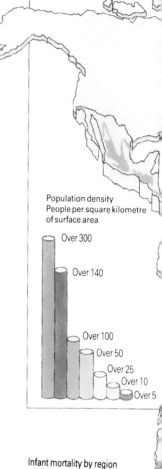

Population density
People per square kilometre
of surface area

Over 300
Over 140
Over 100
Over 50
Over 25
Over 10
Over 5

Infant mortality by region
Over 100 per 1,000
Over 50 per 1,000
Under 50 per 1,000

South and S Asia 2.1% per annum Zero growth about 2085

South America 2.1% per annum Zero growth about 2090

Africa 3% per annum Zero growth about 2100

Death rate

Birth rate

6000 BC   5000 BC   4000 BC

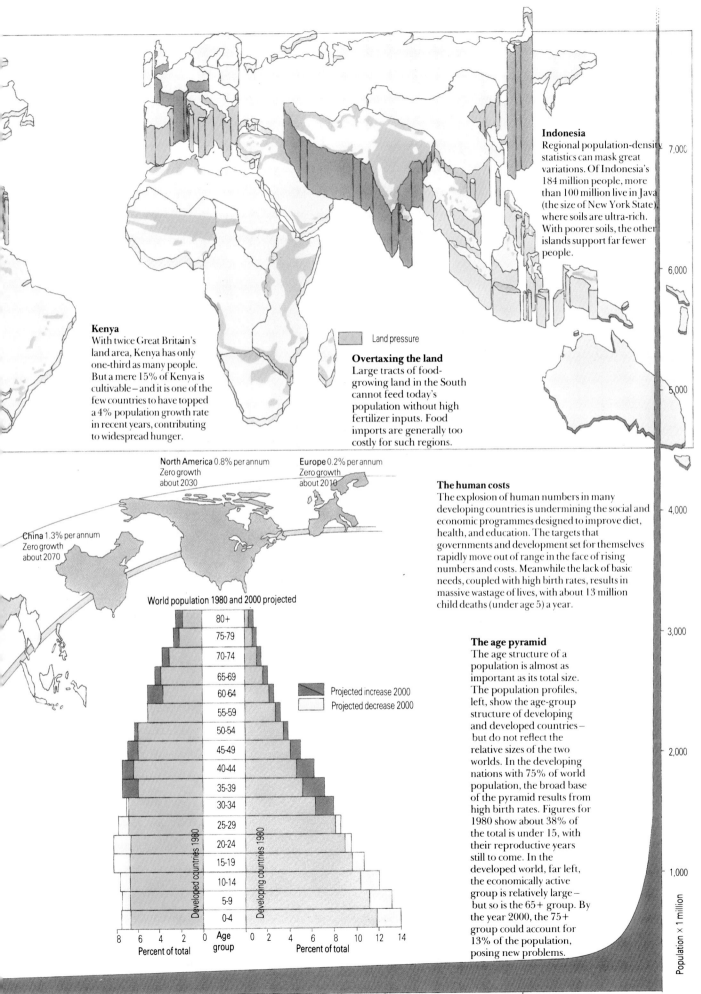

Indonesia
Regional population-density statistics can mask great variations. Of Indonesia's 184 million people, more than 100 million live in Java (the size of New York State), where soils are ultra-rich. With poorer soils, the other islands support far fewer people.

Kenya
With twice Great Britain's land area, Kenya has only one-third as many people. But a mere 15% of Kenya is cultivable – and it is one of the few countries to have topped a 4% population growth rate in recent years, contributing to widespread hunger.

Land pressure

Overtaxing the land
Large tracts of food-growing land in the South cannot feed today's population without high fertilizer inputs. Food imports are generally too costly for such regions.

North America 0.8% per annum
Zero growth about 2030

Europe 0.2% per annum
Zero growth about 2010

China 1.3% per annum
Zero growth about 2070

The human costs
The explosion of human numbers in many developing countries is undermining the social and economic programmes designed to improve diet, health, and education. The targets that governments and development set for themselves rapidly move out of range in the face of rising numbers and costs. Meanwhile the lack of basic needs, coupled with high birth rates, results in massive wastage of lives, with about 13 million child deaths (under age 5) a year.

The age pyramid
The age structure of a population is almost as important as its total size. The population profiles, left, show the age-group structure of developing and developed countries – but do not reflect the relative sizes of the two worlds. In the developing nations with 75% of world population, the broad base of the pyramid results from high birth rates. Figures for 1980 show about 38% of the total is under 15, with their reproductive years still to come. In the developed world, far left, the economically active group is relatively large – but so is the 65+ group. By the year 2000, the 75+ group could account for 13% of the population, posing new problems.

World population 1980 and 2000 projected

Projected increase 2000
Projected decrease 2000

80+
75-79
70-74
65-69
60-64
55-59
50-54
45-49
40-44
35-39
30-34
25-29
20-24
15-19
10-14
5-9
0-4

Developed countries 1980
Developing countries 1980

Age group

8 6 4 2 0
Percent of total

0 2 4 6 8 10 12 14
Percent of total

7,000
6,000
5,000
4,000
3,000
2,000
1,000

Population × 1 million

2000 BC
1000 BC
AD 1
1000
2000

provision of every type convinces parents that they need more children – to ensure that enough survive the rigours of childhood to help work for the household income and to care for their parents in old age. And so the cycle of deprivation and environmental degradation continues.

The employment crisis

Jobs have recently been disappearing in the North at a rate unequalled since the Great Depression, while in the South a grossly inadequate supply of jobs is being severely aggravated by rapid population growth. Politicians in the North wrestle with the social and economic implications of mass unemployment; they now admit that their original goal of full employment is becoming unattainable. While the impact of the work famine differs between regions of the world, it is universally perceived as one of the most pressing of today's crises.

Figures for the OECD countries show that unemployment has recently been growing fast: between 1960 and 1973, the unemployment rate ranged between 4 and 7 percent in North America and between 2 and 3 percent for Western Europe. In 1983, the unemployment rate in the OECD countries peaked at 8.9 percent, and in 1988 the rate was 7.3 percent. The figures for the former Communist countries show full employment, but conceal underemployment.

In the developing countries the problem is not only

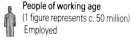

People of working age
(1 figure represents c. 50 million)
Employed

Unemployed or underemployed

Others

Children entering the workforce 1985-2005

GNP per capita by country

Over $10,000

Over $7,000

Over $4,000

Over $1,700

Under $1,700

The work famine

In both North and South the problems of un-and underemployment (unproductive work for little reward) are growing. The ILO's workforce figures cover only 64% of the working-age population in the South. The "others", family labourers for example, added to the ranks of un-and underemployed, outnumber the employees. Worse still, those who will reach working age over the next decade or two threaten to overwhelm opportunities for adequate work.

Northern unemployment

In the North, unemployment grew from 6% in 1975 to 8.9% in 1983. Economic trends and technological change render many skilled workers obsolete and trap the untrained and unskilled in a vicious circle of inexperience and rejection. The resulting sense of frustration and inadequacy may last throughout a person's life.

Odds against employment

Black 14.1% unemployed   7:3   White 6.3% unemployed

Women 7.9% unemployed   8:7   Men 7.2% unemployed

Unemployment and discrimination in the US

In the US, women and blacks are doubly disadvantaged. They find it hard to get work, and the range of jobs they hold is restricted. In the professions, few women reach the top, except in such areas as law and medicine. Some 64% of the 1.5 million "discouraged workers" in the US are women. Discouraged workers are those who know from experience that no jobs are available, many of them failing to register as unemployed. About 30% of those identified are black. Most blacks who do find work are condemned to low-paid blue-collar jobs. Most ethnic minorities suffer such discrimination.

North 80% of working-age population in workforce

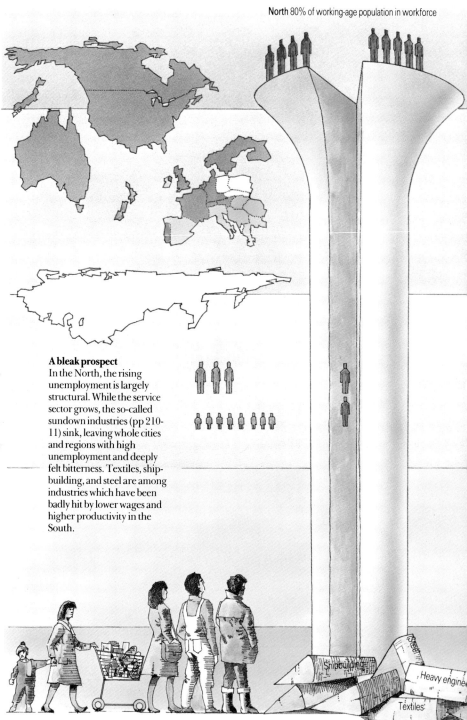

A bleak prospect
In the North, the rising unemployment is largely structural. While the service sector grows, the so-called sundown industries (pp 210-11) sink, leaving whole cities and regions with high unemployment and deeply felt bitterness. Textiles, ship-building, and steel are among industries which have been badly hit by lower wages and higher productivity in the South.

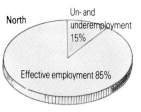

North
Un- and underemployment 15%
Effective employment 85%

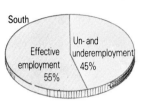

South
Effective employment 55%
Un- and underemployment 45%

lack of an effective waged employment sector, but extremely low incomes generally. Six out of ten people depend on working the land: of the world's one billion absolute poor, half are small farmers and about a fifth are landless peasants searching for work.

Poor rewards are all the worse since workers are a proportionately small part of the population. The under-15 age group accounts for over 30 percent of the total population. Then there are the "others", adults outside the workforce who may be unpaid family labourers or unregistered unemployed people. While 100 workers in the North support an average of 50 dependants, in the South the workers are swamped: in Africa 100 workers support 92 dependants. Millions more join the job market every year. There is a relentless quest for work which encourages a pattern of discrimination.

Women, the young, and racial minorities are particularly disadvantaged, both in the North and the South. In the US, the number of "discouraged workers", those who have given up the hunt for work in despair, is now estimated at 1.5 million. In the South, women suffer exploitation in the unpaid family-labour system, while those who have succeeded in finding a paid job tend to be the first to be hit by recession and upgrading of technology.

But those who suffer most are often the migrants and minorities. The lure of jobs encourages migration from rural, undeveloped areas to urban, industrialized areas – where the streets are seldom

Underemployment in the South

In the South, the most pressing crisis is one of huge underemployment as well as unemployment. Without unemployment benefits and social security, the poor must work if they are to survive at all. But survival often depends on a job with little substance to it apart from low pay and long hours – unprotected by work legislation of any sort. About 60% of workers in the South are self-employed. Government intervention and overseas aid often focus on inappropriate high-technology, capital-intensive projects. They also concentrate on boosting production of cash crops and goods, rather than providing the work which the poor need to buy food and goods. Many respond by moving to the cities.

The "invisible" workforce

The statistics on the male and female workforce in the South suggest that women are less than half the total. In fact, their work is officially invisible, like that of children. Women in the South suffer a "double burden" of duties, bearing domestic responsibilities and acting as unpaid family labourers in the fields (pp 192-3). A 15- to 18-hour day is not uncommon. Where productivity is lowest, children are also engaged in work, in conditions which may well retard their development. Over 98% of almost 100 million child labourers worldwide are found in the South. In Thailand, for example, about 3.5 million aged 11-16 years are in the labour force. Many are at work far younger. In the dry season, poor peasants in the northeast region are often forced to sell their children to child traders. Most go as factory hands and domestic servants, others as child prostitutes.

South hit by North

Employment in the South is hit both by accumulation of wealth by the few and by Northern protectionism. The EEC's agricultural strategy, for example, undermines the sugar industry in the Caribbean. And much Northern investment has gone into job-destroying technology.

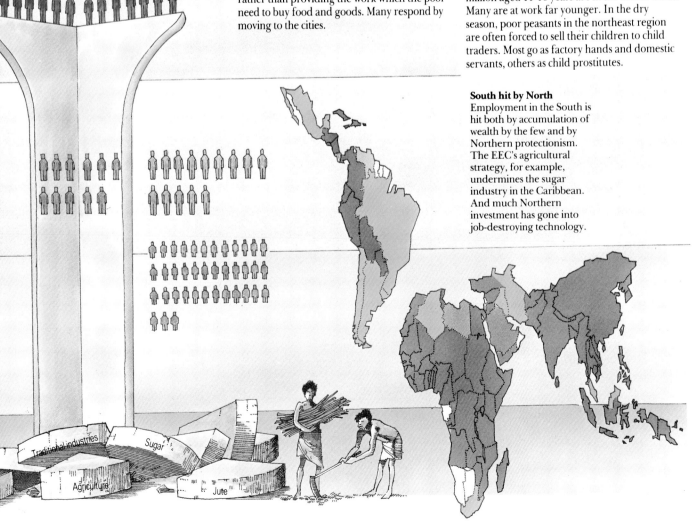

South 64% of working-age population in workforce

Traditional industries   Sugar   Agriculture   Jute

paved with gold. In Western Europe, the post-war economic boom attracted more than 15 million immigrants and "guest workers", who are often treated as scapegoats during a recession.

Crisis in health

Ill-health incapacitates an enormous proportion of the world's 5.5 billion people. As bald statistics are often unintelligible, consider some major world diseases in terms of populations of known size: the equivalent of the entire population of non-Communist Europe partly blind with trachoma; the entire population of Iran sightless with river blindness; everyone in Japan, Malaysia, and the Philippines sweating and shivering with malaria;

and all Americans urinating blood because of bilharzia. These equivalents show only part of the grotesque extent of world sickness.

Despite the fact that 800 million people in the South have no access to medical services, we spend 20 percent more each year on war than on health. A tragedy – or a crime against humanity?

The microbe, meanwhile, continues to display extraordinary ingenuity in outwitting medical science; new health threats are constantly emerging. This century has seen epidemics resulting from lifestyles in the North: coronary heart disease; the cancers; industrial disease; nervous and mental health problems; and the growing toll in road accidents, alcoholism, and drug-addiction deaths.

Sickness and stress

Worldwide, the bulk of health expenditure still goes on curing illnesses rather than on tackling their root causes. High-technology medicine consumes vast budgets. By contrast, primary health care, community schemes, and preventive medicine are often drastically underfunded. And despite major publicity campaigns, the public at large is often ignorant of the links between illness and, for example, diet or cigarette or alcohol consumption.

Ill-health in the North

In the industrialized North, most of the pressing health problems have sprung from the very conditions generally considered to be the hallmarks of progress. The population profile (right) shows a growing proportion of older people, many of them suffering from neglect as well as ill-health; the shaded areas show morbidity. Most seriously, there are the "new deaths": cancers and cardiovascular disease promoted by modern environments and lifestyles. There is also a growing array of minor illnesses, many readily preventable.

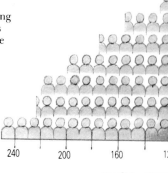

The new outcasts
Increased longevity has created new outcasts in the North – elderly people living alone or in nursing homes and hospitals. This problem will get significantly worse, as the populations of developed countries age. A US survey estimated that between 20 and 40% of elderly people in nursing homes and hospitals could live in the community if only adequate home care was provided. A recent study, right, of the percentage of people aged 60+ living with more than 4 people in a developed country (green) and a developing one (brown) highlights growing isolation in the North.

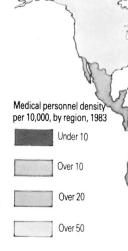

% of people aged 60+ living with more than 4 people

60-4 65-9 70-4 75-9 80-4 85+

The new deaths
Circulatory-system diseases account for half of deaths in developed countries. About 21% of all deaths are due to cancer, a third involving the respiratory system. Smoking is associated with 80% of premature deaths.

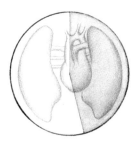

Minds under stress
Spending on tranquillizers in the North matches total public health expenditure in the world's 67 poorest countries. Many drugs used for stress-related illness are addictive. Worldwide, there are over 40 million mentally ill people. More than 1 million Europeans are in mental homes.

Ivory tower medicine
The bulk of health budgets in the North, in some cases as much as two-thirds, goes to "disease palaces" – which only cure about 10-20% of diseases. The emphasis on hospitals and high technology has often locked up resources which might otherwise have been used to prevent illness.

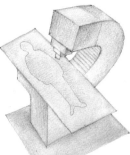

Global health care
The total health budget in the North is 10 times greater than that in the South, even though the developing countries' population is larger and in far greater need. In the South (right), health care often centres on towns and most people lack access to basic services.

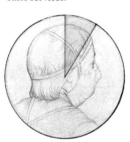

Medical personnel density per 10,000, by region, 1983

Under 10

Over 10

Over 20

Over 50

Deaths from want 1945 – 90

Deaths from war 1945 – 90

Death tolls
Since 1945, there have been 20 times as many deaths from neglect, including lack of food, unsafe water, and poor sanitation, as from wars (1 skull = 10 million deaths).

The developing countries have a double burden to bear. The central battleground for them remains infectious disease, but they also have to provide the infrastructure for health care. Even simple, readily countered threats, such as the micro-organisms that cause diarrhoeal diseases, kill millions of children in the South.

Malnutrition is perhaps the cruellest affliction, with children, young girls, and pregnant and lactating women most at risk. Vitamin-A deficiency, for example, causes over 330,000 infants to go blind every year. It can be tackled – in Bangladesh and Indonesia, the distribution of vitamins to children at risk costs less than 4 cents per child per year. But less than 15 percent of the people in

developing countries live within walking distance of a health facility of any sort.

In the South, a short period of ill-health can push a family into a deepening spiral of poverty, forcing it to sell land, animals, or other possessions which, with high interest rates, it will never be able to buy back. With all these existing problems, the South can ill-afford to import the lifestyle-related diseases of the North. But lung cancer, for example, is now as common among smokers in the South as among their Northern counterparts.

Around the world, too, there is the rising spectre of infirm old age. Even in the North, where the average age of populations is highest, far too little is being done to help the aged cope with such chronic

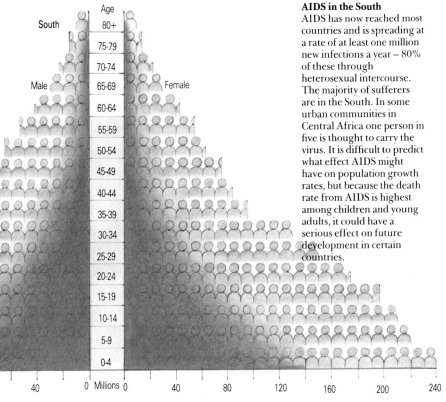

AIDS in the South
AIDS has now reached most countries and is spreading at a rate of at least one million new infections a year – 80% of these through heterosexual intercourse. The majority of sufferers are in the South. In some urban communities in Central Africa one person in five is thought to carry the virus. It is difficult to predict what effect AIDS might have on population growth rates, but because the death rate from AIDS is highest among children and young adults, it could have a serious effect on future development in certain countries.

Sickness in the South
High overall mortality, and very high infant mortality, are the bitter reality in many developing countries. In the first step on the population pyramid, left, 1 in 6 children die; disability and illness then affect a further third of the population. Among the causes are the lack of health workers, the lack of information about drugs, and poor sanitation. Poverty and malnutrition, the most intractable problems of all, are endemic. Moderate to severe malnutrition affects 1 in 5 Latin Americans, 1 in 3 Africans and Asians. There is a complex interplay between all of these killers that frustrates single-minded solutions: poor sanitation causes diarrhoea, which aggravates malnutrition, sapping energy and earning-capacity. UNICEF estimates that approximately one-third of child admissions to hospital in the South are caused by diarrhoea. Several hundred million people suffer from waterborne diseases for which no vaccine exists, notably schistosomiasis (bilharzia) and onchocerciasis (river blindness). Tackling water hazards (pp 118-9, 134-5) has proved a herculean task, aggravated by mis-spending. In the early '80s, for example, the Philippines government spent $50 million on a heart surgery unit which benefits a few hundred people at most.

The child killers
While an average of 12 out of every 1,000 infants (age 0-12 months) die in the North, 71 die in the South – national figures run up to 162. Of all infant deaths between 1975 and 1990, the South suffered 97%. Causes: malnutrition, diarrhoea, and other infectious disease.

Diarrhoea
The deadliest killer of small children in the South is diarrhoea; attacks amount to 1.4 billion a year. In one South American survey, malnutrition associated with diarrhoea caused 57% of deaths. The best cure, ORT (p. 191), is cheap and simple, but fails without education for its use.

Imported drugs
Private spending on health in developing countries outstrips government spending 3 to 4 times, and at least a third of this money goes on useless or harmful drugs. Companies are notoriously slow to remove "contra-indicated" drugs from the Third World market.

health problems as rheumatism and arthritis, while growing numbers of old people are simply sidelined into institutions.

Barriers against progress

Today, there are some 950 million illiterate adults, 920 million of them in developing countries. The predicaments of ill-health and lack of education cannot be examined in isolation. A global map of ill-health would coincide with maps of malnutrition, poverty, and illiteracy. The illiterate person is not only unable to read or write, but he, or more usually she, is also poor, hungry, and highly vulnerable to disease and exploitation.

Many developing nations have made great progress in promoting literacy. Africa, where the greatest recent advances have been made, reports a 50 percent illiteracy rate, Asia 34 percent, and Latin America 15 percent. But the decline in percentage illiteracy rates is often deceptive: the number of people without adequate functional literacy is still increasing in Sub-Saharan Africa, the Arab States, and Southern Asia.

When we focus on sheer numbers, we begin to see the gap between the people and the provisions. Between 1960 and 1980, the number of illiterate men grew by 20 million and the number of illiterate women by 74 million. Financial constraints dictate priorities. The South has only 12.5 percent of the world's education budget. Since the 1970s there has been an alarming plunge in education expenditure (as a percentage of GNP), particularly among developing countries. Venezuela's spending fell from 22.9 to 5.4 percent between 1970 and 1987. Many African countries have dropped beneath the 5 percent line. Over 90 percent of primary-age children in the South were enrolled in school in 1990, but at least 100 million children did not have a school to attend.

Behind the global school-enrolment figures lies a second chasm. Of every 100 children entering primary education in the South, 29 will drop out during their primary school years and a further 21 will drop out before finishing secondary school. So only 71 out of the original 100 complete four years of primary education. And those who fail to learn basic skills by the end of normal school age have little chance of acquiring them as adults.

It is ironic that many developing nations still model their higher education systems along Western lines. The arts are often treated as the art of the European élite, while medical training fosters the expensive, high-technology, essentially curative health care of the North. It is no coincidence that the predominant language in education is often the legacy of the former colonialists, perpetuating the idea of foreign intellectual superiority.

Success in higher education can aggravate problems higher up the scale. In Tanzania, for example, when 1.4 million Tanzanians passed the adult literacy test, further education could offer only

The literacy chasm

School enrolment by country

Primary and secondary < 50%
Primary > 50%; secondary < 50%
Primary and secondary > 50%
Primary > 50%; secondary > 75%

Lost cultures
The history of the dominant European cultures is also a history of the destruction of countless tribal and other diversified cultures. Education can be one of the most powerful tools of suppression.

In the modern nation state, governed by written statutes and fuelled by forms and letters among other written materials, literacy is an increasingly critical skill. Illiterate, you will not get a good job, indeed you may not get a job at all. Illiterate, you are very unlikely to know your statutory rights, and unlikelier still to be able to enforce them. In some countries, you can vote only if you are literate. Illiterate, you may fall victim to any unscrupulous official or confidence trickster. All too often, the most deprived communities, the rural villages and the urban slums, benefit least from national spending on education.

The media chasm
The North/South literacy chasm is mirrored by unequal access to media resources. The North has most of the newsprint, radios, and TVs. The North also dominates markets for news (p. 208). Limited access to radio and TV is a major handicap in developing countries, where there are high proportions of illiterate adults who could benefit from the latest media-based education programmes. Such programmes can also help preserve vanishing cultures, particularly those based on rich oral traditions.

Newsprint distribution
North 81%
South 19%

Radio distribution
North 72%
South 28%

Television distribution
North 74%
South 26%

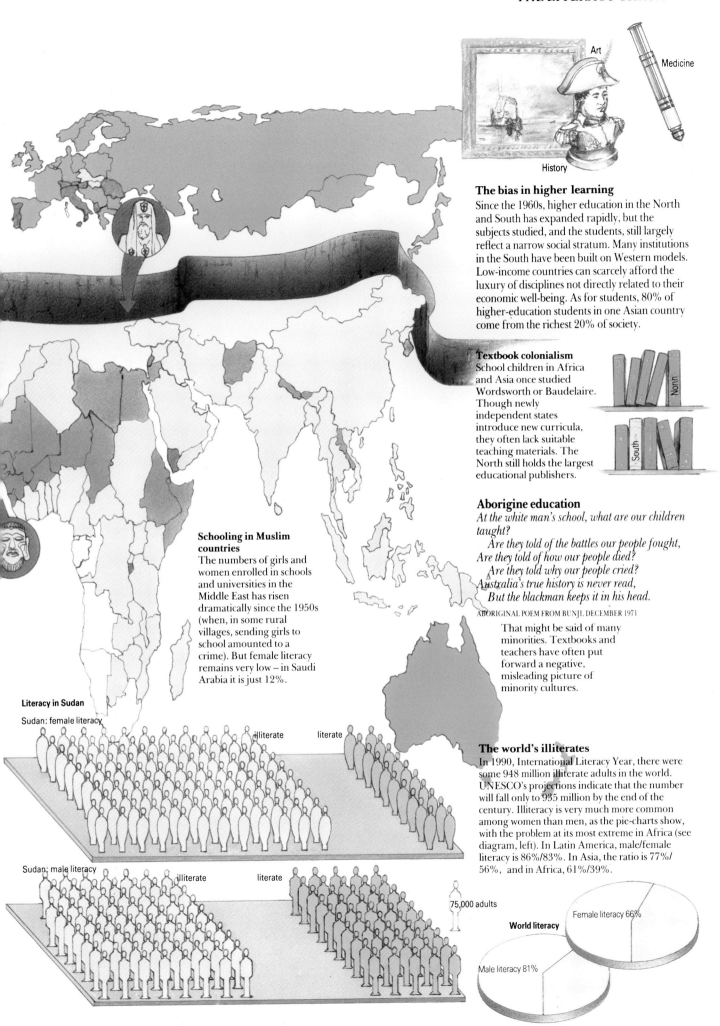

History

Art

Medicine

The bias in higher learning

Since the 1960s, higher education in the North and South has expanded rapidly, but the subjects studied, and the students, still largely reflect a narrow social stratum. Many institutions in the South have been built on Western models. Low-income countries can scarcely afford the luxury of disciplines not directly related to their economic well-being. As for students, 80% of higher-education students in one Asian country come from the richest 20% of society.

Textbook colonialism

School children in Africa and Asia once studied Wordsworth or Baudelaire. Though newly independent states introduce new curricula, they often lack suitable teaching materials. The North still holds the largest educational publishers.

North

South

Aborigine education

At the white man's school, what are our children taught?
Are they told of the battles our people fought,
Are they told of how our people died?
Are they told why our people cried?
Australia's true history is never read,
But the blackman keeps it in his head.

ABORIGINAL POEM FROM BUNJI, DECEMBER 1971

That might be said of many minorities. Textbooks and teachers have often put forward a negative, misleading picture of minority cultures.

Schooling in Muslim countries

The numbers of girls and women enrolled in schools and universities in the Middle East has risen dramatically since the 1950s (when, in some rural villages, sending girls to school amounted to a crime). But female literacy remains very low – in Saudi Arabia it is just 12%.

Literacy in Sudan

Sudan: female literacy

illiterate   literate

Sudan: male literacy

illiterate   literate

The world's illiterates

In 1990, International Literacy Year, there were some 948 million illiterate adults in the world. UNESCO's projections indicate that the number will fall only to 935 million by the end of the century. Illiteracy is very much more common among women than men, as the pie-charts show, with the problem at its most extreme in Africa (see diagram, left). In Latin America, male/female literacy is 86%/83%. In Asia, the ratio is 77%/56%, and in Africa, 61%/39%.

75,000 adults

World literacy

Female literacy 66%

Male literacy 81%

3,400 places. The expectations of those who have succeeded in education are inevitably very high; opportunities for suitable work are very low.

Exile and persecution

The very word *refugee* rings alarm bells; it evokes a sense of immediate crisis, and never more so than today when, according to conservative estimates, there are at least 17 million refugees worldwide. Half of them are children, who may never know their country, and young people of school age. Of the world's refugees, 85 percent live in the South. There is an urgent need for concerted government action worldwide.

A refugee, in the blunt definition of the UN Convention of 1951 and the Protocol of 1967, is a person who cannot return to his or her own country because of a "well-founded fear of persecution for reasons of race, religion, nationality, political association or social grouping". The majority of refugees have been created not by natural disasters, but by political instabilities, including the 160 or so undeclared wars waged since 1945.

The latest statistics of the UN High Commission for Refugees (UNHCR) include more than five million Afghan refugees, now the largest national refugee group in the world. The Palestinians come second: since the creation of the state of Israel, more than 2 million refugees have been created, most of them living in Middle Eastern countries. At the peak of the 1991 Gulf crisis more than 1.5 million people fled Iraq – around a quarter of a million remain in Iran. But many refugees elude the statistics altogether. Of the Boat People from Vietnam, for example, only those who got through were counted.

Some of the countries now hosting considerable numbers of refugees are among those least able to afford such an increased burden. Africa is probably the worst-affected continent, hosting more than four million refugees, most of them displaced by wars in Somalia, Ethiopia, Sudan, Angola, and Mozambique. Approximately one in every hundred Africans is a refugee. There are entire ghettoes of refugees in some cities. Certain governments are adopting restrictive regulations or changing their reception policies. About one-third of all governments in the world now have policies to reduce immigration.

Quite apart from the political exiles and refugees, there are countless social and economic outcasts within countries, including many "internal" refugees – migrants in search of hope. Some are worse off than "official" refugees. Members of the Baha'i religion in Iran have been subject to harassment, persecution, imprisonment, and execution since 1979. Many immigrant workers in Europe suffer less savage, but real discrimination. In tens of countries across the world, ethnic minorities are in danger of being crushed out of existence. The effect of persecution can strengthen a group's identity – but far more often it consigns the victim to deprivation, isolation, and even extinction.

Outcasts and refugees

There is mounting international concern that the number of refugees worldwide has soared since the 1960s. In addition millions more people do not or cannot escape the oppressions that haunt their lives, and some may be even worse off than the exiles. They may be severely persecuted for their race, religion, politics, or nationality, and it may not be within the power of any relief organization to assist them. Or they may depend on one of numerous charitable organizations, who are desperately short of funds. But not all the unwanted are behind bars or in the world's human "reserves". Every country has its "second-class" citizens – people who suffer and are disadvantaged because they are old, sick, homeless, jobless, or because their skin is a different colour.

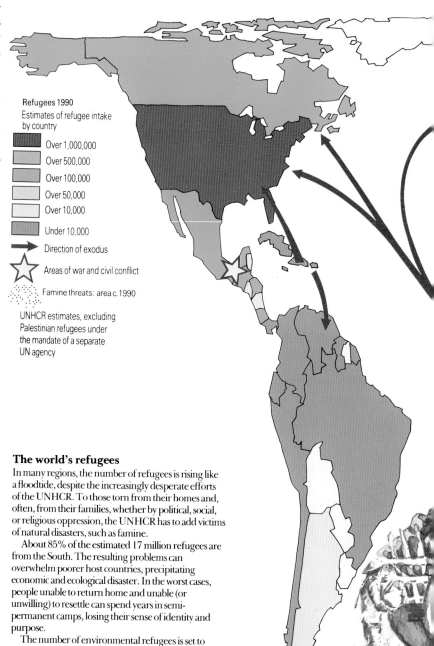

Refugees 1990
Estimates of refugee intake
by country

- Over 1,000,000
- Over 500,000
- Over 100,000
- Over 50,000
- Over 10,000
- Under 10,000
- → Direction of exodus
- ☆ Areas of war and civil conflict
- Famine threats: area c. 1990

UNHCR estimates, excluding
Palestinian refugees under
the mandate of a separate
UN agency

The world's refugees

In many regions, the number of refugees is rising like a floodtide, despite the increasingly desperate efforts of the UNHCR. To those torn from their homes and, often, from their families, whether by political, social, or religious oppression, the UNHCR has to add victims of natural disasters, such as famine.

About 85% of the estimated 17 million refugees are from the South. The resulting problems can overwhelm poorer host countries, precipitating economic and ecological disaster. In the worst cases, people unable to return home and unable (or unwilling) to resettle can spend years in semi-permanent camps, losing their sense of identity and purpose.

The number of environmental refugees is set to increase, also. Desertification is driving many from their homelands, and global warming may throw up further waves of refugees.

The disappearing tribes

Some Latin American Indians are still treated almost as wild animals, or worse. They are "protected" in mean reserves, or they are pushed out far into marginal lands. Even there they may be harassed by new settlers or come into conflict with government or commercial interests. The Yanomame Indians, for instance, now live in the dense forest of the Guyana Shield in Brazil and Venezuela (left), but are threatened by the intrusion one of Brazil's new highways, the Perimetral Norte, which will cut through previously impenetrable Amazonian terrain. Their small world, their distinctive culture, and their ancient history are likely to become fodder for bulldozers.

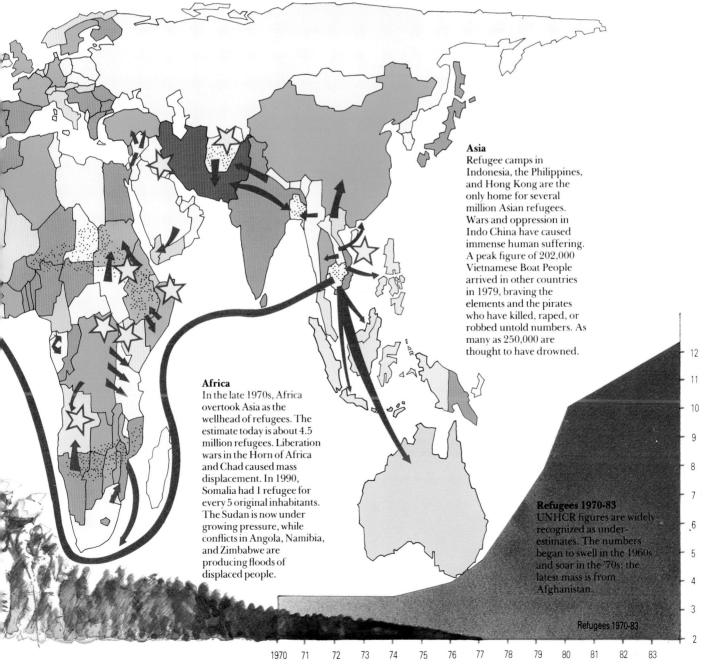

Asia

Refugee camps in Indonesia, the Philippines, and Hong Kong are the only home for several million Asian refugees. Wars and oppression in Indo China have caused immense human suffering. A peak figure of 202,000 Vietnamese Boat People arrived in other countries in 1979, braving the elements and the pirates who have killed, raped, or robbed untold numbers. As many as 250,000 are thought to have drowned.

Africa

In the late 1970s, Africa overtook Asia as the wellhead of refugees. The estimate today is about 4.5 million refugees. Liberation wars in the Horn of Africa and Chad caused mass displacement. In 1990, Somalia had 1 refugee for every 5 original inhabitants. The Sudan is now under growing pressure, while conflicts in Angola, Namibia, and Zimbabwe are producing floods of displaced people.

Refugees 1970-83

UNHCR figures are widely recognized as under-estimates. The numbers began to swell in the 1960s and soar in the '70s; the latest mass is from Afghanistan.

Refugees 1970-83

12
11
10
9
8
7
6
5
4
3
2

1970 71 72 73 74 75 76 77 78 79 80 81 82 83

MANAGING OURSELVES

" 'Well, in *our* country' said Alice . . . 'you'd generally get to somewhere else – if you ran very fast for a long time, as we've been doing.' 'A slow sort of country!' said the Queen. 'Now, *here*, you see, it takes all the running you can do to keep in the same place. If you want to get somewhere else, you must run at least twice as fast as that!' " These words by Lewis Carroll aptly express the plight of many Third World countries in the development race – they have to "run twice as fast" to make any headway. They are handicapped by their rising numbers, by wars, by sickness, and by legacies of inequity. Removing the handicaps is an urgent task, for each nation and for the global community. Fortunately, development is showing signs of moving out into entire communities, to help the women, the poor, the illiterate, the workless, and the rejected minorities to reach the starting line.

What changes birth rate

 Family planning

 Services

 Better health

 Employment

 Later marriages

Education
Improved women's status

More equal incomes

Managing numbers

Ninety-five percent of the world's population lives in countries with family planning services of some description, but, according to the United Nation Population Fund (UNFPA), some 300 million women in developing countries do not have ready access to safe and effective means of contraception.

The forces working against the lowering of fertility rates are deeply rooted in cultural, social, and economic conditions that have prevailed for generations. It is no coincidence that fertility rates are highest in developing countries, where economic deprivation is endemic, and lowest in affluent educated societies with good social provision. The desire for large families is the result of many factors, among them high infant mortality, labour-intensive means of subsistence, and the need for support in old age.

Some aspects of the developmental process, such as improved health, better education, and increased employment opportunities for women, work together with family planning to cut fertility rates. A good measure of national well-being is the Physical Quality of Life Index (PQLI). Birth rates generally vary in inverse proportion to PQLI ratings, though both may be affected by deeper cultural factors.

Physical Quality of Life Index
The PQLI is a useful measure of human progress. Calculated by employing three indices – child mortality, literacy, and life expectancy – it gives equal weight to each factor. Although there is some correlation between high income and high PQLI, there can be significant variations. Sri Lanka, for example, has a low per capita income but a high PQLI. The generally inverse correlation between birth rate and PQLI is shown below.

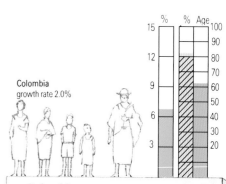

Colombia
growth rate 2.0%

Colombia
With an infant mortality rate less than half India's, Colombia has achieved a dramatic demographic transition – from a 3.2% annual rate of population growth to a current rate of 1.9%. The government's multi-faceted social policy has increased life expectancy, while infant deaths have dropped to 56 per 1,000 live births.

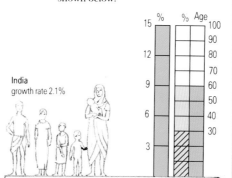

India
growth rate 2.1%

India
India's population has exploded, reaching some 853 million in 1990. Average infant mortality and illiteracy are still high, but the state of Kerala has succeeded in lowering child deaths by improving health services and working conditions.

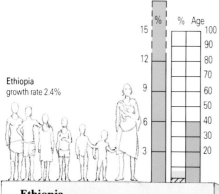

Ethiopia
growth rate 2.4%

Ethiopia
Less than 1.5% of all Ethiopian couples use any form of contraception – and government support for birth control is very recent. The revolution of 1974 has led to a new emphasis on primary health, which will help family planning.

Target: 8 billion
Worldwide family planning services could save about 150,000 abortions a day. Ending absolute poverty, improving health, education, and the status of women will also slow population growth. The aim should be to bring down the projected population by 2050 from 10 billion (UN median) to around 8 billion.

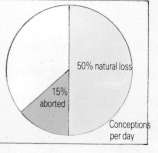

50% natural loss

15% aborted

Conceptions per day

Population management

Population trends are not easy to influence and manage. In a sense, governments face a "chicken and egg" situation – birth control is most effective in a context of improved health care and rising employment and incomes. But it is often lack of birth control that makes these factors so difficult to achieve. However, experience confirms that best results are obtained when modernization, including improved status and education for women, runs in tandem with effective family planning.

Family planning has been making major inroads over the last decade, with countries such as China and India making major cuts in their population growth rates. More recently, Peru, Senegal, and Ethiopia have been taking an interest, as population growth is increasingly seen as a major obstacle to their development.

The costs have often been substantial, the results mixed. In Kerala, India, family planning has been assisted by good health services, the strong economic status of women, high literacy, and excellent communications. Indonesia has introduced effective family planning to the poor of East Java and Bali, thanks to government support for village services. Other successes have been scored in Cuba, Costa Rica, Hong Kong, South Korea, Mauritius, and Taiwan. Family planning has been less successful in

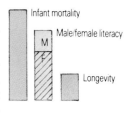

Infant mortality

Male/female literacy

M

F

Longevity

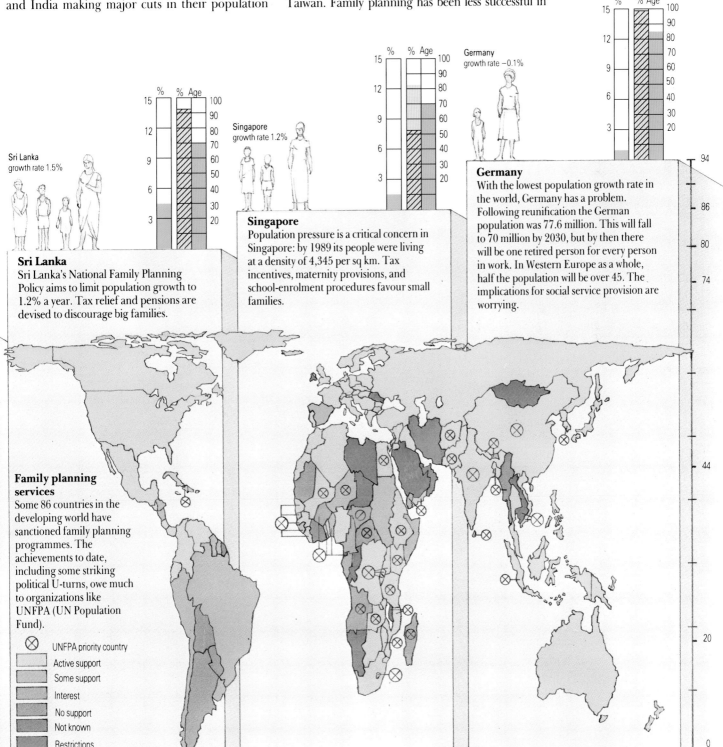

Sri Lanka
growth rate 1.5%

Sri Lanka
Sri Lanka's National Family Planning Policy aims to limit population growth to 1.2% a year. Tax relief and pensions are devised to discourage big families.

Singapore
growth rate 1.2%

Singapore
Population pressure is a critical concern in Singapore: by 1989 its people were living at a density of 4,345 per sq km. Tax incentives, maternity provisions, and school-enrolment procedures favour small families.

Germany
growth rate −0.1%

Germany
With the lowest population growth rate in the world, Germany has a problem. Following reunification the German population was 77.6 million. This will fall to 70 million by 2030, but by then there will be one retired person for every person in work. In Western Europe as a whole, half the population will be over 45. The implications for social service provision are worrying.

Family planning services
Some 86 countries in the developing world have sanctioned family planning programmes. The achievements to date, including some striking political U-turns, owe much to organizations like UNFPA (UN Population Fund).

⊗ UNFPA priority country

Active support

Some support

Interest

No support

Not known

Restrictions

Muslim countries such as Pakistan, underscoring the importance of cultural traditions. Sub-Saharan Africa has been least successful, because of cultural factors and very weak health care.

Equality for women?

The statistics on the contributions which women make to society, and on the share of society's assets they actually own, tell a devastating story. It is a story that dispels cosy notions about a woman's place in the world. No longer can men claim to be the sole providers and bearers of responsibility: a worldwide scan shows that it is women, the often ignored "other half", who sustain families, looking after their health and providing food and care. They also make enormous contributions to domestic and national income.

Think about women today, especially in developing countries, and it is difficult to avoid thinking about oppression. Yet their contribution is remarkable: in Africa as a whole, they do 60 percent of all agricultural work, 50 percent of all animal husbandry, and all the food processing. In Tanzania, women work an average of 2,600 hours a year in agriculture, while the men put in only about 1,800 hours. Most of this work, however, goes unrecognized: women account for less than a third of the official workforce in the South.

In the North, by contrast, one of the most significant socio-economic trends in the post-war period has been the entry of women into the ranks of official workers. Almost 40 percent of the workforce in the US, UK, and Japan today is female. But the unpaid, "invisible" work done by women at home is still substantial: in the US, it is estimated at 40 percent of Gross Domestic Product.

In waged work, discrimination hits women in both North and South, with a strong tendency for them to be regarded as cheap labour. The problem is twin-headed: women often end up in the worst-paid jobs or, where they break out of stereotypical female work, are paid less for comparable work. Female workers are paid 25 percent less than male workers in the UK, 40 percent less in the US.

Illiteracy is a key factor inhibiting women in the South. Two out of three women are illiterate, a far higher proportion than for men. The literacy chasm is closely linked to other indicators of deprivation such as poverty and malnutrition, and it is promoted by segregation of the sexes.

Even today, only a small fraction of women are actively engaged in politics, but the voice of women is increasingly heard. And it is, more and more, a voice in defence of peace, conservation, and humanitarian values. The contemporary women's movement comprises hundreds of international organizations and thousands of pressure groups. It can be described as one of the most global social movements of recent years. Self-help groups are legion. Peace campaigners like the Greenham Common Women, UK, and the tree-huggers of the Chipko Andolan

The voice of women

Despite the growing impact of women's movements, it is still an indisputable fact that women are under-represented at all levels of decision-making. No nation on Earth has a 50:50 balance of the sexes in its legislature – and there are still countries where universal suffrage is denied. Even after 75 years of suffrage in some countries, women have achieved less than 10% national representation.

The high proportion of women in further education in many parts of the world underscores this waste of national resources. A few countries, like Sweden and Cuba, encourage equal sharing of home responsibilities to facilitate power-sharing at the top. Few other nations are tackling this grass-roots inequality. But women's contribution in the workforce is helping to open new doors to power. In such countries as Egypt and Jordan, for example, the growing impact of working women has triggered new laws designed to improve their position.

Getting men to share political power and the breadwinner role can be a major hurdle in male-oriented cultures, but women are making inroads by taking the initiative both in the work sector and in political movements, demonstrating and working for their equality.

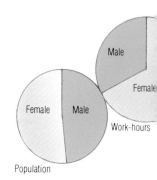

Equal work, equal pay?
The imbalances between female work-hours, income, and ownership (above) have been revealed by a survey made by the UN Decade for Women. This is despite rising numbers at work. Equal-pay laws need to be enforced rigorously and extended. Another key target must be the sharing household work.

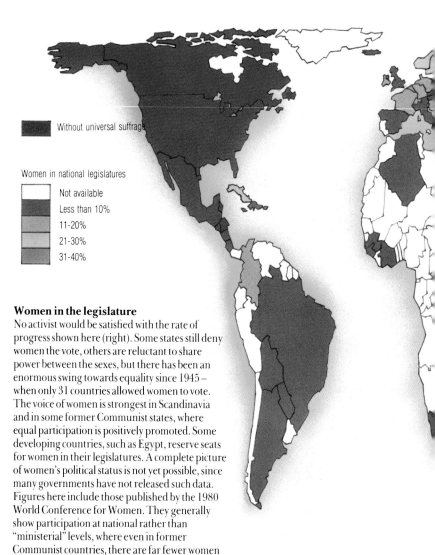

Without universal suffrage

Women in national legislatures

Not available
Less than 10%
11-20%
21-30%
31-40%

Women in the legislature
No activist would be satisfied with the rate of progress shown here (right). Some states still deny women the vote, others are reluctant to share power between the sexes, but there has been an enormous swing towards equality since 1945 – when only 31 countries allowed women to vote. The voice of women is strongest in Scandinavia and in some former Communist states, where equal participation is positively promoted. Some developing countries, such as Egypt, reserve seats for women in their legislatures. A complete picture of women's political status is not yet possible, since many governments have not released such data. Figures here include those published by the 1980 World Conference for Women. They generally show participation at national rather than "ministerial" levels, where even in former Communist countries, there are far fewer women than men.

Female

Male

Income

Female

Male

Ownership

Zambian woman's workday
1 Walking to field
2 Ploughing, planting
3 Collecting firewood
4 Pounding grain or legumes
5 Fetching water
6 Lighting fire, cooking
7 Serving food, eating
8 Cleaning and washing

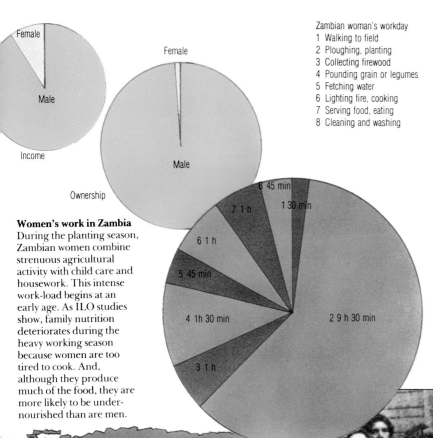

8 45 min
7 1 h
6 1 h
5 45 min
4 1 h 30 min
3 1 h
2 9 h 30 min
1 30 min

Women's work in Zambia

During the planting season, Zambian women combine strenuous agricultural activity with child care and housework. This intense work-load begins at an early age. As ILO studies show, family nutrition deteriorates during the heavy working season because women are too tired to cook. And, although they produce much of the food, they are more likely to be under-nourished than are men.

The Self-Employed Woman's Association

Founded in Gujarat, India, in 1972, this co-operative provides new hope and horizons for some of the country's poorest women. At least 25,000 women have taken part in SEWA – all benefiting from special credit schemes, training programmes, welfare facilities, and negotiated minimum earnings. Members of SEWA range from weavers to vegetable sellers, but their aims are constant: economic and social uplift.

Women in the Muslim world

Some orthodox Muslim states still bar women from voting. But the rising number of women moving into the workforce, particularly in the non-oil states, has begun to modify 1,000-year-old customs.

Despite the unhappy example of Iran, such countries as Jordan, Iraq, and Egypt have forged ahead with government-backed programmes designed to attract more women into the workforce.

Women's protest movements

Demonstrations against tyranny and the threat of war have been increasingly forceful. In Chile (above) and Argentina, the mothers of "the disappeared" protest their loss despite reprisal threats. At Greenham Common in the UK (right), women have kept non-violent permanent vigil since September 1981 to demon-strate their opposition to the installation of Cruise missiles – despite continual harassment. No-one pretends that women have a monopoly in peace protests, but peace has become a unifying theme in women's movements worldwide.

movement in India (p. 53) are part of an upsurge of activity which found international expression in the UN Decade for Women, 1975-85.

Global health care
Physiological causes are by no means the only roots of ill-health. Indeed, the problem of finding management solutions to massive world ill-health is enormously complex. Problems do not hinge only on primary health, vaccines, clean drinking water, and sanitation. They include political decisions on the relative priorities of health and defence; the state of the world economy and the prospects for employment; decent low-cost housing; and the status of women and the entire stratum of under-privileged people.

Freedom from the threat of war is one of the key social objectives of the World Health Organization (WHO) in Europe. It also hopes to achieve equity both between and within the 33 nations of Europe, and to ensure that health services do not favour the few at the expense of the majority.

A critical element in ending global inequality is primary health care, based on a close alliance between local communities and their health workers. Strengthening this sector alleviates the distress of isolation and ignorance, and it promotes prevention through early recognition of problems – an

Health for all

In both North and South, there is a new awareness of the far-reaching advantages of primary health care. Community services critically depend on popular support, and emphasize the preventive role of medicine.

Health care for the North
Community programmes are a fairly recent phenomenon in the North. In the US, Neighbourhood Health Centres reduce infant deaths in particular and hospital admissions in general. In Karelia, Finland, a voluntary campaign promotes healthy lifestyles, lowering the incidence of heart disease. The scope for improvement through motivation is enormous. Self-help groups like Alcoholics Anonymous have come to play an essential role.

Prevention is the key
Provided that there is a safe environment, food for balanced diets, clean water, and reasonable access to housing, work, and education, the ultimate success of preventive medicine is in people's own hands. There has been enormous media support for health-promoting lifestyles in the US, and this trend is increasingly seen in Europe. Information on the risks involved in eating fatty foods, sugar, and salt has influenced dietary habits. Cigarette smoking is declining in the North, reducing the incidence of illnesses.

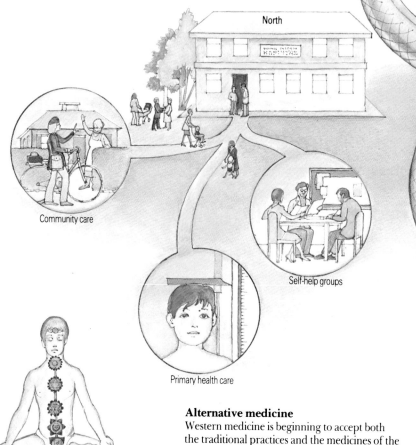

Community care

Self-help groups

Primary health care

Alternative medicine
Western medicine is beginning to accept both the traditional practices and the medicines of the East. Doctors in the North have mostly lost the art of treating the whole person, although many are now rediscovering the effectiveness of a comprehensive approach to health and sickness. Clinical studies have shown that many traditional medicines and therapies can be highly effective. Homeopathy, acupuncture, and Eastern exercises, such as yoga, are gaining widespread recognition.

A healthier Norway
Norway's concern to improve national health has led to a major campaign to achieve better diets – promoted through health education and community programmes. The results include major cuts in deaths from heart disease (a drop of 45-50% was recorded in Oslo), a general decline in teenager tooth decay (fillings were cut by 60% between 1970 and 1979), and benefits in terms of the incidence of some cancers. The campaign has also achieved a rapid increase in breast-feeding.

Heart disease in Oslo 1972-77

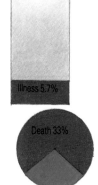

Control group

Illness 5.7%

Death 33%

Intervention group

Illness 3%

Death 16%

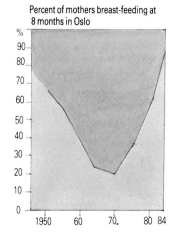

Percent of mothers breast-feeding at 8 months in Oslo

%
90
80
70
60
50
40
30
20
10
0

1950   60   70   80 84

approach which makes a great deal of financial sense. The North could well learn from those nations of the South which, like China and Cuba, have pioneered networks of paramedics.

The battle to achieve health in developing nations is being fought across a far larger field. There is an acute need for adequate sanitation and safe drinking water. Family planning services are not available to many people, while funds for medical training, essential drugs, and health-care technology are very short. Some key approaches, such as vaccination and oral rehydration therapy, have proved enormously cost-effective, but they still need extensive services and political support.

The challenge facing most governments in the South demands new priorities, with city-centred, doctor-oriented administrations converted into extensive networks for community care. Primary health care can cost relatively little. Many resources are often already in existence since traditional healers and midwives can usually be trained in hygiene and primary medicine.

But ultimately, the drive for health comes from people, not simply from governments. Public participation is often the key to success – in planning community services, lobbying for environmental reform, in the introduction of clean water, and preventive campaigns. Health is a function not only

Regional hospitals
The decentralization of health care and hospital services is essential in the developing countries. The provision of regional hospitals helps to bridge the gap between rural and urban health levels.

Health for all in the South
Several developing countries have made enormous strides in community health, increasing life expectancy, banishing endemic diseases such as smallpox, and reducing child mortality. Oral rehydration therapy (ORT) – giving patients a simple mixture of sugar and salts in water – is a major cure for diarrhoea. The benefits of primary health-care schemes have included improvements in educational performance and worker productivity. China spends about $4 per person per year, which is broadly comparable with health expenditure in many developing countries, but has achieved major successes by giving top priority to primary health care and to birth control measures. Other successes have been reported in Colombia, Cuba, S Korea, Sri Lanka, and Tanzania. Key targets include increased breast-feeding, major immunization programmes, and safe water.

Immunization success
There has been a massive increase, to about 80%, in the proportion of the world's children now protected against the major immunizable diseases, compared with only 20% in 1980.

Family planning
Experience in many countries has confirmed that professionally run family planning services can improve the health of mothers and children. Better spacing between births cuts infant deaths and allows for longer breast-feeding. Growth monitoring has proved an extremely effective means of preventive child care.

Barefoot doctors
Official health services in the South have begun to exploit the reservoir of skills found among traditional healers and midwives. China has 15-20 times more homeopathic doctors than it has conventionally trained doctors. In Africa, there is one traditional healer for every 500 people, compared to one doctor for every 28,000.

An adequate diet
The world can easily afford to feed its 800 million ill-fed people. For the cost of a few modern fighter aircraft, millions can be fed. And vitamins and protein supplements which combat serious illness cost under 20 cents per child per year.

A health revolution in Sudan
To provide health care to a scattered population covering an area of 2.5 million sq km, the Sudanese government has established an effective network of health-care units. The units, which are linked to a dispensary, usually cover a radius of about 16 km and serve an average of 4,000 people. These units are backed up by travelling health-care workers serving distant or nomadic communities. In 1970, Sudan had about 0.60 medical staff per 10,000 people – now it has 1.15 per 10,000.

sanitation
safe water
Paramedic

Village health care
One of the most fundamental elements in any Primary Health Care (PHC) strategy is the involvement of the community. This is the key to mobilizing the human, material, and financial resources. There have been notable successes in campaigns in Africa and South America, but they depend on the sustained backing of national health authorities. WHO research has shown that a trained local health worker, equipped with only 15-20 drugs, can effectively treat the majority of common illnesses. Community PHC depends on a multi-stranded approach, from drugs to the provision of sanitation and decent housing.

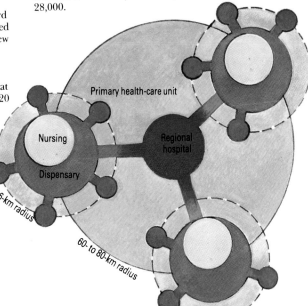
Primary health-care unit
Nursing
Dispensary
Regional hospital
16-km radius
60- to 80-km radius

of medical care, but of the overall integrated development of society. The central objectives of health promotion and health-care programmes should be to improve health, eliminate poverty and inequality, spread education, and enable the poor and underprivileged groups to assert themselves.

Community tools and access to ideas

A growing number of communities are watching angrily as their economic life-blood flows away. In the North, structural change is ripping the economic heart out of entire regions which depend for their livelihood on the sundown industries. In the South, the powerful undertow exerted by the exploding cities is undermining rural communities by draining away their young people.

What can be done? Are these community deaths as inevitable as they seem? Sometimes they are: sometimes communities have established themselves on a single, precarious resource, on a narrow economic base peculiar to a particular location, and are, therefore, left out on a limb when their economic monoculture collapses. But, as often as not, communities find that there are other ways of making a living, of rebuilding a viable community.

Dealing with outright disaster, like a flood or major factory accident, is often easier for a community than coping with a long drawn-out process of decline, yet there are many examples of communities which have gone against the flow of economic decay, refusing to give way to despair; with totally inadequate resources, they have managed to pull themselves up by communal will. Policy makers and planners need to recognize, and support, all these regenerative activities.

In the North, old industrial buildings have been converted into thriving community workshops. Urban "greening" projects have harnessed the enthusiasm, energy, and skills of local residents – which are the real "tools for community". In the UK, for example, towns like Rochdale have started to upgrade derelict areas to attract new industrial employment. Cities and towns across Europe and North America have transformed derelict land and building sites into parks and community gardens, boosting the sense of community both during construction and use.

In the South, the provision of local employment, often based on co-operatives and the use of various forms of appropriate technology, has helped build local economies better able to counter the call of the city. In Sri Lanka, the co-operative movement Sarvodaya Shramadana, which means "the awakening of all through the sharing of time, thought, and labour", draws on "gifts of labour" from villagers and volunteers to help poor Sri Lankans.

And, in both North and South, a growing number of schools have turned themselves into community centres, offering adult education, tutorial help for students undertaking distance learning, parents' groups, and adult literacy support schemes.

Tools and ideas

Unemployment and underemployment are undermining the economic base of an alarming number of communities in both North and South. Technological change, the fluctuating world economy, and the exploding world population are all contributing factors. To turn the tide of people surging into the cities in search of work and better facilities, it is essential that rural and small-town economies should be stimulated, and educational and health care services decentralized.

Community participation has proved to be a vital ingredient in the planning and implementation of public and private-sector development programmes. Given the tools and other resources, communities have pushed through their own programmes. The Naam movement in Burkina Faso (pp 238-9) is a prime example, and there are now thousands of other communities which are pulling together.

Communities in school
Education has to keep pace with changes in work and technology. Schools and colleges are vital tools for communities adapting to new needs, and are also major employers in their own right. In both North and South, good rural schools, used as a resource by all the community, can help stem the flow of people to crowded cities.

Co-operative membership 1981-2
International Co-operative Alliance membership 366,764,185

- 35% consumer societies
- 5% housing
- 2% workers
- 4% miscellaneous
- 1% fisheries
- 18% agricultural
- 35% credit, thrift, and saving

Working ideas
Improved methods and low-cost technologies can revitalize existing work and help create new industry. Groups such as Intermediate Technology Development (ITDG) promote proven systems (e.g. fish farming) and appropriate tools to be used where expensive technology is often irrelevant. Information is vital. One Indian scheme, SITE, used a satellite to broadcast developmental programmes (pp 220-21).

Distance learning
Students learning at home through correspondence, radio, TV, or telephone can help overcome problems of access to education. China and the former USSR reach millions this way, while S American governments have pushed up literacy rates.

Student enrolment in distance-learning institutes by major region
- Over 500,000
- 100,000-500,000
- Under 100,000
- Unavailable
- ● Distance-learning institute with over 100,000 enrolment

Community co-operatives
Worker participation and profit-sharing schemes can both boost productivity and encourage innovation. There have been many ventures throughout Asia, including SEWA (pp 188-9) and, in Sri Lanka, Sarvodaya Shramadana – which has mobilized as many as 3,000 villages for communal road-building, farming, and local industry schemes. The ICA membership (left), covers multitudinous schemes.

Children's TV and radio
The mass-communication media have done much to build up the "global village". A successful television programme like *Sesame Street* can spark off hundreds of editions worldwide, and help children to understand the polycultural world.

Rural regeneration
Most developing countries that are committed to tackling rural poverty have implemented land reforms, notably China, Cuba, South Korea, and Taiwan. Such land distribution promotes employment and rural development. Where there is surplus labour, large-scale regeneration schemes prove highly productive. Examples include reforestation schemes in India and in South Korea (pp 52-3).

New uses in old buildings
Whole cities and towns in the North are blighted by industrial decay. Small workshop schemes, such as those created by British Steel in Clydeside, Scotland, in 1979, have proved fruitful. British Steel (BSC) provided credit and business advice to former employees setting up new firms and co-operatives in obsolete buildings. The "greening" movement is also promoting community gardens and parks in cities.

CIVILIZATION

Introduced by János Vargha

President of ISTER (East European Environmental Research Institute), Hungary, and recipient of the Right Livelihood Award – the alternative Nobel Prize

The evolution of life for the past four billion years has been based on the "technology" of nucleic acids and proteins. Yet, very recently, one species – *Homo sapiens* – started to use other technologies of its own devising to gain ecological advantages over the limitations of the physical environment, over other species of the biosphere, and, eventually, to gain military power against other human populations.

For 10,000 years there has been an accelerating increase of the human population, their chosen domesticated species, and other lifeforms – either parasites or species living in symbiosis with them. In addition, there has been a proliferation of the products of human technology. All of these elements – living things and technological products – constitute a system which can be called the "technosphere". Today, technology dominates the development and behaviour of the technosphere, simply because it provides the power to "rule" over the biosphere. The biological elements of the technosphere, therefore, have to accommodate themselves to the technologies. Lifeforms which are not able to do this become endangered as the technosphere grows. Indeed, they are becoming extinct in growing numbers.

Although extinction has been a natural part of evolution from the beginning of life, the reality today seems to be very different. The pre-human evolution of the biosphere was a result of competition and co-operation between living things, and all life was based on the same biological "technology". Our species could never have become dominant without tasting the apple of knowledge. This was a break in the evolutionary process, rather than its logical continuation. We have started to apply "alien" technologies and so have begun to part from the community of the biosphere. We cut the bands of dependence with the same careless negligence as we would tear into shreds a web in our way. But this particular web is a result of billions of years of evolution.

Our existence is inseparably linked to processes lasting far longer than the lifetime of a human being. But consciousness of this has little effect on our activities. We are only moved to action when our senses suddenly detect that something is wrong. Faced with a dying lake, for example, we are motivated to act only when we become irritated by its stink. Until a disaster is already knocking at our door, we do not begin to use our intellectual gift for analyzing and conceptualizing the world about us, and applying our creative abilities. If we intend to understand our relationship to ecological processes, if we do not want to tear into shreds the sophisticated web of our biosphere, we will have to look at the world in a new way, beyond the limited, short-term perceptions of our everyday senses.

The Braun company advertizes its appliances with the copy line: "A form the human hand has been designed for". For me, this sentence reveals more about our "break" with evolution than about any knowledge and wisdom our civilization has gleaned from technological developments. It reflects the haughtiness of technocracy: technology has arrived at a stage when it does not serve us – it employs us. The following pages offer some guides as to how we might turn the tables back, so the products of our civilization serve us again. If we do not, if we are to be mere employees, we might find ourselves dismissed.

J. Vargha

THE POWER OF CIVILIZATION

For more than 5,000 years, this planet has witnessed the rise and fall of successive civilizations. Each has spread its technology, culture, and beliefs widely; each has declined when its resource base, or its administration, or both, became overstretched, vulnerable to external attack or internal disintegration. The powerhouse behind all these cultures has been the phenomenon of the city – the centre of civilization, reflecting both the best and worst of human aspirations.

We have been building cities ever since we learned to drain and effectively use the fertile river valleys of Egypt, Mesopotamia, and China. The resultant agricultural surplus allowed a division of labour, and the leisure to think and organize, which underpin urban civilization. Over the ages, our towns and cities, often sited strategically on trading junctions, have served as cultural and racial melting pots. At their best, they have poured forth the wealth of art, literature, architecture, scientific discovery, and social, political, and ethical concepts which are humankind's legacy to the future.

But the sheer pressure of human numbers in cities has always created huge problems. Poverty, as well as wealth, is concentrated; crime as well as justice; disease as well as medicine. Today, many Third World cities are ringed by vast shanty towns encompassing much human wretchedness. Some, like Calcutta and Lagos, have become administrative nightmares; everywhere, space is at a premium.

The burgeoning Third World cities demonstrate another high cost of urban civilization: cities are in essence parasitical, having an insatiable appetite for food, energy, raw materials, and human labour. The economy of early cities was linked directly to the productivity of the surrounding countryside; later, colonialism and trade links helped to support their growing populations. Modern cities, by contrast, are nodes on a web of long-range communications, dependent on local, regional, and global markets for their supplies.

Until the Industrial Revolution, only one person in five lived in a settlement of over 10,000 people. In the past 100 years, however, there has been a massive influx to the cities, first in the North, latterly in the developing nations. On present trends, the cities of the Third World will swell to house over half its population by the year 2000, compared to 40 percent now.

What potential will the mega-cities of the South have as centres of the newly emerging 21st-century

The world city

When people congregate in cities, they can specialize to a much greater extent than in rural areas. This specialization becomes more diverse as cities grow until, as with the modern city, the functions of its inhabitants are as different, yet as interdependent, as those of individual cells in the human body.

When cities were relatively isolated, each developed a distinct culture: some became famous for their religious sites, others for their universities, public buildings, textiles, glass, or other artefacts. But today, worldwide cultural diversity is being eroded by mass markets and media. The middle classes in the big cities around the world often wear the same clothes, listen to the same pop-music and watch the same television programmes. We are seeing the emergence of a global "city culture".

The major conurbations are increasingly linked by transport and telecommunications, creating a single "world city" with increasing functional specialization of its individual members. Although separated geographically, they are combined in action: London, Zurich, and Tokyo operate as linked financial centres and the UN, based in New York, has specialized agencies in Paris, Nairobi, and Rome.

Financial control
While economic and political power once coincided, telecommunications have relaxed the link: New York, Rio de Janeiro, and Frankfurt are financial (but not political) capitals.

The changing city
Cities first developed some 4 to 5,000 years ago when a combination of agricultural and technological developments boosted productivity to the point where urban economies could support a range of specialist workers. Early cities were walled for defence – and within these walls craft industry flourished, sustained by the peasantry of the hinterland.

As cities outgrew the productive diversity of the surrounding areas, trade developed in importance, remaining the principal engine of city growth until the Industrial Revolution. It brought the need for banking and other commercial services, and with them, a wealthy and powerful merchant class. With the fall of the Roman Empire, urban life dwindled until the resurgence of European trade in the 11th and 12th centuries. In the 18th century, the mechanization of agriculture and industry released labour from the land and created a growing demand for it in factory cities. Railways and later the internal combustion engine helped reshape cities, the spreading suburbs and expanding new towns draining population from inner city areas. Now another wave of change, based on new technologies, is beginning to shift the world's cities into what has been dubbed the "post-industrial society".

Literature
It is impossible to imagine world literature without Joyce's Dublin, Dicken's London, or Sartre's Paris. Most social, intellectual and spiritual currents which fuel literature still concentrate in the cities.

Architecture
Many cities still bear the imprint of Rome's 1,000-year domination of European civilization. The 20th-century counterpart is New York, with its trendsetting skyscrapers.

3000 BC

200 BC

The "chained" cities
The circles (above) represent stages of development of an idealized city, the chains their growing dependence on trade. Key factors are the financial, industrial and defence sectors.

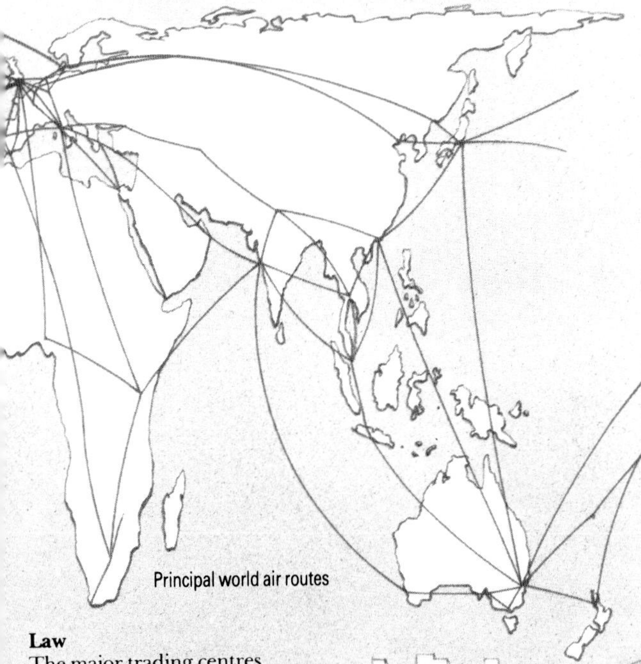

Principal world air routes

Energy
No city could survive without massive energy inputs to drive machines and power transport systems. The harnessing of fossil fuels allowed the increases in agricultural productivity needed to sustain urban growth.

Religion
Our cities have always been centres of worship – their temples, mosques and cathedrals are among our most precious buildings. Rome, Mecca, and Jerusalem are religious capitals for three of the world's largest faiths.

Law
The major trading centres inevitably became legal centres. Cities like Rome, London, Amsterdam, and Paris exported goods and legal codes.

Food
Today's cities have come to rely on ever-increasing trade with remote regions, having mostly long since outstripped the productive capacity of their original hinterlands; and they are more vulnerable to interruptions of supply.

Art
Throughout their history, cities have made unique artistic contributions, many exerting a powerful influence on subsequent, or distant cultures, including the Athens of Pericles, Michelangelo's Florence, and Nigeria's Benin.

State power
Cities are great centres of power – economic, military and political. As commerce grows, however, the military function usually disperses and State power, with its bureaucracy, is often transferred to new capitals.

Mega-city 2000
World population is increasing by over 250,000 people a day. In little more than a decade we will add another 1 billion people to the global population – and would need to build a city the size of Lima (Peru), every 22 days to house them. Instead, we expand existing cities, creating the mega-cities of the future. Northern cities are already merging, while in the South, centres such as Mexico City dwarf even sprawling Los Angeles.

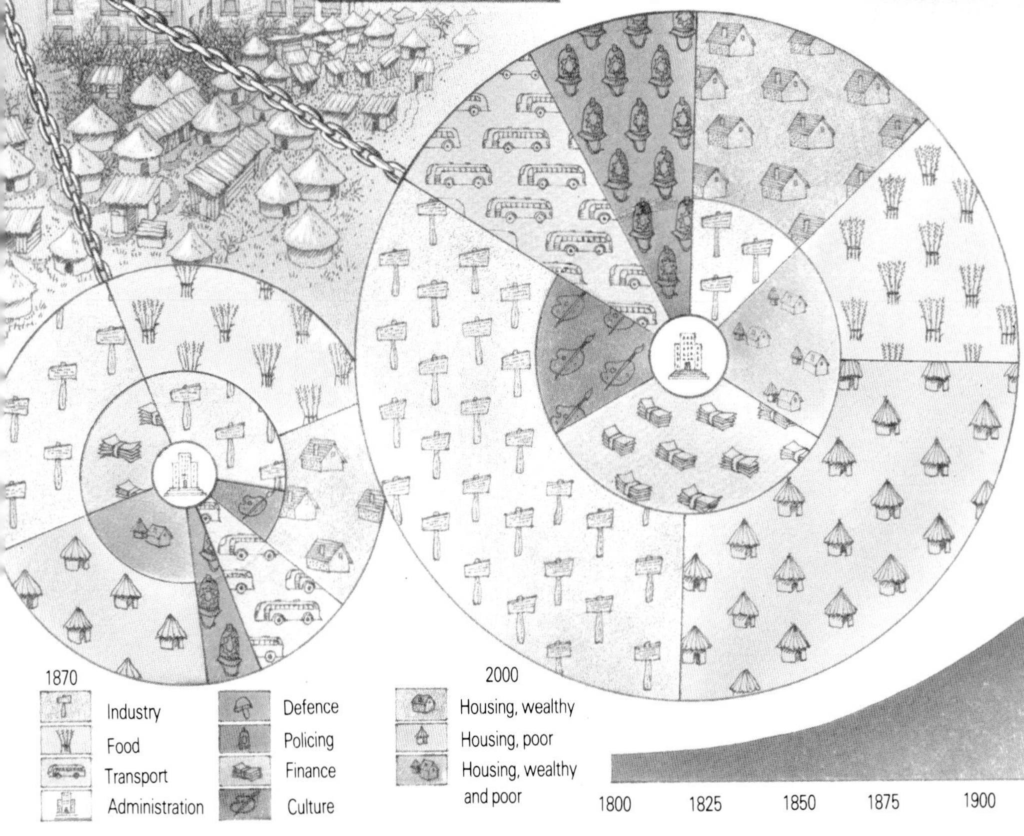

An urbanized world
By the year 2000, 50% of the world's population will be urbanized, compared with 14% in 1900.

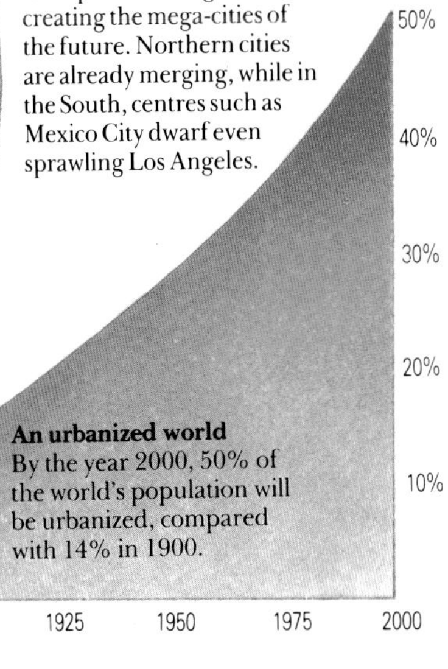

1870

2000

50%

40%

30%

20%

10%

Industry	Defence	Housing, wealthy
Food	Policing	Housing, poor
Transport	Finance	Housing, wealthy
Administration	Culture	and poor

1800   1825   1850   1875   1900   1925   1950   1975   2000

civilization? Will they retain their central role? The upsurge of communications and new information technologies makes it increasingly possible for city functions to be decentralized. It also opens the door to integration of urban and rural sectors in a one-world economic network.

The global factory

No civilization in history has held such technological power in its grasp as does our present generation. And none has allocated to science such a commanding, almost mystical role as the provider of health, wealth, power, freedom, and happiness. But, while technology and mass production have brought enormous rewards, the global factory has proved a double-edged weapon in our battle to improve the human condition.

The world production system has undergone a rapid, increasingly radical change since the optimistic post-war era – a boom period which was fuelled by cheap energy and sustained by the belief that surging growth in the North would boost standards of living in the South.

The outpourings of the global factory spawned the affluent, throwaway society – wasteful, pressurized, dominating the global imagination with its seductive imagery of gleaming automobiles and shining cities. But it also produced cures for major epidemic diseases, doubled food production, and provided mass mobility, communications, and mass media, which effectively shrank the planet into a "global village".

The key technologies of the 1950s and 1960s were based on steel, chemicals, and oil. The bulldozer, tractor, and chainsaw, together with fertilizers and pesticides, revolutionized agriculture. New drugs and surgical techniques transformed medicine. Wave after wave of durable goods, from refrigerators to radios, poured off faster, automated production lines.

And then, during the early 1970s, a series of convulsions ran through the global economic system. Soaring oil prices, depleted natural resources, widespread pollution and health hazards, an escalating and ever-more sophisticated arms race, and worsening world poverty and hunger, dealt hammer blows to our faith in technology.

Government backing for technology and private investment in research is still rising on an unprecedented scale. However, over 97 percent of the world's research and development budget is spent in the North, and global production has become astonishingly concentrated: in 1990, the top ten nations accounted for 82 percent of world production – and just three nations (the US, Japan, and Germany) accounted for around 50 percent. An unquantified but large proportion is in the hands of the transnational corporations, whose power has been escalating since 1945.

More positively, new technologies have emerged which, while they trigger some deep anxieties, do

The world factory

The engine that drives the global factory is commerce, fuelled by research, energy, and innovative ideas. Riding on successive waves of social and technological change, the factory is constantly adjusting to fluctuations in demand, partly created by advertising, and to changes in labour productivity and relations sparked by new technologies. If it seems to promise happiness to many, it also provides power to the few. Both research and production are concentrated in just a few countries of the North; much is in the hands of international megacorporations. This concentration of power often makes it near impossible to develop equitable alternatives to the approaches and technologies promoted by the major industrial nations. But the present order is beginning to alter in the face of profound social and technological changes: we are seeing the birth of a radically different form of global factory.

Technology – for whose benefit?

Fears of mass unemployment and an emerging technocracy beyond the control of ordinary citizens may simply express natural hostility to a new social structure. The new technologies provide the possibility of greater diversity and freedom of work and more consumer control.

However, about $100 billion a year, plus half a million scientists and engineers, feed the research and development needs of the military "albatross", an investment which contributes little to economic and social advancement. The US and former USSR devote about half their R & D budgets to military research, which may well have been a factor in their decline in economic competitiveness.

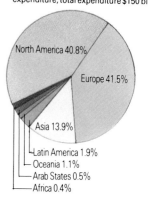

Distribution of research and development expenditure, total expenditure $150 billion

North America 40.8%
Europe 41.5%
Asia 13.9%
Latin America 1.9%
Oceania 1.1%
Arab States 0.5%
Africa 0.4%

The manufacturing base
A major impetus for technological innovation is humanity's constant desire for a greater variety of goods. Growth of the service sector depends on a strong manufacturing base to provide a large range of machines and equipment. Thus manufacturing operations are a key constituent of the world's technological resource. These constitute continuous flow goods, such as chemicals and steel; and batch items, which make up the great mass of consumer goods, such as cars and clothes.

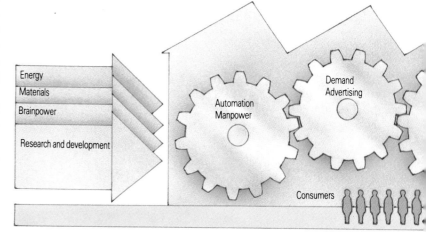

Energy
Materials
Brainpower
Research and development

Automation Manpower
Demand Advertising
Consumers

Technological waves
Major technological innovations are often clumped together and proceed in waves, coinciding with economic growth and recession.

Prosperity
Recession
Revi
Revival
Depression

Steam locomotive
Power loom

Electric locomotive
Electric telegraph
Internal combustion engine

Telephone
Assembly-line fa
Bulldozer

1800   1825   1850   1875

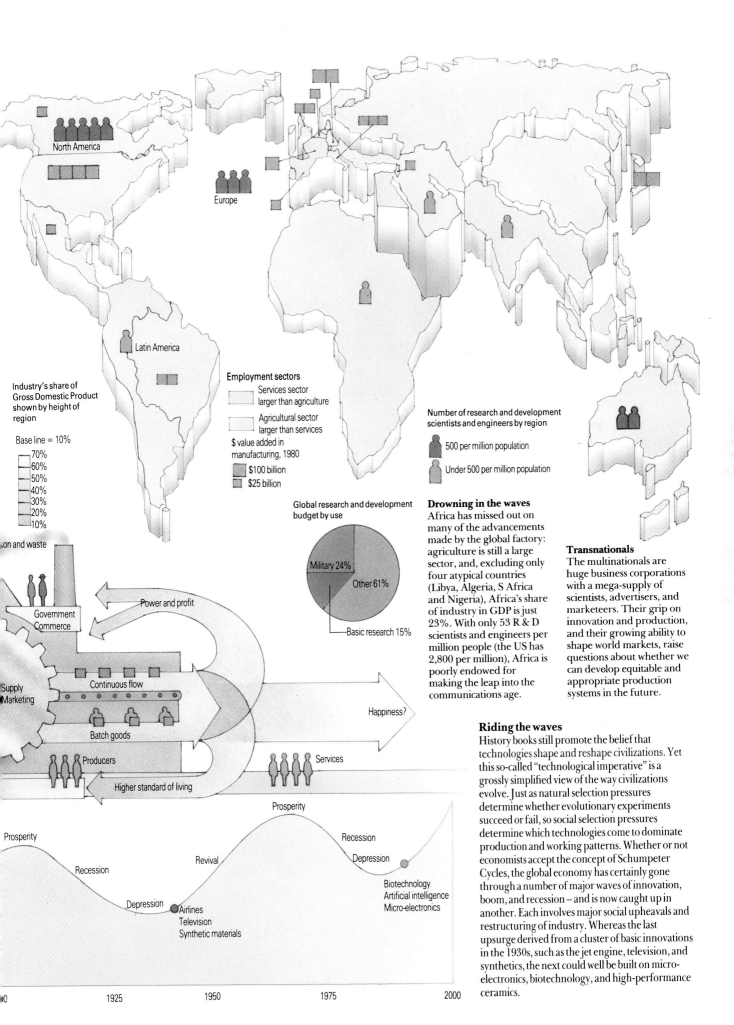

North America

Europe

Latin America

Industry's share of
Gross Domestic Product
shown by height of
region

Base line = 10%

70%
60%
50%
40%
30%
20%
10%

Employment sectors

Services sector
larger than agriculture

Agricultural sector
larger than services

$ value added in
manufacturing, 1980

$100 billion
$25 billion

Number of research and development
scientists and engineers by region

500 per million population

Under 500 per million population

...on and waste

Government
Commerce

Power and profit

Supply
Marketing

Continuous flow

Batch goods

Producers

Higher standard of living

Happiness?

Services

Global research and development
budget by use

Military 24%

Other 61%

Basic research 15%

Drowning in the waves
Africa has missed out on
many of the advancements
made by the global factory:
agriculture is still a large
sector, and, excluding only
four atypical countries
(Libya, Algeria, S Africa
and Nigeria), Africa's share
of industry in GDP is just
23%. With only 53 R & D
scientists and engineers per
million people (the US has
2,800 per million), Africa is
poorly endowed for
making the leap into the
communications age.

Transnationals
The multinationals are
huge business corporations
with a mega-supply of
scientists, advertisers, and
marketeers. Their grip on
innovation and production,
and their growing ability to
shape world markets, raise
questions about whether we
can develop equitable and
appropriate production
systems in the future.

Riding the waves
History books still promote the belief that
technologies shape and reshape civilizations. Yet
this so-called "technological imperative" is a
grossly simplified view of the way civilizations
evolve. Just as natural selection pressures
determine whether evolutionary experiments
succeed or fail, so social selection pressures
determine which technologies come to dominate
production and working patterns. Whether or not
economists accept the concept of Schumpeter
Cycles, the global economy has certainly gone
through a number of major waves of innovation,
boom, and recession – and is now caught up in
another. Each involves major social upheavals and
restructuring of industry. Whereas the last
upsurge derived from a cluster of basic innovations
in the 1930s, such as the jet engine, television, and
synthetics, the next could well be built on micro-
electronics, biotechnology, and high-performance
ceramics.

Prosperity

Prosperity

Recession

Depression

Revival

Recession

Depression

Biotechnology
Artificial intelligence
Micro-electronics

Airlines
Television
Synthetic materials

1925   1950   1975   2000

appear to offer new hope for the developing nations. Micro-electronics, information technology, new materials, and biotechnology are just some of the locally adaptable, low-energy, low-pollution technologies which may provide means for many parts of the South to boost their productivity.

The communications resource

The world is witnessing a communications revolution – a "third wave" of human advance transforming society just as dramatically as did the Industrial Revolution (the "second wave") and the Agricultural Revolution (the "first wave") several thousand years earlier. The interaction of several recent high-technology developments – space satellites, micro-electronics, optic fibres, laser beams, and the computer – is hugely magnifying our ability to store and analyse data, and to communicate and disseminate information, with major consequences for the nature of civilization.

In terms of communications, the world has shrunk rapidly. Today the contents of a dozen volumes of Encyclopaedia Britannica can be transmitted across the world in seconds. There has also been a shift from broadcasting to "narrowcasting". For example, the US cable television network offers people a choice of a hundred or more channels catering to every possible taste and viewpoint; the channels can be two-way, allowing viewers to communicate directly with those planning the service.

Satellites amplify the coverage and flexibility of both television and telephone communications. Indonesia's successful linking of its 3,000 islands by radio and telephone via satellite would not have been practicable by more conventional means. The use of satellites has an enormous potential for serving human needs: in weather forecasting, planning more efficient land use, and the detection and management of the Earth's natural resources. Remote-sensing satellites, like the US's Landsat, can map minerals, forests, and other resources using infra-red frequencies, and also provide early information on cyclones, crop yields, hazards such as blights, and other dangers from acid rain to pollution. Theoretically this should improve planning and enable governments to arrive at better policy decisions.

Computer and telecommunication links are also producing a mounting flood of data transactions. Data flows include financial information of every kind – commodity and share prices, currency rates, debt and credit ratings as well as market supply-and-demand trends, and "sensitive" product and technology information. Governments and businesses have vested interests in controlling information banks, and have a dangerous tendency to resist releasing information.

Of even greater concern is the unequal ownership of communications, both the technology and its information sources. Northern governments and multinational companies own the majority of the

The power of communications

The store of human knowledge has never been higher, nor our ability to communicate it greater. In today's world, split between the affluent North and the developing South, the new communications technologies have huge potential for solving problems – both for decentralizing facilities that have always tied people to towns and cities, for instance, and for speeding up development in the poorer countries of the world. They can also operate at many levels and can be tailored to suit widely varied needs. With literacy rates low in the South, broadcasting, for example, can represent a lifeline to vital information and participation, boosting educational programmes to rural communitites who were formerly without access.

The geostationary orbit
To avoid Earth stations having to track a satellite as it orbits, the satellite's speed can be matched exactly to that of the rotating planet. Such a satellite can send signals to half the globe, but its focus is usually limited to a smaller "footprint."

Language and new technology
For most of human history, languages have diverged and multiplied as human groups have dispersed and developed different identities. The mass media has tended to standardize and to strengthen the official languages. However, we can now support language diversity, through cable TV and community radio stations.

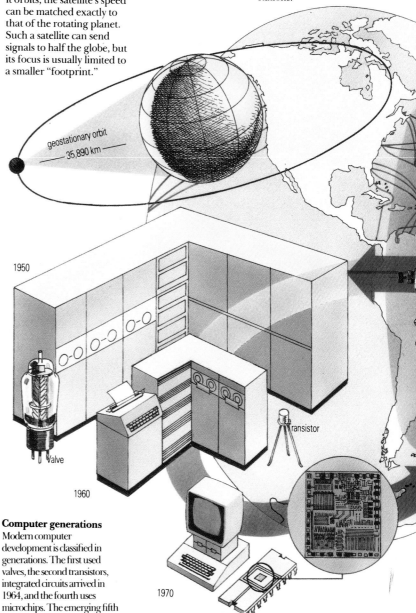

geostationary orbit
35,890 km

1950

Valve

1960

transistor

Computer generations
Modern computer development is classified in generations. The first used valves, the second transistors, integrated circuits arrived in 1964, and the fourth uses microchips. The emerging fifth has "artificial intelligence".

1970

Microchip   1980

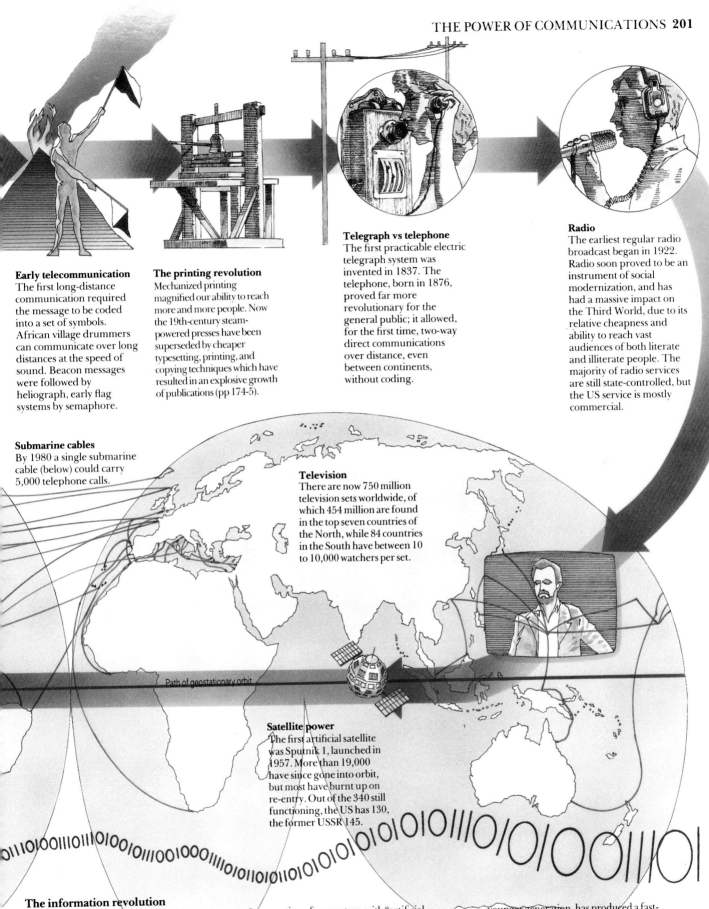

Early telecommunication

The first long-distance communication required the message to be coded into a set of symbols. African village drummers can communicate over long distances at the speed of sound. Beacon messages were followed by heliograph, early flag systems by semaphore.

The printing revolution

Mechanized printing magnified our ability to reach more and more people. Now the 19th-century steam-powered presses have been superseded by cheaper typesetting, printing, and copying techniques which have resulted in an explosive growth of publications (pp 174-5).

Telegraph vs telephone

The first practicable electric telegraph system was invented in 1837. The telephone, born in 1876, proved far more revolutionary for the general public; it allowed, for the first time, two-way direct communications over distance, even between continents, without coding.

Radio

The earliest regular radio broadcast began in 1922. Radio soon proved to be an instrument of social modernization, and has had a massive impact on the Third World, due to its relative cheapness and ability to reach vast audiences of both literate and illiterate people. The majority of radio services are still state-controlled, but the US service is mostly commercial.

Submarine cables

By 1980 a single submarine cable (below) could carry 5,000 telephone calls.

Television

There are now 750 million television sets worldwide, of which 454 million are found in the top seven countries of the North, while 84 countries in the South have between 10 to 10,000 watchers per set.

Path of geostationary orbit

Satellite power

The first artificial satellite was Sputnik 1, launched in 1957. More than 19,000 have since gone into orbit, but most have burnt up on re-entry. Out of the 340 still functioning, the US has 130, the former USSR 145.

The information revolution

The first computers, built in the 1950s, cost enormous sums, and filled several rooms. Now millions of private users have desk-top computers, while commerce, banking, and industry have been revolutionized by advanced information systems and international telecommunications. There are more than 100 million daily international transactions between data users in Europe alone, and the figure is expanding with great rapidity. The emerging fifth generation of computers, with "artificial intelligence" and the ability to use oral and visual inputs, represents another quantum leap in our ability to telecommunicate.

Computers are revolutionizing personal life too – from schooling and working at home to "armchair" banking, holiday booking, and shopping. The advent of word processors, two-way broadcast, and TV computer links, together with the rapid take-up of computer skills by the younger generation, has produced a fast-changing society fuelled by easy access to information and technology. The consequences are still unknown – George Orwell's fear of "Big Brother" control may be applicable to the international economy, manipulated by Northern governments and megacorporations. But mass access makes the breakdown of information monopolies a more likely outcome in the richer nations.

new hardware, and exploit its power, reinforcing their monopoly of global expertise. Many people in the South do not have access to a television (or indeed a radio), let alone a personal computer. Few countries can afford satellite technology. This inequitable sharing of potential benefits is greatly reducing opportunities for development.

Changing trade flows

The huge stepping-up of activity on the global "trading floor" since World War II has profoundly influenced the emerging world economy. By the early 1990s, world output of goods and services exceeded 20 trillion dollars, of which about a fifth was destined for international markets. Trade has been increasing faster than production, and nations exchanging goods and services are ever more interdependent: suppliers in one part of the world rely on buyers in another, and consumers enjoy a growing choice of foreign products.

As the market has expanded, the production of goods has increasingly split up around the world – components and raw materials from one country are shipped overseas for assembly or processing, then returned to their origin, or re-exported to a third nation. It is mainly labour-intensive processes that are shifting Southwards however; technology-based and more profitable industries remain in the North. This "assembly line" approach has been fuelled by the emergence of the multinational corporations, and of newly industrialized countries (NICs) like South Korea or Brazil, where low wage-rates attract Western-based investment. These "new Japans", elbowing their way on to the trading floor, are challenging older industrial countries.

Multinationals have long been involved in production of Third World primary commodities, from minerals to tea, coffee, rubber, palm oil, and bananas. But the growth of their involvement in manufacturing, with subsidiaries around the world taking a sizeable share of global production and trade, is a more modern phenomenon. Of the 100 biggest economic entities on Earth, about half are countries – the other half megacompanies.

These trends are a matter of bitter controversy. Some argue that rapid expansion of trade has been the engine of economic growth, and that market specialization encourages efficiency, so that all prosper. Others are more hostile to the whole trend to interdependence. They hold that the benefits have gone overwhelmingly to rich nations, and that specialization simply means Third world countries stay locked into low-wage, low-technology functions, while the rich countries and multinationals maintain their dominance.

This debate has influenced decisions by the developing countries on whether or not to link their economies to the world market. To follow the process of export orientation means gearing production to satisfying overseas demand, while importing goods needed at home. The alternative,

The world market

The traders in the world market do not bargain on an equal footing; there are huge disparities in their share of trade, and the value of goods they exported. A large part of the trade flows are directed by the megacorporations, mainly based in the rich North, which dominate the market. Snapping at its heels, however, are newly industrialized Third World nations. The OPEC group (Organization of Petroleum Exporting Countries) exploits their high-earning commodity, oil – a temporary bonanza. But the majority of developing nations still export low-value primary products. Often unable to buy the North's manufacturing exports, their share of world trade is small – only 3% for non-OPEC Africa. If the rapid growth of the world market is to provide prosperity for all, much depends on restructuring the trading flow, to make room for the "new Japans", and to increase the earnings of the poor majority.

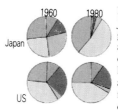

Market leaders: Japan, US
Japan's manufacturing growth has surpassed all non-Communist countries. The US is the leading industrial nation, also accounting for one-fifth of world agricultural exports.

Structure of merchandise imports
Food ─┐
Fuel ─┼→ Primary commodities
Other ─┘
Machinery and transport equipment
Other manufacturers

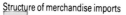

Structure of merchandise exports
Fuels, metals and minerals ──┐── Primary commodities
Other ──┘
Textiles and clothing
Machinery and transport equipment
Other manufactures

On the face of it, the rich industrial countries have all the cards stacked in their favour. They account for 65% of all export earnings, have benefited greatly from the boom in manufacturing value, and are able to import raw materials at a price advantage. But the older sundown industries, such as steel, are facing very tough competition from cheaper Third World products. Recession, high oil prices, and weakening markets in the poorer nations are all a threat to the viability of Northern consumer goods production and heavy industries.

Rich industrial countries

Value of exports, 1990 $2,700 billion

OPEC

Value of exports, 1990 $164 billion

Share of world oil production

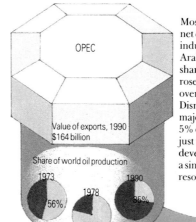

1973 56%
1978 76%
1990 35%

Most oil nations are major net exporters, but have little industry. After the 1973 Arab-Israeli war, OPEC's share of world oil production rose to 76% (by 1978); overall, prices rose ten-fold. Disruptions to trade were major. Today, OPEC earns 5% of world export income, just under a third of the developing world total, from a single, non-renewable resource.

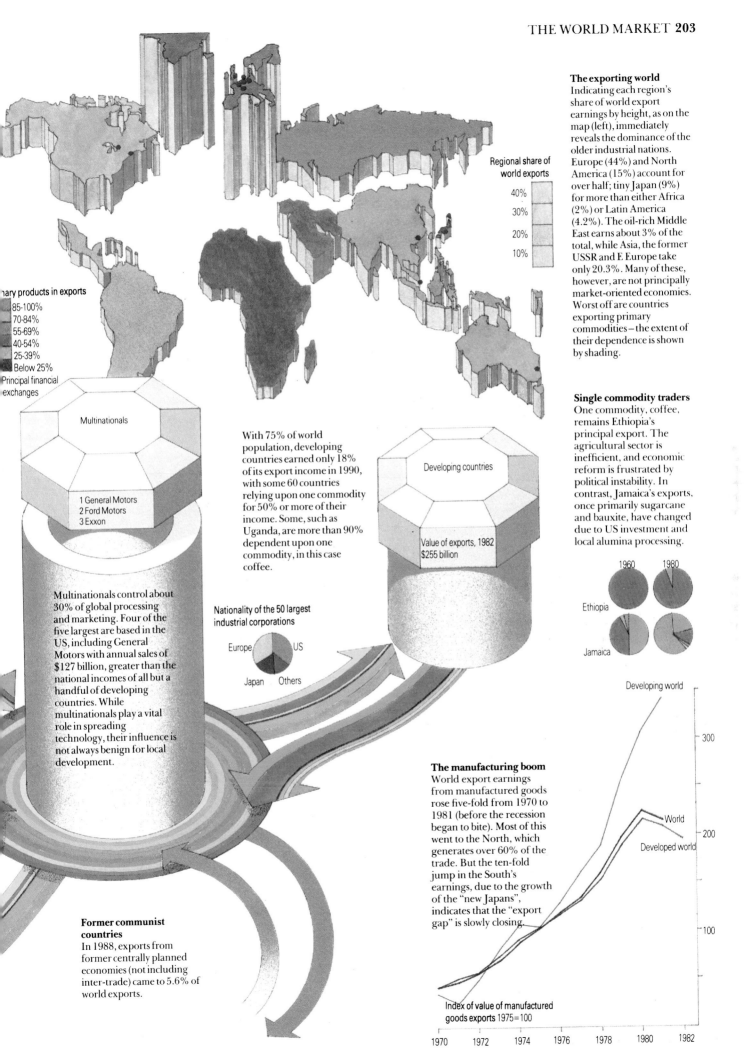

Regional share of
world exports

40%

30%

20%

10%

Primary products in exports

85-100%
70-84%
55-69%
40-54%
25-39%
Below 25%
Principal financial
exchanges

The exporting world
Indicating each region's
share of world export
earnings by height, as on the
map (left), immediately
reveals the dominance of the
older industrial nations.
Europe (44%) and North
America (15%) account for
over half; tiny Japan (9%)
for more than either Africa
(2%) or Latin America
(4.2%). The oil-rich Middle
East earns about 3% of the
total, while Asia, the former
USSR and E Europe take
only 20.3%. Many of these,
however, are not principally
market-oriented economies.
Worst off are countries
exporting primary
commodities – the extent of
their dependence is shown
by shading.

Single commodity traders
One commodity, coffee,
remains Ethiopia's
principal export. The
agricultural sector is
inefficient, and economic
reform is frustrated by
political instability. In
contrast, Jamaica's exports,
once primarily sugarcane
and bauxite, have changed
due to US investment and
local alumina processing.

1960   1980

Ethiopia

Jamaica

Multinationals

1 General Motors
2 Ford Motors
3 Exxon

With 75% of world
population, developing
countries earned only 18%
of its export income in 1990,
with some 60 countries
relying upon one commodity
for 50% or more of their
income. Some, such as
Uganda, are more than 90%
dependent upon one
commodity, in this case
coffee.

Developing countries

Value of exports, 1982
$255 billion

Multinationals control about
30% of global processing
and marketing. Four of the
five largest are based in the
US, including General
Motors with annual sales of
$127 billion, greater than the
national incomes of all but a
handful of developing
countries. While
multinationals play a vital
role in spreading
technology, their influence is
not always benign for local
development.

Nationality of the 50 largest
industrial corporations

Europe   US

Japan   Others

The manufacturing boom
World export earnings
from manufactured goods
rose five-fold from 1970 to
1981 (before the recession
began to bite). Most of this
went to the North, which
generates over 60% of the
trade. But the ten-fold
jump in the South's
earnings, due to the growth
of the "new Japans",
indicates that the "export
gap" is slowly closing.

Developing world

300

World

Developed world

200

100

**Former communist
countries**
In 1988, exports from
former centrally planned
economies (not including
inter-trade) came to 5.6% of
world exports.

Index of value of manufactured
goods exports 1975=100

1970   1972   1974   1976   1978   1980   1982

import substitution, means self-sufficiency through gearing production to domestic needs. Of the countries that are trying self-sufficiency, a few manage above-average economic growth, despite the attendant problems. Some export-oriented nations prosper; but their dependence on fluctuating markets and prices, especially those of raw commodities, makes their prospects uncertain.

Assets and earning power

The wealth of the planet is enormous. Even our human share of it – including the monumental assets of civilization – is considerable. What we have made of this resource, however, is clearly a very different matter.

Today, the First World, with a fifth of the world's population, enjoys nearly 75 percent of world income. The three-quarters who live in developing countries have only 12%. The balance goes to the former Eastern bloc.

These stark indicators of world inequality do not tell us much about relative affluence and poverty, or about potential. Income, usually referring to Gross National Product (GNP), is the conventional yardstick for assessing a nation's prosperity. We need to know too about national *wealth,* whether in the form of railways, factories, major rivers, soil and unexploited natural resources, or workforce skills and the health and education of a nation's children.

This kind of wealth is, like the proverbial talents, a stock of assets that can be increased or depleted, from which income can be derived. In this sense, a poor country in the South, such as Zaire, with vast but unexploited natural resources, is potentially very rich. By contrast, high-income Japan is rich through the ingenuity of its people, plus their capacity and enthusiasm for hard work.

A high GNP tends to bring power and influence. GNP determines a nation's shareholdings in the International Monetary Fund (IMF), for example, and the size of its shareholdings in turn determines its votes and borrowing rights. The rich countries do not borrow much. But they do decide policy, and so ensure that the international economic system functions in the way that they prescribe.

Exploitable natural wealth can provide counter-bargaining power however – witness the disruption of global economic patterns caused by OPEC's oil bonanza, or the emerging value of genetic resources in the South (pp 156-7).

But the problem for poorer countries is that, whatever their resources, their development costs money. They must either save the little income they earn, or they must borrow. During the 1970s, banks were the major source of capital, largely because of vast, virtually unspendable sums of petro-dollars that were deposited in banks – which then went looking for credit-worthy customers in the South. But now, bank lending has almost dried up as world recession and high interest rates have turned the resulting debts into almost unsupportable burdens.

The world's wealth

No nation's wealth is independent of world power structures and priorities. Income is increasingly dependent on market factors, and development on world rankings for aid, credit-worthiness, or investment. Since the 1944 Bretton Woods agreement on the World Bank and the IMF, the flow of wealth across the world and the ground rules for who gets what, on what terms, have been controlled by a group of largely Northern-based institutions comprising the financial community. Properly managed, international wealth flows *can* help poorer nations realize their potential. Wealth however, tends to flow towards wealth rather than where it is most needed.

Credit and investment

A nation's ability to raise credit or attract multinational investment is more often a measure of its strength than its need. The sources of credit are private banks, plus two major sister institutions: the World Bank, which provides investment capital, and the IMF, which concentrates on short-term balance-of-payments assistance for countries in difficulty.

Credit-worthiness hinges not only on performance indicators (some countries like Singapore and Brazil have brought good returns through their rapid economic growth), but also on political stability and "preferred" economic policies. A handful of rich, industrialized nations control the voting rights at the IMF (p.229).

Wealth vs income

For individuals, as for nations, income and wealth do not necessarily go together. The table below compares the situations of an Amazonian smallholder, with 50 ha of land and four cows, with that of a young, salaried European, making down-payments on an apartment.

	Small farmer Latin America	European professional
Wealth ($)	8,000	6,000
Annual income ($)		
cash	780	18,000
kind	460	–
Total income ($)	1,240	18,000

Technology and aid

Access to technology and skills for development costs money. Increasingly, developing countries rely on commercial enterprise or aid – multilateral funds from agencies like UNICEF, or bilateral agreements between countries. OECD and OPEC aid rose to $49 billion in 1989. By contrast, by the mid 1980s, transnational corporations were taking profits of $13 billion a year from 93 developing countries. Many transnational investments in the South lock recipient countries into buying unwanted goods and services.

IMF, credit agencies, and commercial banks

Terms of credit

$8,642

$119 billion

Total balance-of-payments deficit for developing countries, 19

$827

Terms of trade

Mean per capita GNP, 1981 ($)
Developing countries
Developed countries

Share of technology

Technology

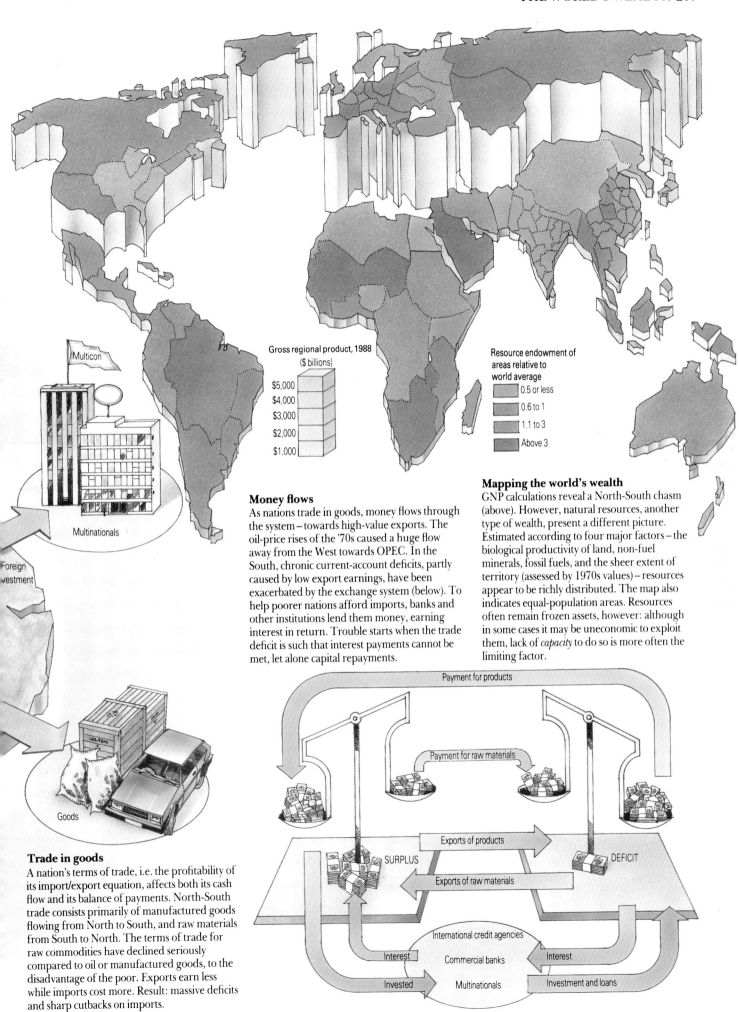

Gross regional product, 1988
($ billions)

$5,000
$4,000
$3,000
$2,000
$1,000

Resource endowment of
areas relative to
world average

0.5 or less
0.6 to 1
1.1 to 3
Above 3

Money flows

As nations trade in goods, money flows through
the system – towards high-value exports. The
oil-price rises of the '70s caused a huge flow
away from the West towards OPEC. In the
South, chronic current-account deficits, partly
caused by low export earnings, have been
exacerbated by the exchange system (below). To
help poorer nations afford imports, banks and
other institutions lend them money, earning
interest in return. Trouble starts when the trade
deficit is such that interest payments cannot be
met, let alone capital repayments.

Mapping the world's wealth

GNP calculations reveal a North-South chasm
(above). However, natural resources, another
type of wealth, present a different picture.
Estimated according to four major factors – the
biological productivity of land, non-fuel
minerals, fossil fuels, and the sheer extent of
territory (assessed by 1970s values) – resources
appear to be richly distributed. The map also
indicates equal-population areas. Resources
often remain frozen assets, however: although
in some cases it may be uneconomic to exploit
them, lack of *capacity* to do so is more often the
limiting factor.

Trade in goods

A nation's terms of trade, i.e. the profitability of
its import/export equation, affects both its cash
flow and its balance of payments. North-South
trade consists primarily of manufactured goods
flowing from North to South, and raw materials
from South to North. The terms of trade for
raw commodities have declined seriously
compared to oil or manufactured goods, to the
disadvantage of the poor. Exports earn less
while imports cost more. Result: massive deficits
and sharp cutbacks on imports.

Multicon

Multinationals

Foreign
vestment

Goods

Payment for products

Payment for raw materials

Exports of products

SURPLUS

Exports of raw materials

DEFICIT

International credit agencies

Commercial banks

Interest

Multinationals

Interest

Invested

Investment and loans

CRISIS: THE DIVIDED WORLD

Despite three decades of growth and development planning, the numbers of absolute poor have steadily increased. With present growth rates, Brazil, a leader of Third World development, would take 362 years to close its income gap with the rich world; Mauritania, one of the poorer nations, would need 3,224 years – if such equality were ever possible. The gap between rich and poor threatens the very roots of our global civilization.

Cities in crisis

Cities in the developing world are exploding under the pressure of their populations. Their surging growth is similar to that of the 19th-century industrializing North, but on a massive scale. Every day an estimated 75,000 poor people, hoping for work, stream in from rural areas to overwhelm city services and administrations. The majority of the newcomers head for the *barrios, favelas,* and shanty towns – made from corrugated iron, plastic sheets, and packing cases – which now ring the outskirts of most cities. Such squatter settlements house as much as 75 percent of the entire population of cities such as Calcutta (India).

Many millions live in the most degrading and menacing environment. In 1990, about 244 million people in developing world cities did not have safe drinking water and 377 million were without adequate sanitation – many millions more than in 1980. Widespread diarrhoea, dysentery, and typhoid are inevitable. Food and fuel are scarce.

With known unemployment frequently between 20 and 30 percent, the urban poor are trapped in a vicious circle. Those lucky enough to have a job suffer long hours, low wages, exposure to chemicals and dust, excess noise, and dangerous machinery.

Administrations stretched beyond capacity by the swelling slumland have developed harsh policies. The most draconian are straight eviction for those on illegally occupied land, and bulldozing of settlements. Many city governments also refuse to extend basic services and infrastructure, fearing that any such improvements will attract more newcomers.

Even the most workless and derelict post-industrial cities of the North do not compare with such misery; but life is also bleak. Instead of outer-city growth, Northern cities are afflicted by inner-city decay. As industries decline, and the rich move out to suburbs, those least able to move – the old, the ethnic minorities, and the poorly educated – are left behind. This results in declining revenues for local

Chaos in the cities

Most major Third World cities are really two cities: the inner city of the rich elite, which mimics the forms and lifestyles of the affluent North, and the largely self-built outer city of the poor. Urban incomes in the South average three times rural incomes, and modern services such as doctors, teachers, sanitation, clean water, and electricity are at least within reach. Thus the city acts as a "honeypot" for the rural poor; appalling though conditions are, they can be much better than in the countryside. Unable to afford ready-built housing, immigrants have no choice but to swell the shanty towns, where populations are now growing at four times the overall city growth rate. Many are already unmanageable. In Cairo, for instance, a water and sanitation system built to service a population of 2 million is collapsing under the burden of 11 million people.

Proportion of urban population

By 2000, more than 50% of the world population will be urban; this population will include 57 mega-cities (42 in the developing world), with populations in excess of 5 million.

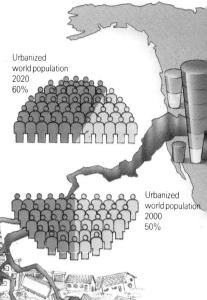

Urbanized world population 2020 60%

Urbanized world population 2000 50%

The older Northern conurbations, by contrast, are often struggling through a long, drawn-out, post-industrial crisis. Instead of a massive influx of the poor, they have suffered instead from an exodus of the rich. Lacking new inputs, the city structure – buildings, sewers, and roads – has increasingly fallen into disrepair. Instead of overcrowding, hunger, lack of water, and poor sanitation, the streets of Harlem (New York), Watts (Los Angeles) or Toxteth (Liverpool) offer loneliness, drug-addiction, and violence. Some city halls face bankruptcy, their falling revenues unable to meet the upkeep of the decaying physical and social fabric around them – notably New York, and Liverpool, UK.

Unemployment, homelessness, and tension between social and racial groups are exacerbated by economic recession and competition for scarce jobs. The elderly, in particular, are vulnerable – to neglect, to isolation, and to the mounting petty crime which outstrips police control. A UK study showed the British housing stock in need of over $35 billion of repairs – at present rates of funding, one estimate suggested that renovation might take 400 years.

Northern city degradation

The Northern post-industrial society no longer needs to concentrate employment and resources in the inner cities. Chicago, for instance, lost 212,000 jobs from 1969 to 1979, its suburbs gaining 220,000. The work remaining is mainly administrative and "white-collar", but the wealthier citizens who hold these positions are moving to the suburbs. In Hartford, Connecticut, suburbanites hold 90% of all jobs paying $15,000 or more. The same tale could be told in much of Europe.

Many inner cities have become "reservations" for the worst-off members of society, unable to provide the economic development to move people up the income ladder. Bootle (right), near Liverpool, typifies this urban decay.

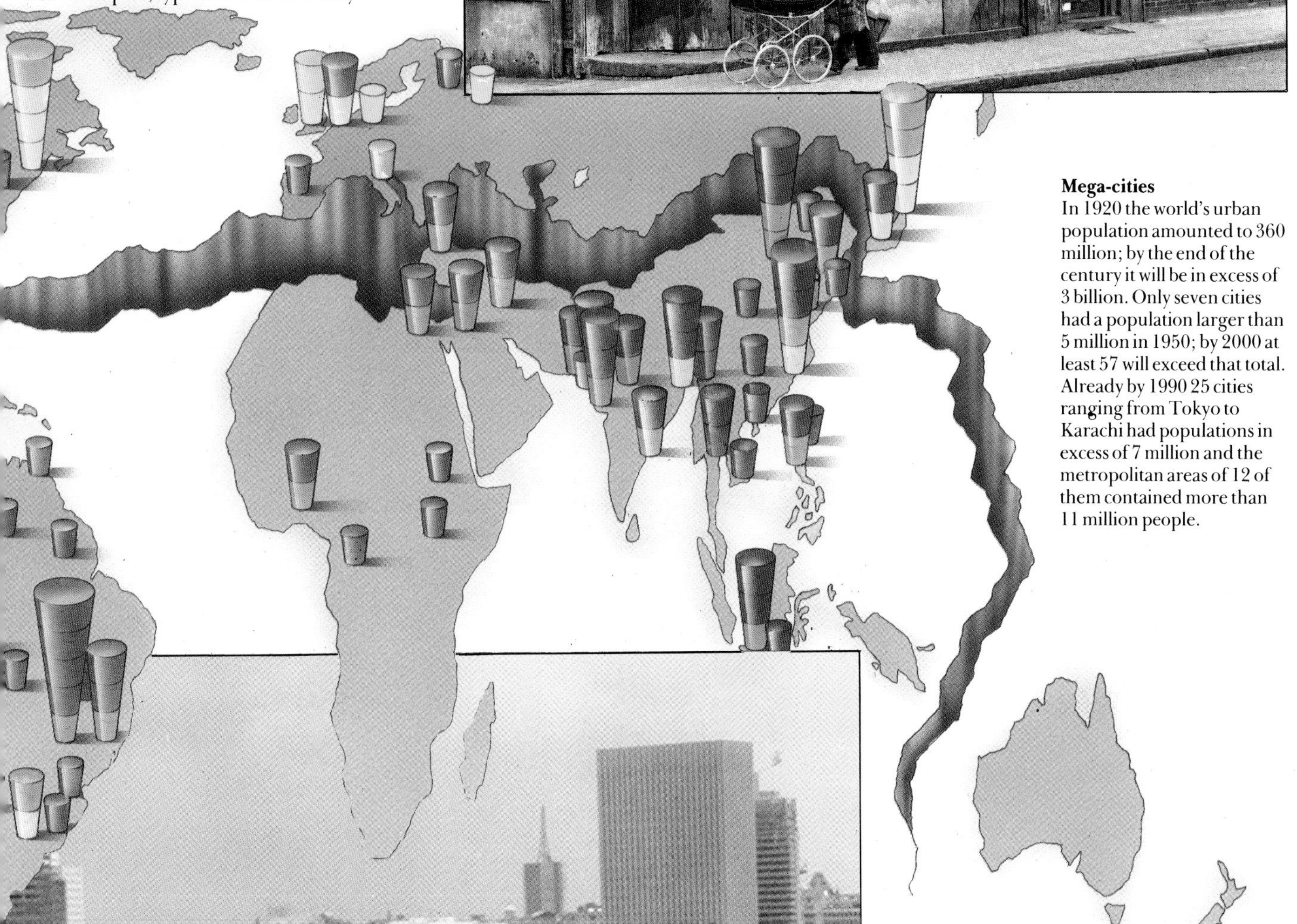

Mega-cities

In 1920 the world's urban population amounted to 360 million; by the end of the century it will be in excess of 3 billion. Only seven cities had a population larger than 5 million in 1950; by 2000 at least 57 will exceed that total. Already by 1990 25 cities ranging from Tokyo to Karachi had populations in excess of 7 million and the metropolitan areas of 12 of them contained more than 11 million people.

Bangkok squatters

Klong Toey is a community in the metropolitan district of Krung Thep Maha Nakhon, better known as Bangkok. More than 30,000 people live as squatters in houses they have built themselves over a swamp. Most of the inhabitants are recent migrants from Thailand's rural areas. An average of 6 people live in each dwelling, often as many as 4 to a room. Only 3 in 100 people have direct access to a water supply, most people draw water from the river or buy it. Lack of sewage connection or rubbish disposal (both virtually non-existent) is a constant threat to health. A third of the school-age children cannot attend school. Klong Toey is well-off compared to many other Southern urban communities.

authorities, which respond by cutting services. The growth of crime that often accompanies this pattern encourages more of the wealthier people to leave the city, concentrating poverty into ever-deeper pockets. Slums lie side by side with affluence even in cities like New York, which have moved well into the modern communications age.

Communications in crisis

Information is power, more valuable than oil, more precious than gold. And most of it is created, stored, and distributed in the rich countries. A few Western nations are the arbiters of taste and cultural values throughout much of the world. Their news media have something close to a monopoly in interpretation, judging of news value, and dissemination of ideas. Even stories about the South, published in the South, will often have passed through a Western filter. At the same time, government censorship in the Third World (sometimes in the belief that this will help nation-building, but more often to preserve personal power) has created a thirst for foreign newspapers, magazines, films, television, and radio programmes – thus perpetuating the West's values and its economic interests.

The Western monopoly could be further strengthened by technology. Take satellites: the smaller, cheaper types of satellite, currently using the lower

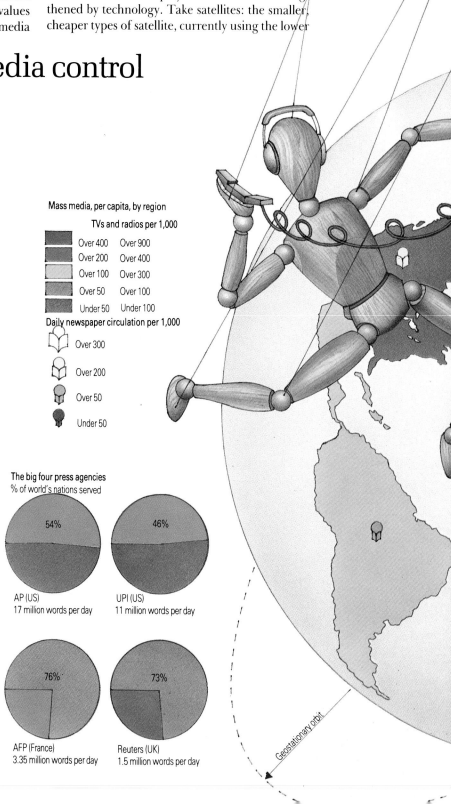

Governments

Monopolies in media control

The first public perception of the huge power of Information Technology (IT) came through George Orwell's novel *Nineteen eighty-four*. Orwell foresaw state control, thought control, and total invasion of privacy through mass media and censorship. The picture today, though less extreme, is disturbing. The mass media are in the control of state authorities and a few transnational companies. Censorship is widespread, particularly in one-party states and the dictatorships of the South. Almost everywhere, privacy is invaded through computerization of personal files.

The IT revolution is strongly centred in the North, and it proceeds apace, far faster than regulations to prevent its abuse. With few exceptions, the South is outside the network. And the Third World lacks many of the benefits of basic communications: 9 out of 100 people have a radio; 1 in 30 has a daily newspaper; and 1 in 500 a TV.

Mass media, per capita, by region

TVs and radios per 1,000

Over 400	Over 900
Over 200	Over 400
Over 100	Over 300
Over 50	Over 100
Under 50	Under 100

Daily newspaper circulation per 1,000

Over 300

Over 200

Over 50

Under 50

Broadcast "invasion"

US commercial TV has a huge export market. Neighbouring Canada, in a 1970s survey, imported nearly half of its TV shows. The Canadian authorities complain that their people are swamped by US culture. The same process of "invasion" applies between the US and Europe, US and Mexico, and within Europe.

Global press control

The press plays a critical role in forming public opinion. But 90% of foreign news published in the world's newspapers comes from just four Western news agencies – Reuters, UPI, AP, and AFP (right). About a quarter of their output is concerned with the Third World.

Telephone tapping

Modern society is highly vulnerable to disruption and invasion of privacy. During struggles with Solidarity in Gdansk, the Polish government was able to cut internal and external telecommunication. Even in the UK the government was taken to the European Court of Human Rights in 1983 after the exposure of a secret telephone-tapping centre.

The big four press agencies
% of world's nations served

54%

AP (US)
17 million words per day

46%

UPI (US)
11 million words per day

76%

AFP (France)
3.35 million words per day

73%

Reuters (UK)
1.5 million words per day

Geostationary orbit

frequency bands, are potentially of most benefit to developing countries. But their use is mainly controlled by the technologically advanced countries "in the interests of efficiency". The North can already gather more information about the South – its mineral potential, its crop yields, its untapped water resources – than these countries know about themselves. For example, a commodity speculator in New York is more likely to know about a coffee tree blight in Brazil before a Brazilian farmer is informed. The commercial, and to a lesser extent, the military advantage that this gives to the North is considerable. The Third World cannot afford to let the First World increase its stranglehold.

Similar questions arise in relation to cross-border data flows between computers. Critical information about a country's markets and companies can be collected and transmitted abroad, without ever being available to the country of origin, undermining national control. Even Canada has had to devise a bank by-law to prevent its banks from sending data to the US for processing. Brazil has already prevented local subsidiaries of Western multinationals from linking their Brazilian customers, via satellite, to overseas databases. These companies have been asked to replicate their foreign databases locally. The revolution in data transmission also affects the global financial systems.

Multinational companies

Data-processing monopoly
In 1989, Japan, Western Europe, and the US accounted for 91% of the world data-processing industry. This monopoly effectively stifles the growth of a data-processing industry in the developing world.

Satellites
☐◻○◻☐ 10 functioning satellites
◻◻●◻◻ US
▨▨●▨▨ Former USSR
◻◻○◻◻ Others: Japan, China, France Italy, India, UK, European Space Agency

Satellite control
The North controls most of the geostationary orbit (pp 200-201). The orbit and the radio spectrum are finite resources, and crowding has caused conflict.

Who pulls the strings?
As long as people have sought power – whether political or economic – they have sought to exercise control over communication channels and information. The greater the extent of mass media, and the more powerful the development of Information Technology (IT), the more people can be manipulated by the controllers.

With mass media, the crudest control is straight censorship. More subtle ways are the licensing of printing presses or newspapers (which is widespread) – and even the restrictive allocation of typewriters, paper, and copying machines (as in Eastern Europe). Government control over radio and TV in nearly all nations has gone far beyond what is actually necessary for state security. Legal controls like the libel laws and secrets acts back up the state's control. People languish in prison because of what they have written, or what they have said. And where the state can manipulate the media without opposition, what is printed or broadcast bears little resemblance to objective truth.

The transnationals have tight control in the production of electronic components, computers, and telecommunications technology. As national economies become more dependent on these technologies, the power of the companies will increase. They are creating an "electronic brain-drain", gathering business from the South.

Multinationals, using international banks to access their own funds, become their own bankers and send massive sums of money between countries. The banks may be riding "electronic tigers". Computers can be programmed to buy and sell stocks and shares automatically when prices reach predetermined levels, which can on occasion lead to wild swings in the marketplace.

Looming economic crisis

The world market is in deep trouble. As the world economy began to deteriorate in the 1970s, the climate for international trade became steadily bleaker. Growth declined, unemployment in industrialized countries rose, and inflation rose too. Over-production in some industries contrasted with idle factories in others. The price of money rose steadily, thrusting developing countries further into debt, and the exchange rate of key currencies became badly misaligned. Between 1985 and 1989, global GNP per capita grew only by an average of 1.6 percent a year.

Alongside these malfunctions, a new trade protectionism emerged in the West. The transfer of some production processes to low-cost Third World areas and the rise of the "new Japans" now began to appear as a threat to jobs in Europe and North America. Workers and employers started to demand protection and the rationing of imports – often by artificial "agreements" which exporters in Japan and the South were forced to accept.

Protectionism and unstable pricing make trade and production everywhere less efficient. They have a serious additional consequence for the North. If the developing world's earnings are restricted, so too is its spending, and a deflationary bias enters the world economy. According to World Bank estimates, if the industrialized countries increase protection by the equivalent of a 15 percent rise in import taxes, the cost to the GDP of middle-income developing countries would be 3.4 percent. The cost to GDP in the North is put at 3.3 percent.

The developing countries themselves also practise protectionism. Barriers are erected around infant industries (especially those involved in advanced technology) that would otherwise fall in the face of powerful competition from Japan and the West.

The Third World debt problem, which exploded in 1982, has exacerbated the problem. Many countries have been forced to slash imports. In 1983 the seven major debtors in Latin America cut imports to more than 49 percent below 1981 levels. Desperate to earn money to meet debt bills, some also resorted to subsidies to help their exports, fuelling the North's demand for protectionism.

While protectionism afflicts the industrializing nations, declining terms of trade, especially in relation to oil and manufactured imports, damage countries exporting raw materials and foodstuffs (though they too face some trade barriers). Between 1980 and 1988 the purchasing power of

Market tremors

As Japan and the newly industrializing countries (NICs) clash with the dominant West, major shifts in the market lead to a rash of protectionist barriers – which does no one much good in the long term. World trade slows down, economic distortions grow, and the gap between nations widens. The height of trading stations (right) indicates percentage increases in trade value over 10 years; dials at the top show the shift in their share of the market. The West, particularly the US, has lost heavily to Japan, OPEC, and the NICs. The remarkable progress of Japan and West Germany may in part be due to their lack of a "military albatross" around their necks – long prohibited from arms expenditure, their non-military industrial growth has soared.

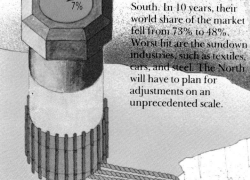

Newly industrialized countries (NICs)

Moving from 7% to 17% of the market in 10 years, the NICs sell goods from textiles to computers. To the first wave of Asian, Latin American and Mediterranean NICs is now added a second, a dozen or so, ranging from Colombia to Sri Lanka. The inability of the market system to accommodate these new Japans is a serious threat. What will happen as their numbers grow, and they challenge high-technology markets?

Developed countries (DCs)
Older industrial nations face severe problems: conflicts in the relationship with successful Japan; strains posed by the NICs; and reduced markets in the South. In 10 years, their world share of the market fell from 73% to 48%. Worst hit are the sundown industries, such as textiles, cars, and steel. The North will have to plan for adjustments on an unprecedented scale.

Development distortion
Export-led growth does not benefit all equally. Lured by Western affluence, many developing nations have neglected fundamental development issues such as agricultural reform, rural development, and health, in favour of prestige projects from airports and hospitals to dams and model farms, and concentrated on cash exports to pay for them. As the market fails, they are now reaping a bitter reward. Some multinationals have been deeply involved in this development distortion, while intermediaries cream off up to a third of the profits. Bilateral aid deals tied to purchasing Northern manufactured goods have often not helped. And the Green Revolution often exacerbated divisions between rich and poor (pp 56-7). Burdened with debt, or locked in the commodity trap with rising poverty, many nations face a radical rethink of their whole development policies.

Least developed countries
Despite an increase in trade (largely imports), the poorest nations' overall share of the market is still below 1%. Northern protectionism hurts. So does nearly stagnant world demand, which reduces income from exports. In real terms, African per capita incomes have hardly improved since 1960 and in some cases have deteriorated.

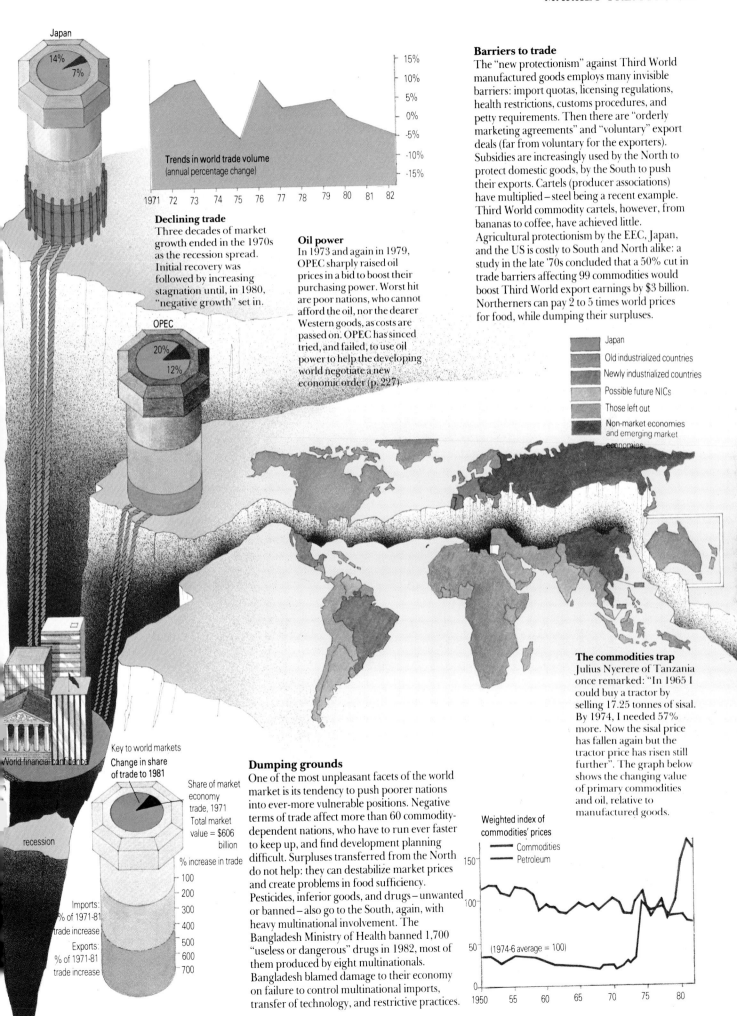

Japan
14%
7%

Trends in world trade volume
(annual percentage change)

15%
10%
5%
0%
-5%
-10%
-15%

1971 72 73 74 75 76 77 78 79 80 81 82

Declining trade
Three decades of market growth ended in the 1970s as the recession spread. Initial recovery was followed by increasing stagnation until, in 1980, "negative growth" set in.

OPEC
20%
12%

Oil power
In 1973 and again in 1979, OPEC sharply raised oil prices in a bid to boost their purchasing power. Worst hit are poor nations, who cannot afford the oil, nor the dearer Western goods, as costs are passed on. OPEC has sinced tried, and failed, to use oil power to help the developing world negotiate a new economic order (p. 227).

Barriers to trade
The "new protectionism" against Third World manufactured goods employs many invisible barriers: import quotas, licensing regulations, health restrictions, customs procedures, and petty requirements. Then there are "orderly marketing agreements" and "voluntary" export deals (far from voluntary for the exporters). Subsidies are increasingly used by the North to protect domestic goods, by the South to push their exports. Cartels (producer associations) have multiplied – steel being a recent example. Third World commodity cartels, however, from bananas to coffee, have achieved little. Agricultural protectionism by the EEC, Japan, and the US is costly to South and North alike: a study in the late '70s concluded that a 50% cut in trade barriers affecting 99 commodities would boost Third World export earnings by $3 billion. Northerners can pay 2 to 5 times world prices for food, while dumping their surpluses.

Japan
Old industrialized countries
Newly industrialized countries
Possible future NICs
Those left out
Non-market economies and emerging market economies

The commodities trap
Julius Nyerere of Tanzania once remarked: "In 1965 I could buy a tractor by selling 17.25 tonnes of sisal. By 1974, I needed 57% more. Now the sisal price has fallen again but the tractor price has risen still further". The graph below shows the changing value of primary commodities and oil, relative to manufactured goods.

World financial confidence

recession

Key to world markets
Change in share of trade to 1981

Share of market economy trade, 1971
Total market value = $606 billion

% increase in trade
100
200
300
400
500
600
700

Imports: % of 1971-81 trade increase
Exports: % of 1971-81 trade increase

Dumping grounds
One of the most unpleasant facets of the world market is its tendency to push poorer nations into ever-more vulnerable positions. Negative terms of trade affect more than 60 commodity-dependent nations, who have to run ever faster to keep up, and find development planning difficult. Surpluses transferred from the North do not help: they can destabilize market prices and create problems in food sufficiency. Pesticides, inferior goods, and drugs – unwanted or banned – also go to the South, again, with heavy multinational involvement. The Bangladesh Ministry of Health banned 1,700 "useless or dangerous" drugs in 1982, most of them produced by eight multinationals. Bangladesh blamed damage to their economy on failure to control multinational imports, transfer of technology, and restrictive practices.

Weighted index of commodities' prices

Commodities
Petroleum

150

100

50

(1974-6 average = 100)

0
1950 55 60 65 70 75 80

exports against the price of imports for developing countries fell by 30 percent. This deterioration in the terms of trade meant a loss to Sub-Saharan Africa alone of $50 billion in the second half of the 1980s. This decline is a major factor behind the crisis of development Southern exporting countries now face.

North and South: the divided planet

For the last two centuries there has been a widening gap between the rich countries of the North and much of Africa, Asia, and Latin America. By the middle of the last century, the difference in per capita income was about two to one. In this century, the divergence has increased, particularly since the 1950s. Today, GNP per capita in Western Europe, North America, and Japan is, on average, some 58 times greater than in the low income countries, and six times higher than even in the fast-developing countries such as Brazil. And this evidence of a divided world conceals ever deeper divisions at levels other than those of national economies.

The source of Northern affluence is much disputed – it is variously attributed to industrialization, to climatic factors and cultural values, and to colonialism, which transferred resources from the South to Europe and the US. In recent years, that transfer has become a prominent feature again. When Western banks found themselves with billions of petro-dollars to invest in the 1970s, the Third World seemed to offer a good prospect. But no sooner had the loans been spent than commodity prices fell and interest rates rose. The result is that today the South owes some $1,300 billion in international debt. To pay interest on the loans, there is a massive transfer of resources from the South to the North – $1,345 billion between 1982 and 1990.

A few newly industrialized countries have managed to make substantial progress. But for a sizeable sector of humanity, progress is mostly non-existent. Some are even going backward; most will take hundreds of years to close the gap with the North (if its consumerist lifestyle is thought desirable or ecologically sustainable), if they ever can. One in five of humankind falls below the income threshold necessary for the most meagre existence. The overall number of poor still rises.

Another factor in Southern poverty is gross inequality in income distribution. Latin America exhibits greater inequality than almost any other region, mainly because of the inequitable distribution of landholdings and political elitism.

Many individual countries have pursued economic growth first and foremost, in the belief that everyone benefits from this in the long run, through a "trickle-down" process – they reject redistribution of wealth as being "anti-growth", despite evidence that the two can go hand in hand. "Trickle down" has advanced the richer sectors of Third World

Haves and have-nots

The wedge driving the North and South apart has two components: slightly different growth rates, starting from a considerable GNP gap; and deepening developing world debt, exacerbated by soaring interest rates and falling export earnings. Every 1% rise in US interest rates adds about $4 billion to the debt bill. Latin America owes $410 billion; Brazil, Mexico, Venezuela, and Argentina alone over $305 billion. The "conditionality" factor – stringent economic adjustments, aimed at cutting consumption and boosting exports, attached to IMF emergency loans – often worsens the situation of the poorest. Per capita incomes in Latin America and Africa have fallen dramatically. The commonly-held assumption is that the rich North is transferring vast sums of money to the poor South. Nothing could be further from the truth. Between 1982 and 1990, total resource flows to the developing world amounted to $927 billion. Over the same period, these poor countries paid out $1,345 billion in debt service alone. There is a net transfer of funds from South to North, reaching over $50 billion in 1989.

The technology gap

Development, and hence wealth, demand access to advanced technology – almost all of which originates in the industrialized countries. The transfer of technology to the South via multinational companies is mostly concentrated in a limited number of countries which offer "suitable" economic environments. The rest depend largely on the handouts of aid.

For some nations, lack of access to much-needed technology rules out much hope of competing in the modernized world; even for those who are favoured, lack of control over planning, choice, and management of technology can have equally severe results. Debt-ridden Paraguay and Brazil, for instance, have spent $18 billion on the Itaipu Dam, the world's largest hydroelectric power project – eventually it will produce excess electricity.

Many bilateral aid agencies place their programmes in the hands of Northern experts, who may be insufficiently aware of local needs and can easily perpetuate the receiving country's dependence. One of the saddest examples is that of Egypt's Aswan Dam power scheme (pp 132-3), which has diminished the productivity of agricultural land and fisheries, thus leaving Egypt more dependent on foreign food imports. Capital-intensive Western technology is inappropriate for many developing nations, with their massive labour surplus and their chronic shortage of capital and foreign exchange.

Developed countries' GNP per capita $12,015 (in 1987 $)

Developing countries' GNP per capita $689 (in 1987 $)

1970   1972

$90 billion
Developing countries' debt (1971)

Brazil's burden
Brazil is the biggest debtor in the world. By 1990, it already owed $111 billion. For millions of poverty-stricken Brazilians looking for jobs, in rural areas or crowded into the notorious shanty towns, the nation's economic crisis is a harsh, unremitting reality. Brazil argues that its debt mountain has been imposed by external factors: its huge borrowing in the 1970s was justified by rapid economic growth. But since then it has been dealt a series of body blows; first the OPEC price rise, then the terrifying rise of interest rates, and finally Northern recession and protectionism halting its export boom.

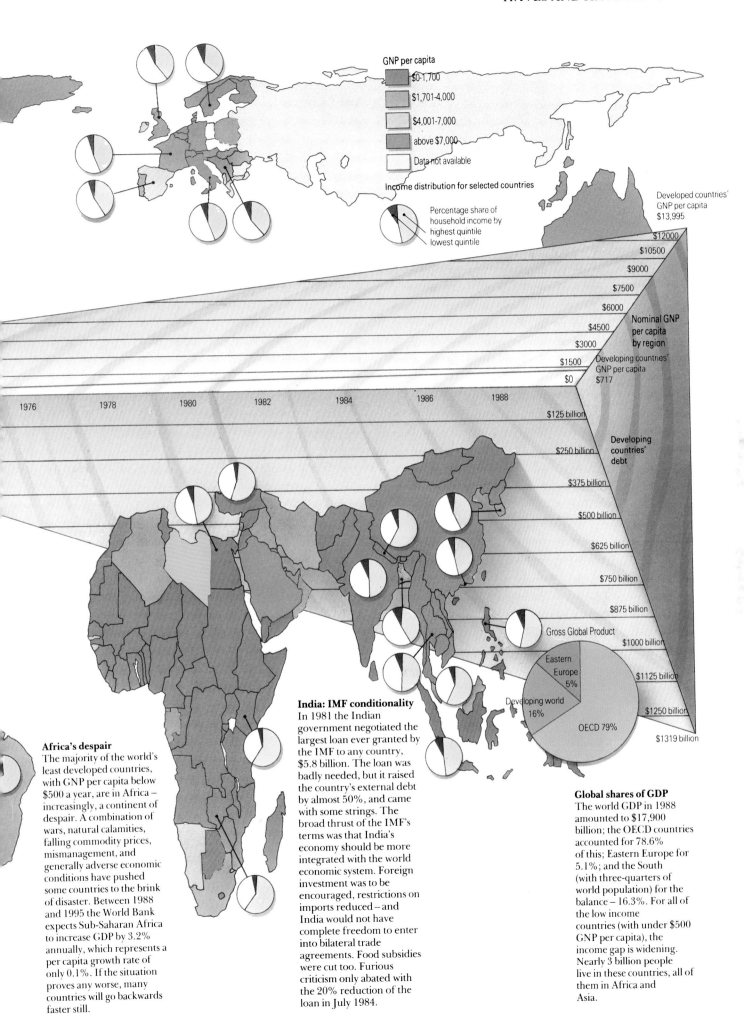

GNP per capita

$0-1,700
$1,701-4,000
$4,001-7,000
above $7,000
Data not available

Income distribution for selected countries

Percentage share of
household income by
highest quintile
lowest quintile

Developed countries'
GNP per capita
$13,995

$12000
$10500
$9000
$7500
$6000
$4500

Nominal GNP
per capita
by region

$3000
$1500

Developing countries'
GNP per capita
$717

$0

1976   1978   1980   1982   1984   1986   1988

$125 billion

Developing
countries'
debt

$250 billion
$375 billion
$500 billion
$625 billion
$750 billion
$875 billion

Gross Global Product

$1000 billion

Eastern
Europe
5%

$1125 billion

Developing world
16%

OECD 79%

$1250 billion

$1319 billion

Africa's despair
The majority of the world's
least developed countries,
with GNP per capita below
$500 a year, are in Africa –
increasingly, a continent of
despair. A combination of
wars, natural calamities,
falling commodity prices,
mismanagement, and
generally adverse economic
conditions have pushed
some countries to the brink
of disaster. Between 1988
and 1995 the World Bank
expects Sub-Saharan Africa
to increase GDP by 3.2%
annually, which represents a
per capita growth rate of
only 0.1%. If the situation
proves any worse, many
countries will go backwards
faster still.

India: IMF conditionality
In 1981 the Indian
government negotiated the
largest loan ever granted by
the IMF to any country,
$5.8 billion. The loan was
badly needed, but it raised
the country's external debt
by almost 50%, and came
with some strings. The
broad thrust of the IMF's
terms was that India's
economy should be more
integrated with the world
economic system. Foreign
investment was to be
encouraged, restrictions on
imports reduced – and
India would not have
complete freedom to enter
into bilateral trade
agreements. Food subsidies
were cut too. Furious
criticism only abated with
the 20% reduction of the
loan in July 1984.

Global shares of GDP
The world GDP in 1988
amounted to $17,900
billion; the OECD countries
accounted for 78.6%
of this; Eastern Europe for
5.1%; and the South
(with three-quarters of
world population) for the
balance – 16.3%. For all of
the low income
countries (with under $500
GNP per capita), the
income gap is widening.
Nearly 3 billion people
live in these countries, all of
them in Africa and
Asia.

society, at the expense of the poorer. The global community, with its emphasis on growth, has fallen into much the same trap.

Trade is an environmental issue

Many deep and pervasive environmental problems are associated with trade and the debt crisis. The need for foreign exchange to service debts fuels efforts to expand exports, and these add to the many other pressures on natural resources. As people are pushed into poverty, in part by government measures to deal with debt, and as prices rise faster than incomes, the daily struggle for survival forces the poorest to exploit ecologically-fragile environments.

The central tenet of today's economics is that developing countries can export their way into recovery. Indeed, lending institutions have effectively devoted much of their efforts to promoting export-led growth. Freer trade, it is held, could give poor nations better access to world markets, thus raising income to protect the environment and alleviate poverty. This, at any rate, is the view of officials from the General Agreement on Tariffs and Trade (GATT), the international organization that lays down the rules for almost all world trade. The GATT aims to smooth the path of free trade and remove import and export controls. But critics argue that this could also increase the pace at which nations plunder natural resources, especially

Environmental trade-off?

Trade rules and agreements are a major determinant of how natural resources are used, what pressures are put on the environment, and who ultimately benefits from the $3 trillion that crosses borders annually with the exchange of goods. Major international accords such as the General Agreement on Tariffs and Trade are intended to achieve trade liberalization – by reducing or eliminating regulations that may interfere with the free flow of goods and resources between nations. Yet this may be a double-edged weapon where poverty and environmental degradation are concerned. It may promise developing nations better access to world markets. But it also has the potential to increase the rate of environmental destruction.

The interactions between trade and the environment are increasingly widely recognized. Energy-intensive, trade-based growth in the North has caused most of the global pollution in terms of ozone-depleting and greenhouse gas emissions, along with hazardous waste creation. In the South, debt-fuelled export expansion continues to play its part in the destruction of the natural resource base.

if it caused the prices of commodities to fall even further.

With import and export controls removed, there is no allowance for any subsidies or restrictions to promote more sustainable forms of production. Indonesia, for example, has been accused by the EC and Japan of violating the GATT by imposing export restrictions on unprocessed timber – a measure introduced to increase value-added processing, and conserve forests. As well as becoming "illegal" for countries to protect their markets and environment, it may become increasingly difficult to protect the global commons. In 1991, the GATT overruled a US ban on imports of Mexican tuna which had been imposed because of the large numbers of dolphins killed in Mexican tuna fishing operations. A blow to marine conservationists, the implications of this ruling threaten conservation measures worldwide. Trade sanctions – the most effective non-military weapon available – might no longer be relied upon to enforce non-compliant states to respect international treaties, such as the Montreal Protocol and CITES.

Trade is, of course, a double-edged sword. The overturning of the tuna ban would be a blow to the conservation of marine resources, but a boost to Mexico's economy. Furthermore, developing world governments are aware of a new form of "green imperialism" as a threat to their development efforts. People in the South stress that any concern

GATT and the environment
The GATT highlights the incompatibilities between the two contemporary trends of liberalizing free trade, and safeguarding the environment. In the North, for example, the drive for less-polluting products may add to industry's production costs. But GATT would forbid the setting of import surcharges to allow such products to compete with those produced in countries with lax pollution laws. And if governments do bite the bullet and legislate against polluters, the industries will be more inclined, and able, to relocate to countries with poorer legislation; mostly in the South.

The trend in trade
Previously, multilateral trade negotiations have focused on trade barriers for goods. The trend now is to include, for the first time, trade in services, foreign investment, and intellectual property – areas where the North has a "comparative advantage" and seeks to protect its service industries, expand investment opportunities, and guard its technologies from imitation.

The GATT tuna ruling
In the eastern Pacific various fishing fleets catch huge quantities of tuna. The Mexican fleet, in particular, kills large numbers of dolphins which get entangled in their nets. US law allows for import restrictions on countries which so kill large numbers of dolphins, but when these were enacted in 1991 Mexico claimed the action was contrary to GATT rules. In the first test case of whether environmental considerations beyond a country's border can be a factor in restricting imports, the GATT ruled in Mexico's favour.

Maquiladoras and free trade
There has been an economic boom on the Mexican border with the US. Over the past few years, about 1,500 American-owned companies have set up assembly plants – known as *maquiladoras* – including General Motors and IBM. Special trade agreements currently exempt these companies from US tariffs and other restrictions. They can supply the home market with goods produced at a fraction of the cost. Relocation also avoids expensive anti-pollution equipment needed to meet US health, safety, and environmental standards. Pollution and living conditions in the shanty towns around the *maquiladoras* have turned the region into a "sinkhole of abysmal living conditions and environmental degradation" according to one foreign aid worker. Will this be the shape of things to come with the free trade agreement between the US, Canada, and Mexico?

for the global environment necessitates the tackling of North/South inequality. If free trade is to be of environmental benefit it must first be *fair* trade. If debt burdens were eased and commodity prices improved (and adjusted to reflect the real value of natural resources), then pressure to over-exploit the environment would be reduced. Tackling poverty in the South would benefit both the global environment *and* world trade.

One billion poor

Absolute poverty has been described by Robert McNamara, a former President of the World Bank, as a "condition of life so characterized by malnutrition, illiteracy, disease, high infant mortality and low life expectancy as to be beneath any reasonable definition of human decency". The engine of growth seems to be forcing more and more people into economic vulnerability. They have been systematically marginalized, their ability to supply their own basic needs unrelentingly reduced.

Of the one billion absolute poor, four-fifths live in the countryside (in Latin America half live in urban slums). Most of them are landless labourers, marginal farmers, or unskilled labourers in cities. Most countries with a high proportion of absolute poor (Bangladesh, India, or Indonesia), are found contiguously in two "poverty belts". One extends across the middle of Africa, from the Sahara to Lake Malawi. The other, beginning with the two Yemens and Afghanistan, stretches eastwards across South Asia and some East Asian countries.

These nations frequently exist in a fragile tropical environment, upset by the pressure of expanding populations, and afflicted by droughts, floods, soil erosion, and the encroachment of deserts. Not only has their economic performance been inadequate in recent years, but the changing character of national environment may make supporting their present populations impossible.

There is rarely any system of state benefits to ease the plight of the poor. They are most likely to be without the medical facilities to cope with sickness and disability, without schools to raise the levels of literacy, and without a hope of better employment. On top of physical isolation, they also suffer the social isolation of being unwanted; women, in particular, find it impossible to escape their burdens (pp 188-9). Furthermore, the poor household may belong to a socially inferior class, forced by custom to accept the lowest paid and most menial tasks, and physically segregated. Their children will be condemned to the same lowly status.

The pressures of poverty have often caused explosions in the past, and today still more people are driven beyond the limits of endurance. In Latin America, for example, there have been poverty outbursts, such as the looting of hundreds of food shops in Rio de Janeiro in 1983, and in the Dominican Republic in 1984 where 186 people died.

The poverty bomb

Poverty is self-sustaining and self-generatng, a trap that holds about one billion people, nearly a fifth of the world's population. In general, the poor in the North can at least survive, through a safety net of social services (although not all-embracing). Their poverty can be seen as "extreme social deprivation". In the South, however, the term "absolute poverty" is the only one applicable. Even if they survive a malnourished childhood, hundreds of millions will never have the opportunity to realize their full human potential. All the routes out of the trap are firmly shut, because of lack of education, technical aid or credit, employment, sanitation or safe water, access to health services, transport, or communication.

Most Southern governments are unlikely to implement wide-ranging reforms: resources are too scarce and they lack political will. So the poor are left with the greatest toll of death and misery – and they are wide open to exploitation by landlords and merchants, petty officials and police, and any employer. Insecurity and powerlessness are their lot.

Power
Franchise may be denied to the homeless and the illiterate. Blacks in South Africa cannot vote. The poor are powerless to represent themselves.

Income
The majority of people in developing countries manage to survive on an income of under $100 per annum – most on less.

Credit
A few reasonable credit schemes have been set up for the urban poor. There is little hope for those in remote regions.

Fuel
Over 2 billion people rely on wood for household fuel – but the supply, for 70% of them, is insecure.

Percentage of national populations
in absolute poverty, 1980

	65% and above
	45 – 64%
	25 – 44%
	5 – 24%
	Below 5%
	Unavailable
	Countries with under $400 GNP per capita, 1980

Measures of poverty
Zambia: child mortality is 11 times higher than in Germany.
Mali: average life expectancy is 30 years less than in Sweden.
Ethiopia: only one in seven can read.
Tanzania: one doctor per 17,500 citizens.
India: average calorie deficiency of 23%.

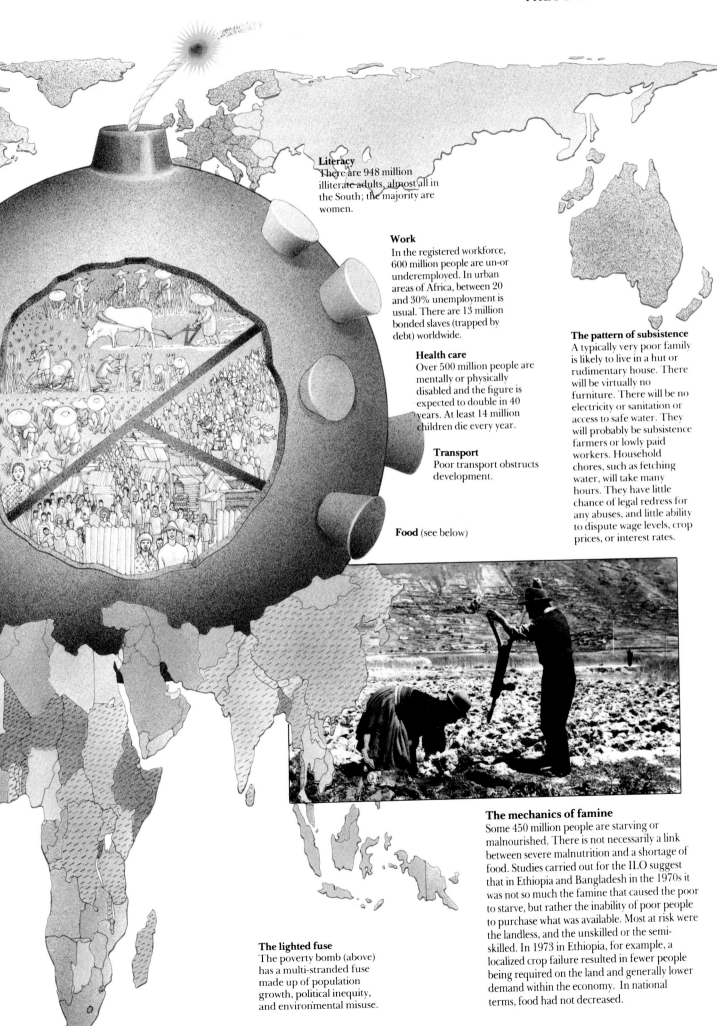

Literacy
There are 948 million illiterate adults, almost all in the South; the majority are women.

Work
In the registered workforce, 600 million people are un-or underemployed. In urban areas of Africa, between 20 and 30% unemployment is usual. There are 13 million bonded slaves (trapped by debt) worldwide.

Health care
Over 500 million people are mentally or physically disabled and the figure is expected to double in 40 years. At least 14 million children die every year.

Transport
Poor transport obstructs development.

Food (see below)

The pattern of subsistence
A typically very poor family is likely to live in a hut or rudimentary house. There will be virtually no furniture. There will be no electricity or sanitation or access to safe water. They will probably be subsistence farmers or lowly paid workers. Household chores, such as fetching water, will take many hours. They have little chance of legal redress for any abuses, and little ability to dispute wage levels, crop prices, or interest rates.

The mechanics of famine
Some 450 million people are starving or malnourished. There is not necessarily a link between severe malnutrition and a shortage of food. Studies carried out for the ILO suggest that in Ethiopia and Bangladesh in the 1970s it was not so much the famine that caused the poor to starve, but rather the inability of poor people to purchase what was available. Most at risk were the landless, and the unskilled or the semi-skilled. In 1973 in Ethiopia, for example, a localized crop failure resulted in fewer people being required on the land and generally lower demand within the economy. In national terms, food had not decreased.

The lighted fuse
The poverty bomb (above) has a multi-stranded fuse made up of population growth, political inequity, and environmental misuse.

MANAGING OUR CIVILIZATION

The increasing division of wealth and power in the world is inimical to all our interests. The North needs the South, if only because a large share of its export markets are located there: half Japan's and 40 percent of the US's. The South, equally, depends on the North. Interdependence applies not only to our global systems, but also within nations, and between countryside and cities. Lop-sided rural and urban development, with cities monopolizing government attention, has already proved a disastrous recipe. Reliant on each other, our management objectives and resources must be shared universally between nations, and between peoples and their governments.

Coping with city chaos

The management of city problems cannot be separated from wider issues – of income distribution (both between social groups and between nations), the international economy, sustainable development, and human values. There may well be many innovative schemes to improve life in cities, but they nearly all hinge on economic strength, on cities having the resources, and the will, to pay for their infrastructure and services.

What the poor of developing world shanty towns lack most is a voice in city management, and security of tenure for inhabitants who live in perpetual fear of eviction. Even the poorest settlement is bursting with human energy: provision of cheap materials and tools would enable families to build and improve houses. Above all, safe water and sanitation are critical, plus environmental controls to prevent pollution and uncontrolled dumping of waste.

Much of this can be done only with government backing, and within the framework of national development, since plans must strike a reasonable balance between the interest of the city and the countryside. Holding down food prices in towns, for example, may upset the rural economy; then even greater numbers of rural poor flock to the cities, swelling the shanty towns. Improving city health care, schools, transport, and other services must not be achieved at the expense of the countryside. In some Third World states, rural land reform and agricultural development have helped slow down the stampede to cities, and improved food supply.

Ultimately, much depends on national priorities – whose interests do the governments serve? The question applies just as much in the North, where financial constraints are far less acute. Northern

Urban regeneration

The science-fiction city of the future – Le Corbusier's grand schemes, or Niemeyer's Brasilia – seems ever less likely to replace our decaying Northern cities and sprawling Southern slums. A few new "open field" towns may be built as satellites to our biggest cities to fulfil this dream, like Reston or Columbia in the US; gleaming towers may rise in the South; but neither will house the millions of urban poor. The new urban strategies are aimed at mobilizing local communities and stretching scarce resources to cope with massive problems. Just as it is vital that the North responds to troubles in the South, so must more affluent citizens help the inner or outer city poor.

Liveable cities in the North

The massive slum clearance and building boom of the '60s and '70s are over and reaction has set in. The human costs of uprooting communities, to rehouse them in socially and constructionally disastrous high-rise blocks, are all too evident. Gradual renewal of our decaying city centres is now under way, through re-use of existing structures plus more sensitive new architecture. The renewal is often community-based (as in Watts, Los Angeles), and many small agencies have sprung up to help municipal and private efforts. Nothing, however, can replace major long-term investment by governments, to deal with obsolescence and disrepair.

Community participation

Encouraging a dialogue between city officials and shanty town dwellers can produce more effective initiatives than top-down planning. Redirected, local skills and organization can carry out low-cost schemes on a large scale, as in El Salvador. Establishing local administrative centres helps to focus community spirit, and allows a degree of self-management.

Urban employment

Current thinking aims at providing incentives to employers, through aid and technical advice, and the provision of small workshops. The World Bank now funds many such schemes. The huge informal economy of many of the South's largest cities (between 40% and 60% in Jakarta, Bombay, and Lima, for example) is a major provider of jobs, and at present receives negligible support through government credit. There is a great demand for loans and sites for small businesses, for the establishment of credit union savings – and vocational training to help the urban poor.

Targeting the inner city

Bringing the decaying areas of older cities back to life typically requires a multi-pronged approach, with inputs of private, voluntary, and official funding and effort. Rehabilitation of usable existing housing is matched by small "in-fill" schemes of new homes, harmonizing with existing street architecture; conversion and re-use of obsolete industrial buildings provide space for small workshops and businesses, or for commercial or community centres, from markets to sports halls. Local employment projects and light-industry units help to revitalize the community, which also needs new open spaces, communal facilities, and environmental improvements, from pedestrian or limited car access schemes, to tree planting and gardens. Based on these approaches, the "village in the city" is emerging all over Europe, and in US cities from Philadelphia to San Francisco.

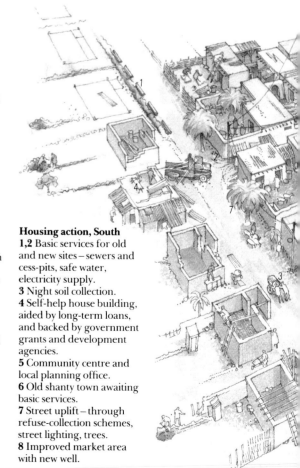

Housing action, South

1,2 Basic services for old and new sites – sewers and cess-pits, safe water, electricity supply.
3 Night soil collection.
4 Self-help house building, aided by long-term loans, and backed by government grants and development agencies.
5 Community centre and local planning office.
6 Old shanty town awaiting basic services.
7 Street uplift – through refuse-collection schemes, street lighting, trees.
8 Improved market area with new well.

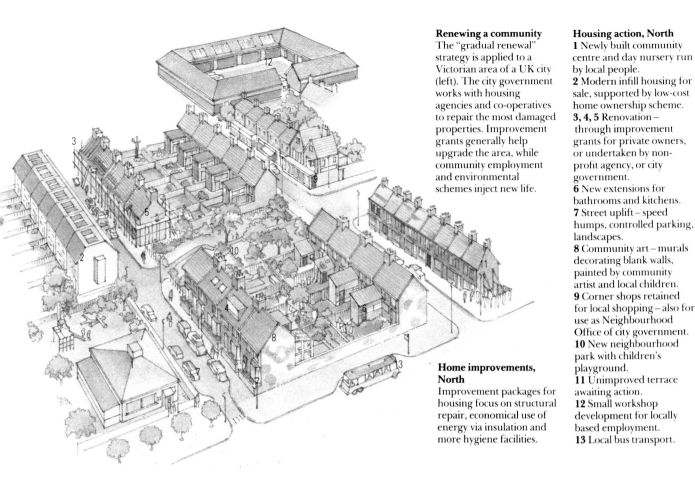

Renewing a community
The "gradual renewal" strategy is applied to a Victorian area of a UK city (left). The city government works with housing agencies and co-operatives to repair the most damaged properties. Improvement grants generally help upgrade the area, while community employment and environmental schemes inject new life.

Housing action, North
1 Newly built community centre and day nursery run by local people.
2 Modern infill housing for sale, supported by low-cost home ownership scheme.
3, 4, 5 Renovation – through improvement grants for private owners, or undertaken by non-profit agency, or city government.
6 New extensions for bathrooms and kitchens.
7 Street uplift – speed humps, controlled parking, landscapes.
8 Community art – murals decorating blank walls, painted by community artist and local children.
9 Corner shops retained for local shopping – also for use as Neighbourhood Office of city government.
10 New neighbourhood park with children's playground.
11 Unimproved terrace awaiting action.
12 Small workshop development for locally based employment.
13 Local bus transport.

Home improvements, North
Improvement packages for housing focus on structural repair, economical use of energy via insulation and more hygiene facilities.

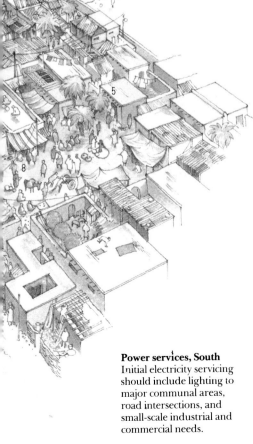

Power services, South
Initial electricity servicing should include lighting to major communal areas, road intersections, and small-scale industrial and commercial needs.

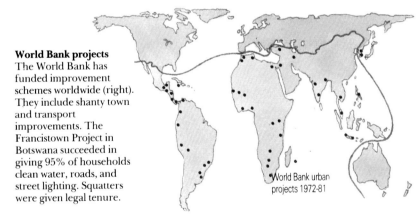

World Bank projects
The World Bank has funded improvement schemes worldwide (right). They include shanty town and transport improvements. The Francistown Project in Botswana succeeded in giving 95% of households clean water, roads, and street lighting. Squatters were given legal tenure.

World Bank urban projects 1972-81

Southern city regeneration
Whenever cityward migrations have reached unusual proportions, conventional housing and infrastructure services have been hard-pressed to cope. The many millions of poor people now crowding into Southern slums and squatter settlements cannot afford even the simplest permanent housing schemes (half Bombay's 11 million population live in slums, 350,000 sleep on the streets). Authorities are being forced to take a different line, tackling only the most basic provision themselves, and letting the settlers do the rest, with minimal aid. Just one of their intractable problems is that of water supply which is often privately owned – and very scarce. The most urgent need, however, is for greater rural investment, to slow the flood to the cities.

In the last decade "sites and services" schemes have concentrated on providing water, sanitation, street foundations, and power, but leaving construction of housing to individual occupants. This policy has evolved into "upgrading" of existing slums and shanties. One project in Lusaka, Zambia, in the '70s, tackled the upgrading and servicing of 31,000 plots, bringing basic needs to about 30% of people.

As in the North, community involvement and leadership are critical. El Salvador, prior to heightened civil unrest in the early '80s, boasted an almost model scheme – a local non-profit making group concentrating on low-cost housing, operating through long-term repayment, appropriate technology, and communal self-help.

But all these efforts must go hand-in-hand with better employment opportunities to generate income. Credit can be provided by quasi-governmental agencies; donor agencies can also play a useful role by helping to provide capital for small-scale industry.

cities too require investment for development, plus grants, loan-credit schemes, and enterprise zones to attract new business.

Strategies for communication

The power of advanced communications technology for assisting development is immense. It has been used to study and control locust damage in Africa; for the better management of rangelands in Kenya; in the study of the humid tropical forests of Peru. It is helping in conservation, the design of road and rail networks, mineral prospecting, and weather forecasting. Apart from land and resource management, satellite technology helps in education and can transmit medical, nutritional, and agricultural advice to large numbers of people, often otherwise isolated.

If the communications revolution is to benefit everyone, however, ways must be found of making the technology widely available, ensuring universal access to information, establishing a diversity of information sources, and achieving maximum participation in the transmission of ideas.

Advances in satellite technology help reduce the costs involved. Countries that cannot launch their own satellite may be able to pay others to launch them. They can also rent communications services from the West's Intelsat or the Russian Intersputnik systems for telephone, telex, data, and video transmissions. Regional co-operation is the most effective way that Third World nations can utilize the space potential. Colombia, Mexico, and Brazil have plans to launch their own satellite and provide a telephone and television link with remote rural areas. It is suggested that the UN should create a Centre for Outer Space, which would provide help and services to member states.

To make full use of the information gathered by satellite, developing countries must have the capacity to process and analyse it. In the field of data processing, international rules for the regulation and control of data flows must be devised. These should encourage multinationals to share skills, technology, and data with the developing countries. Huge libraries of information made available at the touch of a few keys could help reduce the information gap between South and North; but it is just as important to obtain the appropriate information, and develop the capacity to analyse and apply it.

The communications revolution could break the information monopolies. Sweden has shown the way by extending its Freedom of Information laws to ensure private citizens' access to the databanks of government departments. But many democratic countries, notably the UK, resist such reforms. New information technologies are also creating a far greater diversity in the media: cable television, video cassettes, home computers, and the proliferation of software programmes will provide an immense choice, like books on a library shelf. For the future, the entire role of communications must also change,

The will to communicate

The importance of the free communication of ideas and information cannot be over-emphasized: it is a vital key to the management of our global crises. At present many voices – of minorities and of people outside the communications network – go unheard. It may take a single determined journalist reporting on a crisis – such as the situation in Kampuchea in 1982 – to release massive funds for aid. Or an investigative newspaper reporter to uncover governmental or business corruption, such as the Watergate scandal in 1974.

Getting access to the medium, as well as the message, is particularly important in the South. Developing countries need more satellites to multiply TV broadcasts, and allow telephone communication in remote areas. They must also be able to benefit from the abundance of information collected about Southern resources from Northern satellites.

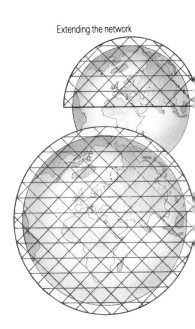
Extending the network

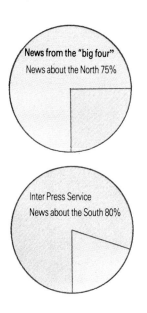

News from the "big four"
News about the North 75%

Inter Press Service
News about the South 80%

Reversing the flow

From a global viewpoint, news coverage and analysis of events in the South are as important to the North as vice versa. Since 1986, the Panos Institute has worked to help journalists ask the right questions about sustainable development, producing syndicated newspaper features and photographs, as well as books, magazines, videos, and radio programmes. Panos has established a regional partnership programme to strengthen the skills of media and private groups in the South. One developing world news agency attempting to break the barriers that prevent full North-South information flow is the Inter Press Service. Started in 1964 by a group of journalists writing about problems in Latin America, it has grown, with the aid of a 30% grant from UNESCO, into an organization with 90 offices issuing 100,000 words per day about the South to 26 national agencies worldwide. It gives a voice to the South, and within the South to rural areas and especially to women.

Satellites for education in India

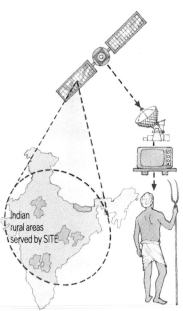
Indian rural areas served by SITE

Broadcasts by satellite reached 24,000 villages in six different states of India in 1975-76. The joint Indian-American Satellite Instructional Television Experiment (SITE) used the ATS-6 satellite provided by NASA, and demonstrated that India could manufacture and maintain both the required Earth stations and the community receiving sets in far-off villages. The communities, many of whom had never seen TV, showed large gains in knowledge about health and hygiene, family planning and political awareness from the educational and developmental programmes. The experiment also gave experience in designing and producing relevant programmes for widely spread areas with different languages and problems. Satellite TV broadcasts are immensely valuable to developing countries as they can reach rural communities, and cost a third of the price for ground-based programmes.

Breaking the monopolies

At present the information pool is held chiefly by governments and multinational companies, giving them massive power. They decide what data the satellites should collect, and they maintain control of the media. The increasing scope of their data pool makes reform an urgent priority. Properly managed, communications can prevent abuse and encourage public participation in government processes. Opening the sluice gates of the monopolies' dam (inset below) means pressing governments and companies to recognize the Freedom of Information movement, and encouraging the sharing of information technology.

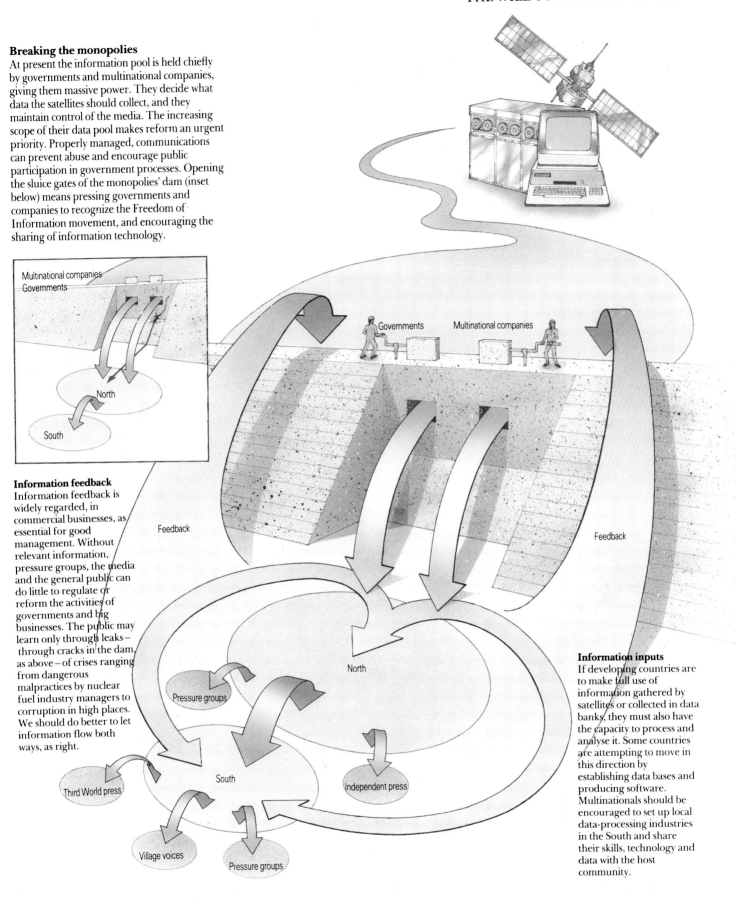

Information feedback

Information feedback is widely regarded, in commercial businesses, as essential for good management. Without relevant information, pressure groups, the media and the general public can do little to regulate or reform the activities of governments and big businesses. The public may learn only through leaks – through cracks in the dam, as above – of crises ranging from dangerous malpractices by nuclear fuel industry managers to corruption in high places. We should do better to let information flow both ways, as right.

Information inputs

If developing countries are to make full use of information gathered by satellites or collected in data banks, they must also have the capacity to process and analyse it. Some countries are attempting to move in this direction by establishing data bases and producing software. Multinationals should be encouraged to set up local data-processing industries in the South and share their skills, technology and data with the host community.

Independent media

Today the advent of citizens' band radio and cable TV promotes, in Alvin Toffler's phrase, the "demassification of society", discouraging tight control, while broadcasting minority views. Ghanaian national unity was noticeably fostered by different ethnic groups sharing the various media in the 1960s.

Village access

At grass-roots level, communications need not now demand high technology. A small 500-watt transmitter powered by two heavy-duty car batteries can broadcast over a range of about 20 km. Such a service provides information of all kinds, and stimulates communications skills. Local newspapers can be launched cheaply too.

Access to Information Technology

Sharing technology regionally offers most independence, while reducing costs. Some Arab states are planning regional telecommunications and the exchange of TV programmes. PADIS, the Pan-African-Documentation and Information System, pools data for the use of African technicians, planners, and others.

from a one-way flow, from North to South, to full dialogue between North and South, so that the current information advantage of the rich societies can be shared with the rest of the world.

Towards new technologies

The technologies developed by the North since the Industrial Revolution may sometimes seem like a Pandoran box of mischief and sorrow, accompanied by a forlorn spirit of human hope. Many of the latest fruits of our civilization fly in the face of conventional wisdom and humanitarian concern – for instance, the latest refinements in missiles of destruction, 'destructive nuclear experiments, the debate over the use of human embryos in scientific experiments – all raise alarming questions about science and the needs of humankind. Modern technology may even be seen as a means of magnifying the chasm between the world's élite and the poor majority – enabling Northern governments and companies to exercise "world control".

It is a truism, however, that no technology is intrinsically evil – like information, it depends on who pulls the strings. Undoubtedly, scientific research and development have produced stunning solutions to disease, food production, and problems of communication. The benefits of satellites, for instance, have already been applied to specific communications problems in the South (pp 220-21). The immediate issue is not only the nature of some

The technology circuit
Technology no longer flows in one direction; the current may be reversed. Technology transfer is becoming a dialogue. As developing world governments and tribal communities discover the value of their knowledge, it will become a potent commodity. The knowledge held in shamen's huts and in farming communities may be as vital to planetary survival as that held in Northern laboratories.

Technology transfer

Technology is transmitted in many different ways, through publications, personal exchanges, development aid, or outright piracy. Most, however, is transferred commercially. Far more attention needs to be paid to the usefulness of the technologies transferred to the South, to ensure that they are not, unwittingly, the 20th-century equivalent of the Trojan Horse – bringing unsuspected political, social and environmental problems in their wake. At present the rate of transfer from the North and from transnationals is lamentably slow. We need to find ways to accelerate transfer, build up Southern research and development, nurture new skills, and lift those controls on information that impede the process.

Southern technology impetus

Targeting the South's "technofamine" requires massive input from both North and South. Barely 1% of current Northern research and development (R & D) spending is focused on problems in the South (while 32% is devoted to defence and space research). In order to build up R & D capacity, by the South, for the South, governments and Southern businesses must persuade megacorporations to play a significantly responsible role in North-South relations. Technological aid efforts should be extended, and Southern universities must be strengthened. When curricula reflect specifically Southern needs, they can expand existing schemes for bringing technology to the village (such as India's Peoples' Science movements). However informal the platform, education to help leap the current technology gap is vital.

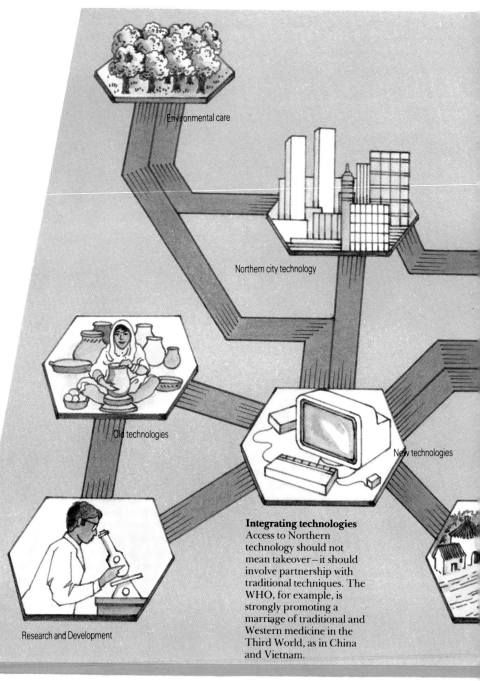

Environmental care

Northern city technology

Old technologies

New technologies

Research and Development

Integrating technologies
Access to Northern technology should not mean takeover – it should involve partnership with traditional techniques. The WHO, for example, is strongly promoting a marriage of traditional and Western medicine in the Third World, as in China and Vietnam.

of our "advances", but the sharing of technological control. As the Brandt Commission concluded, "it can even be argued that their (the developing countries') principal weakness is the lack of access to technology, or command of it".

Many of the technologies currently transferred, including those in the nuclear power and weapons fields, reflect this imbalance, and have proved expensive, almost useless, or positively harmful in the South – despite the myriad efforts of organizations which develop and transfer appropriate technologies. International organizations, however, are taking a growing interest in the social, economic, and environmental implications of technology transfer. The ILO, for example, has been conducting a worldwide pro-

gramme of research on the blending of emerging technologies, such as solar energy, micro-electronics, biotechnology, and new materials, with the traditional technologies of the South. A range of pioneer projects explore technological fusions that could boost productivity and competitiveness, without causing massive unemployment or any other dislocation in the process.

We should be *en route* for a polytechnical world, with space-age know-how partnering traditional technologies, with high and low technologies operating together, and labour-intensive techniques and automation taking their appropriate places. But all developments must be controlled and conditioned by a planetwide perception of our technologies as

Advances in the North
Having experienced some of the worst ill-effects of industrialization, Northern governments are just beginning to lead the way in environmental protection, pollution control, decentralizing information technologies, and the exploration of renewable energy sources – such as solar, water, and wind power. In a better world, such advances, and certain lessons of the Northern industrial experience, should be transferred to the South.

Transnational transfer
Transnationals should be obliged to share their expertise with the developing countries where, at present, some have a record of excessive profiteering and exploitation. They must help their host nations build up their own technology bases, both in fostering pure research and projects of applied technology.

"Outposts" in the North
The African Centre for Technology Studies has established a Biopolicy Institute in Maastrict in the Netherlands to carry out searches in the fields of biotechnology and biodiversity conservation. "Outposts" like these, created by the South but based in the North, should help developing countries search out and acquire environmentally-sound technologies instead of relying on the North finding and supplying them.

NORTH

llution control

Agriculture and countryside

Multinationals

Patent release
Patents, protecting new products and plans against direct imitation, are mostly registered in the North. The World Intellectual Property Organization aims to develop model patent laws, tipping the balance back towards the South by limiting the duration of patent cover on some types of product.

Environment

SOUTH

age technology

City technology

Agriculture and environment

Appropriate technology
Northern technology, based on Northern infrastructures, social patterns, and education, is not always immediately relevant to Southern needs – particularly to local productivity and employment. In the 1960s, the late E. F. Schumacher, trying to envisage a more practical aid, came up with the ideal of appropriate technology. Despite criticisms – that the South needs more than recipes for small-scale enterprise – the fruits of appropriate technology have been widely accepted. The Intermediate Technology Development Group (ITDG) helps foster hundreds of schemes, from local power generators to all kinds of low-cost equipment for small-scale use.

double-edged swords, constructive or destructive. We need to concentrate on resource efficiency, pollution control, and halting and repairing environmental degradation. We need a Gaian technology that is sustainable, energy-efficient, diverse, non-toxic, peaceful, and people-centred.

Trade and development

Nearly every attempt to redress the world's many interlocking crises is threatened by the current market crisis. Any solution to the growing global economic malaise must involve a recognition of the mutual interdependence of North and South, both in establishing demand and in sharing the means of supplying it. Confusingly, an enormous range of possible approaches to this end are being proposed, three of which merit immediate discussion. They can be summarized as "the market knows best", "the market must change", and "the South must delink its markets from those of the North".

As far as the "marketeers" are concerned, the basic problem is seen to be excessive government interference, which has prevented the proper functioning of the market. Price controls, protectionism, and subsidies have given the wrong signals to entrepreneurs, leading to the erosion of incentives and the misallocation of resources. Marketeers think in terms of a development "continuum", running from the poorest to the wealthiest countries – with the implication that the poorest can pull themselves up by their bootstraps.

In contrast, the "reformers", whose views have been convincingly stated in the two reports of the Brandt Commission, argue that nations do not compete as equals – and that global measures are needed to mobilize the weaker nations. Even if the free market could bring benefits to the poor, the reformers stress that this would take too long.

The "delinkers" are essentially a coalition of those who have been disillusioned by failures to win reforms and those who believe that the international capitalist system, whether reformed or not, is inimical to the interests of the developing countries. Some delinkers would go for a total delinking of the South from the rich economies of the North, while others would prefer to see a greater stimulation of South-South trade and co-operation. Some also believe that this approach would ultimately benefit the North, since, while the Northern share of trade would be proportionately smaller, the global economic "cake" would be bigger.

Until the early 1980s, the reformers probably represented the biggest constituency, although the centre of gravity has shifted towards market-orientated solutions. The prospect of continuing economic problems, with erratic growth rates in the North and protracted debt problems in the South, suggests that the plight of the South could well deteriorate further in the near future. There is a growing consensus that any attempt to manage the global market in its present state must

Growing interdependence

Some Northerners believe that they can get along without the South. They are dangerously mistaken. The North badly needs the fuel, minerals, and foodstuffs it imports from the South, as well as the cash it earns from exporting goods and skills. Some 40% of all North American exports of manufactured goods go to developing countries, 50% of Japan's. If demand falls in the South, the effects are soon felt further North. North and South often like to imagine that they are separate, with complete freedom of movement, yet nothing could be further from the truth. Self-sufficiency may be appropriate in some cases, but at the global level we must recognize growing interdependence.

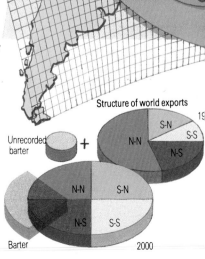

Scales showing percentage change in regions' shares of world exports (1970-82)

-20%   0%
-15%   +50%
-10%   +100%
-5%   +150%
              +200%
0%   +250%

Fairer trading
Tariffs, "invisible" protectionist barriers, subsidies, adverse terms of trade, unstable pricing – all are interlinked. They are both the symptoms and the prime causes of the struggle for shares of a shrinking cake. Dumping, and North/South export/import ties, also act as disincentives to domestic food production. Before the world cake can grow, fairer trading must become a common goal: we must roll back trade barriers, index-link commodity prices to manufactured goods, realign the industrial base – and above all, encourage domestic development and food sufficiency.

Sharing the bigger cake
Present world trade includes a little known sector – barter, which, broadly defined, may now account for over 20% of world trade. Rubber, for instance, is exchanged for cocoa beans, timber for transport equipment, and just about everything for oil. Most of this trade takes place between developing countries, who lack the foreign exchange to employ conventional trade methods. This suggests that emphasizing barter may be one way to stimulate greater South-South trade. However, the very nature of barter agreements makes them difficult to quantify, or plan. The pie-charts (right), compare the present shares of the export "cake" with a possible future division: by increasing South-South and South-North trade and boosting the barter economy, the whole trade flow could be increased.

Structure of world exports

Unrecorded barter

Barter
2000

Trade in the 1980s

During the 1980s, the industrial countries' share of world trade steadily increased. North America's share of exports fell again in the mid '80s before increasing slightly; the former Eastern bloc continued its downward path as Japan's share rose to a peak of almost 10% in 1986. Remarkable turnarounds were seen by Western Europe and the Middle East. Europe increased its share by almost 20%; while the major oil exporters saw their share of world exports crash from 16.4% in 1980 to 5.3% in 1989. The other regions of the developing world also witnessed falls in their share of world exports: by 30% in Latin America; 15% in Asia; and a disturbing 60% in Africa.

Delinking the South from the North

The traditional market structure established during the colonial period is changing fast – hence the rash of "new protectionism". Some economists, however, argue that the North, in order to maintain its position, must keep the poorer South in a state of "economic servitude" – and will never voluntarily reform the biased market. The only hope for the South, some believe, is a complete "delinking" of its economy from that of the North, and freedom from the stranglehold of international capitalism.

"Economic miracles"

Japan is the most obvious example of explosive economic growth; Germany, some OPEC nations, and a few new Japans are also contenders. Such "economic miracle" countries invade established markets and cause much disruption. They must also take a growing share of world responsibilities.

Former USSR and Eastern Europe

Western Europe

Middle East

Africa

Asia

JAPAN

Japan

Percentage share of world exports (1989)

New NICs

Old NICs

Self-sufficiency

Barter

Oceania

The new industrial nations

In Hong Kong and South Korea, 32% and 27% of the workforce, respectively, is in manufacturing. wage rates are rising, and the service sector is growing. A second wave of low-wage nations is taking over the labour-intensive products, among them India and China. They may realize new, more equitable, economic directions.

Self-reliance and South-South trade

Despite recent increases, South-South trade remains at a lower level than common sense would dictate. In 1980, it accounted for 30% of developing countries' total imports and 25% of their exports. Among the reasons: prejudice against so-called "second-best" products made by other developing countries; and poor transport and communications links – the roads built by colonial powers lead to coastal ports, rather than to neighbouring nations. Worse, many developing nations tend to produce competing, rather than complementary,

products, while the rich markets of the North seem to offer a better return. On the whole, regional groupings – such as ECOWAS in West Africa and Andean in South America – have proved disappointing. A system of generalized tariff preferences would help compensate for Northern trade barriers. A new "South" bank is also proposed, to offer buyers extended credit, and so help boost trade. Coupled with increasing technical co-operation between developing countries, such measures could help foster Southern self-reliance and expertise.

Safety nets for the South

There have been many international commodity agreements, but they have done little to stabilize prices. Arranging compensation for developing countries' lost export earnings, when prices fall, is one approach which has been tried by the Lomé trade/aid convention (p. 229). A more ambitious scheme might provide a safety net for all poorer countries, offering a guaranteed minimum income. A Common Fund for the South is also mooted: producers and consumers of the 18 top commodities exported by low-income countries could contribute to a central financing facility, buying up surpluses, and so maintain prices at an agreed stable level.

take a somewhat pragmatic approach – combining market solutions, Brandt-style reforms, and greater co-operation between developing countries. Success will increasingly depend on a willingness to dispense with economic dogma and experiment with alternative management approaches, learning from failures and building on success stories.

Managing the world's wealth

"No society can surely be flourishing and happy, of which the far greater part of the members are poor and miserable."

ADAM SMITH, THE WEALTH OF NATIONS

The stark contrast between the world's haves and have-nots, together with the short-fused "poverty bomb", are symptoms of a number of interlocking, mutually reinforcing crises which have already proved largely immune to a wide range of management solutions. The key question: how can we begin to close the gap between the rich countries of the North and the poor countries of the South?

In the 1960s, the emphasis was on rapid growth, in the belief that the benefits would "trickle down" to the poor. By the early 1970s, however, it was clear that this would achieve too little, too late. A seminal speech in 1973 by Robert McNamara, then President of the World Bank, signalled a switch to "growth with redistribution". The new objective was to ensure that the poor shared directly in the benefits of development through land reform, improved access to credit facilities, and projects providing jobs and adequate incomes to the landless and unskilled.

A key element in this new approach was the focusing of assistance on education, health care, and training programmes for the poor. Other concepts which emerged included the "basic needs" approach and, later, integrated rural development – but they were all overtaken by the world depression of the early 1980s. As one poor country after another ran into debt and balance-of-payments problems, so they were forced to cut social programmes, real wages, and imports, switching resources instead to their exporting sectors.

Clearly, if there is to be any hope of reducing world poverty, a massive effort is required across a very broad front. It must involve reforms at the international level, in trade and financial systems, coupled with far-reaching policies promoting income-distribution and a full-scale attack on poverty at the most localized level.

The international economic system must be redesigned to provide poorer nations with a fair return on their exports and the finance to invest in the sustainable development of their national resources, and this must be backed by aid programmes geared to more equitable development and the abolition of poverty. Voting power within the International Monetary Fund needs to reflect more fairly the interests of both North and South, to allow discussion and reform of loan conditions

Closing the gap

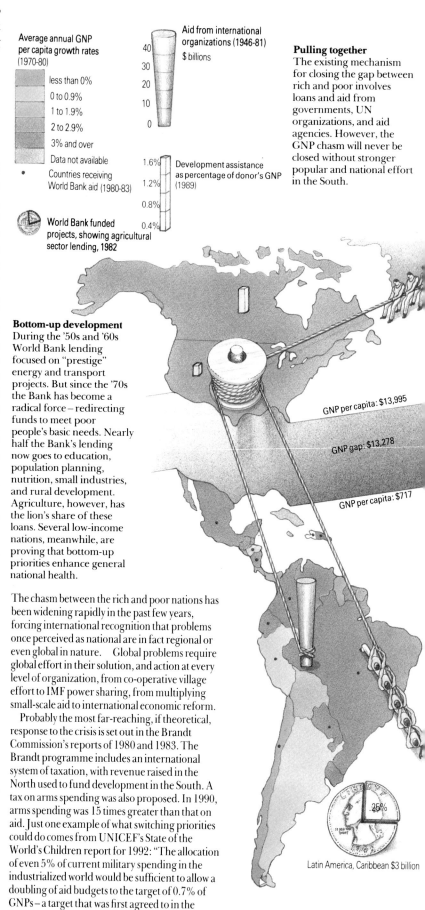

Average annual GNP per capita growth rates (1970-80)

- less than 0%
- 0 to 0.9%
- 1 to 1.9%
- 2 to 2.9%
- 3% and over
- Data not available
- Countries receiving World Bank aid (1980-83)

World Bank funded projects, showing agricultural sector lending, 1982

Aid from international organizations (1946-81)
$ billions
40
30
20
10
0

Development assistance as percentage of donor's GNP (1989)
1.6%
1.2%
0.8%
0.4%

Pulling together
The existing mechanism for closing the gap between rich and poor involves loans and aid from governments, UN organizations, and aid agencies. However, the GNP chasm will never be closed without stronger popular and national effort in the South.

GNP per capita: $13,995

GNP gap: $13,278

GNP per capita: $717

25%

Latin America, Caribbean $3 billion

Bottom-up development
During the '50s and '60s World Bank lending focused on "prestige" energy and transport projects. But since the '70s the Bank has become a radical force – redirecting funds to meet poor people's basic needs. Nearly half the Bank's lending now goes to education, population planning, nutrition, small industries, and rural development. Agriculture, however, has the lion's share of these loans. Several low-income nations, meanwhile, are proving that bottom-up priorities enhance general national health.

The chasm between the rich and poor nations has been widening rapidly in the past few years, forcing international recognition that problems once perceived as national are in fact regional or even global in nature. Global problems require global effort in their solution, and action at every level of organization, from co-operative village effort to IMF power sharing, from multiplying small-scale aid to international economic reform.

Probably the most far-reaching, if theoretical, response to the crisis is set out in the Brandt Commission's reports of 1980 and 1983. The Brandt programme includes an international system of taxation, with revenue raised in the North used to fund development in the South. A tax on arms spending was also proposed. In 1990, arms spending was 15 times greater than that on aid. Just one example of what switching priorities could do comes from UNICEF's State of the World's Children report for 1992: "The allocation of even 5% of current military spending in the industrialized world would be sufficient to allow a doubling of aid budgets to the target of 0.7% of GNPs – a target that was first agreed to in the 1960s".

A New International Economic Order

Through the 1970s and early 1980s, international organizations came up with a variety of blueprints for the future, spearheaded by the call for the New International Economic Order (NIEO). The NIEO comprises the ideal of international redistribution, provisions for basic needs and services, and the sustainable management of our natural resource base. It was adopted by the UN General Assembly in 1974, and by the Group of 77 (which includes 122 developing countries) in 1976; details include a fairer deal on commodity prices, a massive increase in Third World manufacturing, and a call for the financial resources necessary to effect the transition. In effect, the NIEO would be an international welfare state. But since the recession the North, with few exceptions, has rejected nearly all the Third World's proposals.

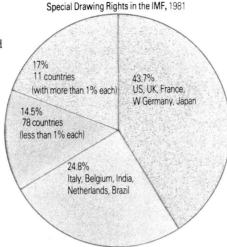

Special Drawing Rights in the IMF, 1981

- 43.7% US, UK, France, W Germany, Japan
- 17% 11 countries (with more than 1% each)
- 14.5% 78 countries (less than 1% each)
- 24.8% Italy, Belgium, India, Netherlands, Brazil

Reforming the IMF

Member nations of the IMF contribute to the Fund on the basis of their GNP, which also determines their "special drawing rights" (SDRs), left, and their *voting power*. Since the rich North dominates policy, bitter controversies have arisen. Third World borrowers complain that the "medicine" applied to their economies now reflects Northern monetarism and market-oriented interests – instead of development strategy. Shortage of foreign exchange, plus "conditionality", have spurred a Third World preference for private financial assistance. Whatever we make of the IMF's policies, it is plain that they should be arrived at by a fairer voting structure. Otherwise, the South's proposals to form their own Third World financing arrangements may become reality – dividing the world even further.

The Lomé negotiations

Through the four Lome Conventions, the most recent of which came into effect in 1991, 69 countries from Africa, the Caribbean, and the Pacific (ACP) have special trade/aid links with the European Community. Lomés objectives were to increase trade through reduced tariffs; the Stabex system aimed to stabilize export earnings; and the European Development Fund gives aid for specific projects. The first Convention was seen as a major step towards the NIEO. But aid funds have proved very low, and the ACP countries have been alarmed by the conditions set. World recession has resulted in a significant fallback from the original ideals.

23%

E Asia and Pacific $2.72 billion

18%

Europe, Middle East, N Africa $2.38 billion

30%

W Africa $1.1 billion

28%

E Africa $0.7 billion

34%

S Asia $3.12 billion

Targeting poverty

Direct action is top priority. Departments of the UN have specific projects – such as WHO's Expanded Programme of Immunization, which has immunized about 80% of all children against the major immunizable diseases, and the UN's target of Health for All (including clean water and adequate sanitation) by the year 2000. Charities such as Band Aid have also made an impact. But they all need more funds. Adjustments in government priorities, with health, education, employment, and food regarded as universal rights, may prove the best long-term solution.

"Sweat equity"

Raising the level of the rural poor is an essential prerequisite of any real achievement. Small farmers and village industries are only too willing to offer muscle power or "sweat equity", and apply new ideas – but they need fair terms and appropriate advice.

Defusing the debt bomb

Economic recovery and sustainable development in the South will not happen unless debt burdens are substantially reduced. The writing-off of some debts, as envisioned by the UK-proposed "Trinidad terms" is one way forward.

and the fund needs to be greatly increased. In an interdependent world, we cannot afford to "let the Devil take the hindmost".

New economics

"Trying to run a complex society on narrow indicators like "Gross National Product" is like trying to fly a Boeing 747 aircraft with nothing on the instrument panel but a single oil pressure gauge!"

HAZEL HENDERSON

The Greek root of the word "economics" essentially means "household management" – an appropriate meaning given our new awareness of the need to preserve the integrity of our one-Earth home. But our old Northern-dominated economic structure seems to many to be destroying, rather than managing, our "home", laying waste not only the environment, but whole societies of people.

Yet still we hanker after traditionally defined "growth", reinforcing the supposed link between economic production and welfare, and measuring development in terms of GNP and other material indices.

Such conventional economic indicators, however, cannot measure the effectiveness or stability of growth, nor who is benefiting. There is no recognition of the degree of freedom in society, or of access to media, or of income distribution. Neither do they take account of desertification, acid rain, or the hole in the ozone layer. For example, a country that has cut down all its trees and gambled away the money would appear from its national accounts to have got richer in terms of per capita GNP.

New economics aims to balance the present accounting system by attaching economic value to these "nonmarket" needs of people and planet. As the Brundtland Commission said: "No business can survive without a capital account. Neither can the planet".

A key premise of the new economics is to set economic growth at a level that is environmentally sustainable. In such a system, the cost of goods and services will bear a direct relationship to their environmental impact in production and disposal. Central also is the objective of eliminating the trap of poverty. The new economics aims to create a system which allows the poor to satisfy their needs, and reduces the gap between the haves and have-nots, both within societies as well as between societies.

Under a new economics, money should reflect wealth that is of real benefit to human welfare, and act as a stimulator of sustainable economic activity. What money most certainly is not, is wealth itself.

We need a new economics: one that is enabling, and allows people to enlarge their ability to support themselves sustainably, and control their economic lives. Its prime concern should be the well-being, not of corporations, or even nations, but of people and planet.

New rules of the game

If the world economy and its impact on people and planet can be considered a game, then a new set of rules for playing and scoring are long overdue. Indeed, even the concept of "winning" and "losing" requires a radical rethink. One of the main methods of scoring economic performance is the Gross National Product (GNP). This measure basically reflects the prices for which goods and services are sold. An increase in GNP is regarded as a good thing, since it indicates economic growth. What it fails entirely to account for is the environmental implications of that growth. As an example, in 1989 the *Exxon Valdez* tanker ran aground in Alaska, releasing 11 million gallons of crude oil and devastating one of the world's few remaining wilderness areas. The approximately $2 billion spent on detergents, labour, equipment, and transportation to clean up the spill counted, perversely, as an increase in GNP. Any rational system of accounting would score this event down heavily. New indicators are emerging, however. The UNDP's Human Development Index is one way of measuring growth and well-being, based on such factors as life expectancy, literacy, and purchasing power; World Bank economist Herman Daly's Index of Sustainable Economic Welfare accounts for environmental deterioration of air and water, cropland and wetland losses, and also elements such as the cost of commuting, car accidents, and income equality.

START
In the context of new economics, our game of Snakes and Ladders is played for high stakes – the sustainable existence of humankind. Landing on the tail of a snake leads us down the route of environmental and human degradation, while landing at the foot of a ladder points in the direction of new accounting strategies that recognize and endeavour to repair the true costs of our activities.

FINISH
Learning the rules of the new economics is essential. Fundamental to its success is acceptance that we are a single, global community; that inequality affects not only the poor but also eventually rebounds on the rich; and that all of us are dependent on the continuing health of Gaia. New economics is not the finish, it is indeed a new start.

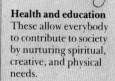

Social decay
Crime figures point to the decay of community life. In the US, homicides have doubled since 1960, as have rapes since 1970, while the drug trade is over $100 billion a year.

Health and education
These allow everybody to contribute to society by nurturing spiritual, creative, and physical needs.

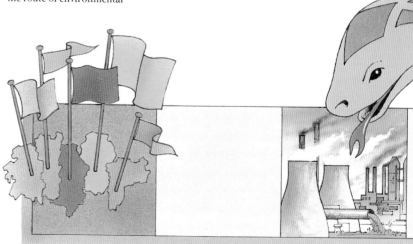

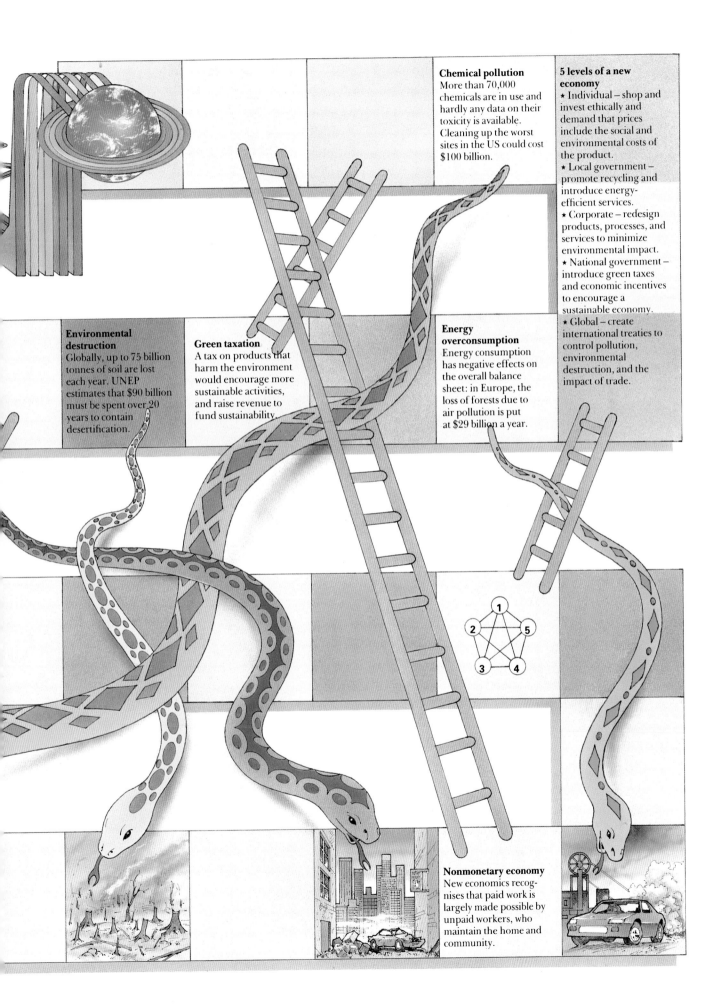

Chemical pollution
More than 70,000 chemicals are in use and hardly any data on their toxicity is available. Cleaning up the worst sites in the US could cost $100 billion.

5 levels of a new economy
★ Individual – shop and invest ethically and demand that prices include the social and environmental costs of the product.
★ Local government – promote recycling and introduce energy-efficient services.
★ Corporate – redesign products, processes, and services to minimize environmental impact.
★ National government – introduce green taxes and economic incentives to encourage a sustainable economy.
★ Global – create international treaties to control pollution, environmental destruction, and the impact of trade.

Environmental destruction
Globally, up to 75 billion tonnes of soil are lost each year. UNEP estimates that $90 billion must be spent over 20 years to contain desertification.

Green taxation
A tax on products that harm the environment would encourage more sustainable activities, and raise revenue to fund sustainability.

Energy overconsumption
Energy consumption has negative effects on the overall balance sheet: in Europe, the loss of forests due to air pollution is put at $29 billion a year.

Nonmonetary economy
New economics recognises that paid work is largely made possible by unpaid workers, who maintain the home and community.

MANAGEMENT

Introduced by Lois Gibbs

*Executive Director of the Citizen's Clearinghouse
for Hazardous Waste (CCHW), USA*

In 1970, people all across the United States joined in celebrating the first ever "Earth Day". In city after city, people marched, protested, celebrated, and spoke out in their own communities against the destruction of the environment. Denis Hayes, one of the organizers of that event, told a crowd in Washington DC to press government and industry to clean up their act. "We will not appeal any more to the conscience of institutions because institutions have no conscience. If we want them to do what is right, we must make them do what is right".

A surge in public concern led to many improvements. New laws were passed and strengthened, such as the Clean Water Act, and the Clean Air Act. Many new environmental agencies and groups were formed. The Environmental Protection Agency, the Environmental Defence Fund, Natural Resources Defence Council, and the Sierra Club Legal Defence Fund to name just a few.

Looking back now, it all seems rather impressive. But a careful analysis would have to include a few other observations. For all the new laws that were passed, several basic flaws remained. The government agencies began to permit and then regulate pollution. The new environmental movement quickly stagnated and grew sedentary. The passage of the new laws led these groups to replace direct action with arcane policy debates. As a result, a gap grew between the groups' grass-roots constituents and their leadership.

More than two decades later, a new environmental movement is growing. Today, more than 7,500 local community groups in the US alone are working to win "Environmental Justice". This is a movement of people that is defining the issues and setting the agenda in their own backyards, through democratically-based community organizations that are usually catalyzed by a local environmental threat.

It all started at Love Canal. This was a toxic landfill created by the Hooker Chemical Company who sold the land to the city of Niagara Falls in 1954. The city built a residential neighbourhood around the landfill, and a school on top of it. In the 1970s I lived in that neighbourhood and my concern for the health of the children led me to go door to door, asking people to help me build a group that could force action on the toxic site. From those early efforts, the Love Canal Homeowners Association was born. The group forced action by naming names and applying political pressure. For example, when New York Governor Carey announced that he would come to the canal and evacuate children under the age of two, we met him at the news conference and put three, four, and five year olds at his feet. With the cameras rolling, Carey was asked if he intended to leave these young kids at the site to die. With that, he agreed to evacuate more people.

Today, this movement of people continues to grow. Groups in the inner cities of the United States fight to stop a garbage incinerator, a radioactive dump, or a landfill. In East Europe, ecology groups played a major role in the democracy movements. Grass-roots action groups protest against the Narmada dam project in India. Many other examples are given in this book. Victories occur on a regular basis and carry one predominant theme. You can't fight city hall because it is a building, so name names and hit them where it hurts. Fight politically and don't let the opposition have one inch of ground. The health of the people is not open to negotiation.

Lois Marie Gibbs

THE MANAGEMENT POTENTIAL

We live in a world dominated by nations, about 200 of them, from micro-states to superpowers. And whether we like it or not, the nation-state remains the principal actor in international relations, and will remain so for some time to come. A few nations are old: Egypt has been a stable nation for 500 years. Most are new: some 150 are less than 40 years old.

A nation is typically a large group of people inhabiting a defined territory, ruled by one government, and united mainly by common descent, history, culture, language, or religion. In many nations, however, unity is no more than skin deep. Large minority groups, often called ethnic groups, effectively form nations within nations, but without adequate representation in the government.

Just how thin the veneer of unity within nations can be is shown by events in what was Communist East Europe and the USSR. Communism effectively kept the lid on nationalism in these countries. The collapse of these authoritarian regimes and the disintegration of the Soviet Union into independent republics, not only brought new freedoms, but released a tide of nationalist fervour that is a major source of conflict. Nationalistic tensions in Europe and a number of other regions are taking us into a highly-volatile period of history.

Nonetheless, at the international level the nation state has proved an effective unit, and an excellent vehicle for trade, commerce, and communications. No other available management structure rivals it as a basis for collective living.

Like any other family, it suffers from squabbles, rifts, and outright conflict. In the wrong hands, the nation state can be a deadly instrument of divisiveness. But however much national leaders may proclaim their rights of sovereign independence, their rhetoric is very often overtaken by their day-to-day dealings with one another. By virtue of trade flows, monetary patterns, inflation linkages, and myriad other relationships, nations are becoming ever more involved in each other's affairs.

Modern nations face many predicaments that can be tackled effectively only through collaborative effort and swapping independence for interdependence. These problems range from threats of nuclear war and terrorism, through energy and food-supply issues, to environmental problems such as acid rain and carbon dioxide build-up.

The process of the formation of new nations is not yet over. Ethnic conflicts will fracture existing nations and federations into smaller nation states. But

Phoenix nation states

We live in a time of revolutionary change, a time of much hope and opportunity. We are witnessing the rebirth of the nation state, like a phoenix, from the ashes of the Cold War. Old ideologies are swept away in a heady rush for freedom. Across the world, people are demanding a fairer, more equitable future, and seek this through self-determination, individual liberty, and the development of democratic institutions. Even totalitarian governments can no longer ignore popular uprisings. Yet the opportunity may be transitory. In many areas, persistent turmoil works against moves to democracy and encourages intervention by outsiders. Building a more stable, new family of nations will be very difficult. Governments seldom generate innovation, they tend to only react to pressure – from democratic process or revolution. The sudden implosion of an empire so huge as the former USSR is unprecedented. When this is accompanied by the chaos produced by the rapid transformation of a massive non-market economy to a free-market one, the resulting privation, internal economic crises, ethnic conflict, and large-scale disorder may well lead to the emergence of xenophobic and aggressive governments.

National time bombs
Ethnic conflict is the greatest source of tension in a number of regions and is hampering the growth of democracy. All of the countries in East Europe have significant populations of minorities. Yugoslavia, with no clearly dominant ethnic group, has fractured bloodily into a morass of independent states. In addition, there are Hungarian minorities in the Czech and Slovak republics and Romania, a large Turkish minority in Bulgaria, and over one million ethnic Poles are scattered throughout the former USSR. Ethnic violence has marred the birth of independent republics out of the former Soviet Union – as, for example, in Nagorny Karabakh, Georgia, and Kurdistan. Domestic ethnic conflicts in East Europe and the former USSR states are bound to lead to disputes between nations, some of which will almost certainly flare up into armed conflict.

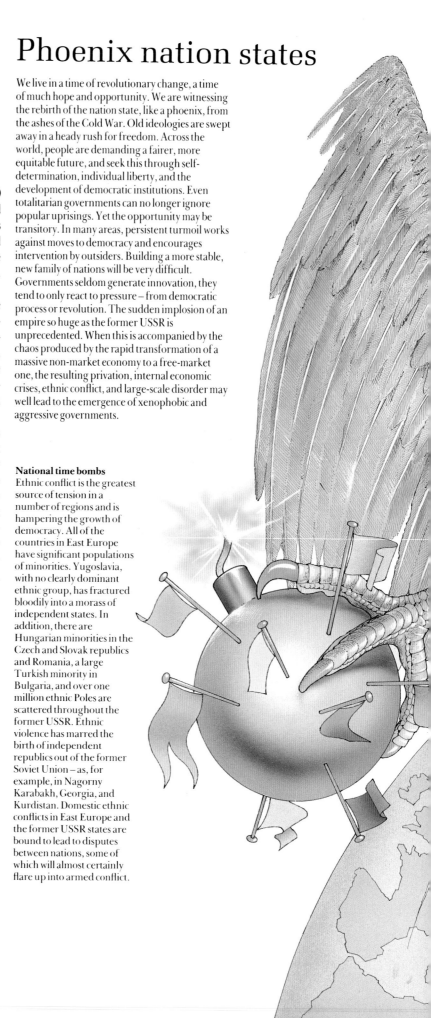

Local, regional, global
The nation state plays a
central role – the conduit
between local, regional, and
global levels of interest and
power. We need to ensure a
system of debate and
decision making with the
closest possible involvement
of those who will be affected.
Global problems need global
forums; but local matters call
for town or community
meetings rather than edicts
from a remote central
bureaucracy. The Swiss, for
example, have 25 cantons,
each with its own
constitution, and central
government acts as an
administration, consulting
the population by popular
referendum.

Emerging democracies
Authoritarian control is
declining. Across the world,
people are demanding
peace, democracy, human
rights, and effective
economic systems. A decade
ago 90% of Latin America
lived under authoritarian
governments. Now more
than 90% lives under
democratically-elected
governments. Communism
in East Europe and the
former USSR has collapsed.
Germany has been
peacefully reunified.
Elsewhere, communist
regimes in China, Cuba,
North Korea, and Vietnam
are under pressure to
change. Trends to
democracy are, however,
often vulnerable to social
pressures generated by
moves from statist to free-
market economic systems.

at the same time, governments are ceding some sovereignty to supra-national bodies. Both internal fracturing and external agglomerating are, perhaps, growing pains in the transition towards some kind of global system.

International and regional organizations

Despite the upsurge in the number of nations and in nationalist sentiment during the past four decades, there has been a simultaneous growth in internationalism. We now have a plethora of organizations whose sole purpose is to take care of our international needs.

Indeed, we can take heart from this growing web of international bodies, reflecting as they do the increasing interdependence of nations. No nation can afford to go totally its own way, not even the superpowers. Disputes between and within nations often depend for their outcome on world opinion – a fact of which national governments are all too well aware. Good public relations may win more battles than effective firepower.

Just as what Washington decides impinges upon London, Tokyo, Jakarta, and Rio de Janeiro, so the way that Riyadh acts affects Zurich, Nairobi, Moscow, and New York. This fast-growing interdependence of the community of nations is one of the most striking revolutions of our time. Far from feeling daunted by the increasing complexity of the global community, we should view it as a remarkably diverse resource to meet our ever more complex needs, with its individual pieces more important than the whole.

True, the two major attempts to establish an international order have stemmed from periods of acute crisis. World War I produced the League of Nations, World War II the United Nations. Today we face an even greater challenge: we need to produce a dramatically more capable system for collective endeavour *ahead* of any future crisis.

The omens are by no means all bad. Despite outbursts of hostility, the community of nations displays no small capacity for co-operation. Virtually every nation willingly operates within a network of international relationships. Already we have an array of regional bodies, from the Rhine River Commission to the Organization of American States (OAS), the Association of Southeast Asian Nations (ASEAN), the Organization of African Unity (OAU), and the Arab League. We have broader-scale consultative bodies, such as the Organization for Economic Co-operation and Development (OECD), and more cohesive groupings which begin to override national sovereignty, like NATO and OPEC. We have North-South organizations like the Commonwealth. And, at the global level, we have the United Nations, with some 30 specialized agencies: a system of near-universal membership for near-universal concerns.

Some of the elements of the one-Earth, one-people society are there. The question is how can we

Reluctant internationalists

The world is walking a tightrope: it needs a safety net. We risk not just war but serious environmental collapse, so long as we continue to view the major "organs" of the biosphere in terms of national self-interest. We have established safety nets before: the League of Nations, for example, emerged from the unparalleled bloodshed of World War I, the United Nations from World War II. But we cannot afford to wait for the nuclear convulsions of a possible World War III to put an improved safety net in place.

We already have many strands of co-operation from which this net can be woven. A central strand must be the UN, together with its myriad specialized agencies. Another possible strand is made up of the many geopolitical organizations which have emerged since 1945, including the EEC, OAU, OPEC, Comecon, and ASEAN, which need to be set into the new planetary pattern.

Every day, fine threads of global activity help pull us together, from international collaboration in space to the posting of an airmail letter. But our new safety net is far from complete – and in many areas already looks frayed. New strands must be added soon: we are a long way out on the tightrope.

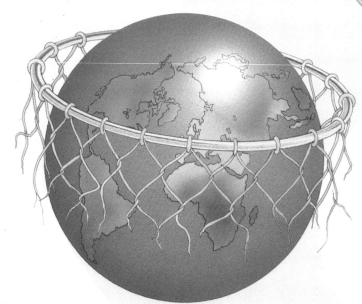

Fragile nets
Safety nets of the past have relied too heavily on fragile strands of political and diplomatic accords. The Congress system, fashioned by diplomats in Vienna in the wake of the Napoleonic Wars, held together an unsteady peace until 1914, when it was torn asunder. The League of Nations introduced some harder-wearing strands based on functional relationships, but the central strands were still predominantly political and diplomatic. This safety net was shredded by World War II. The third major attempt has focused much more on functional relationships, such as those tied together within such UN agencies as FAO, UNESCO, UNEP, and WHO. These strands help hold the half-completed net in place.

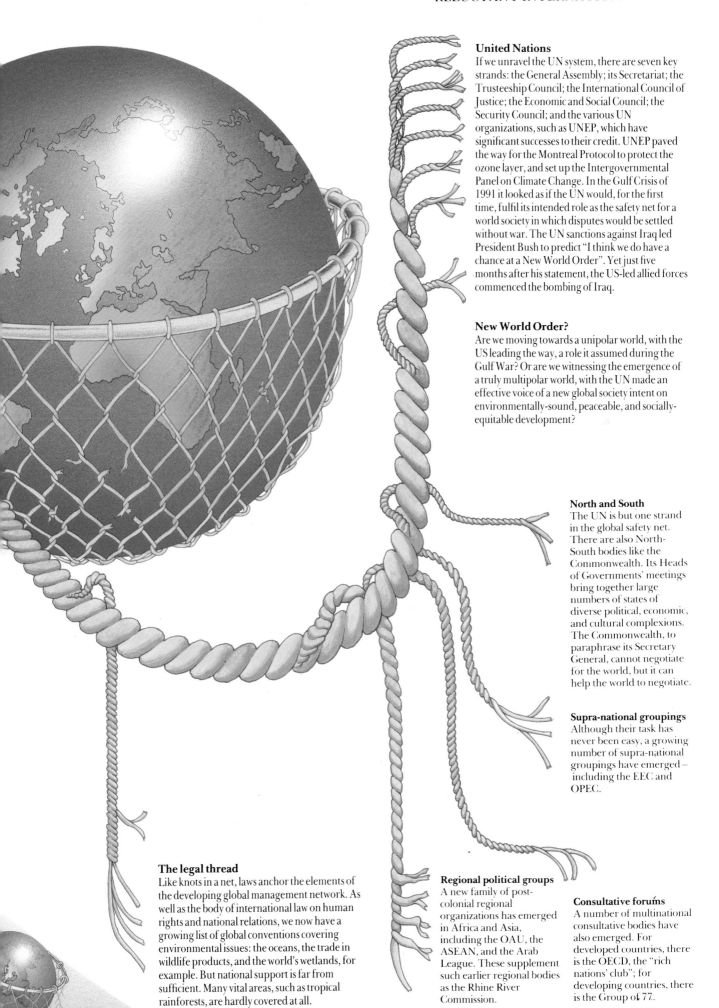

United Nations
If we unravel the UN system, there are seven key strands: the General Assembly; its Secretariat; the Trusteeship Council; the International Council of Justice; the Economic and Social Council; the Security Council; and the various UN organizations, such as UNEP, which have significant successes to their credit. UNEP paved the way for the Montreal Protocol to protect the ozone layer, and set up the Intergovernmental Panel on Climate Change. In the Gulf Crisis of 1991 it looked as if the UN would, for the first time, fulfil its intended role as the safety net for a world society in which disputes would be settled without war. The UN sanctions against Iraq led President Bush to predict "I think we do have a chance at a New World Order". Yet just five months after his statement, the US-led allied forces commenced the bombing of Iraq.

New World Order?
Are we moving towards a unipolar world, with the US leading the way, a role it assumed during the Gulf War? Or are we witnessing the emergence of a truly multipolar world, with the UN made an effective voice of a new global society intent on environmentally-sound, peaceable, and socially-equitable development?

North and South
The UN is but one strand in the global safety net. There are also North-South bodies like the Commonwealth. Its Heads of Governments' meetings bring together large numbers of states of diverse political, economic, and cultural complexions. The Commonwealth, to paraphrase its Secretary General, cannot negotiate for the world, but it can help the world to negotiate.

Supra-national groupings
Although their task has never been easy, a growing number of supra-national groupings have emerged – including the EEC and OPEC.

The legal thread
Like knots in a net, laws anchor the elements of the developing global management network. As well as the body of international law on human rights and national relations, we now have a growing list of global conventions covering environmental issues: the oceans, the trade in wildlife products, and the world's wetlands, for example. But national support is far from sufficient. Many vital areas, such as tropical rainforests, are hardly covered at all.

Regional political groups
A new family of post-colonial regional organizations has emerged in Africa and Asia, including the OAU, the ASEAN, and the Arab League. These supplement such earlier regional bodies as the Rhine River Commission.

Consultative forums
A number of multinational consultative bodies have also emerged. For developed countries, there is the OECD, the "rich nations' club"; for developing countries, there is the Group of 77.

expand the array and put together the parts of the jigsaw in a way which makes sense in both economic and ecological terms?

Twelve thousand alternatives

Impatient of many of its old hierarchies, the world has made a radical shift in the way it organizes its affairs. There has been an explosive growth in recent decades of small and large pressure groups, advisory agencies, research groups, technology and information specialists, aid charities, "network" associations, and professional bodies – the list seems endless. We now have about 400 significant intergovernmental bodies in the international field, and upwards of 4,000 other bodies, dubbed "non-governmental organizations" (NGOs), with memberships and budgets spanning several countries.

Some NGOs are groups of professionals which set international codes of practice, and some are religious in orientation, but the groups which have caught the world's imagination are those which have campaigned on specific issues - such as whaling, intermediate technology, Third World aid, and corporate responsibility. They range from development groups like OXFAM and CARE, to conservation-minded groups such as Greenpeace.

All such groups serve as rallying points for common concerns, disseminating information among their members, totalling many millions. They engage in "networking" activities, which allow them to operate as pressure groups, seeking to influence governments and international agencies. Their strength reflects their single-minded devotion to highly focused campaigns, backed up with specialist skills and information.

NGOs have had a strong impact. Environmental NGOs mobilized the public concern which resulted in the 1972 Stockholm Conference on the Human Environment. This led to the formation of UNEP, which has mounted many highly successful campaigns. Until about 20 years ago, NGOs were largely developed-world phenomena. Today, the developing world has thousands of NGOs and, in response to their lobbying, many developing world governments have set up environmental agencies.

NGOs also had a high profile at the 1992 Earth Summit in Rio de Janeiro. The NGOs attended their own conference, the Global Forum, where about 1,500 groups from 163 countries met. Events included meetings, workshops, debates, networking, and seminars. The Global Forum addressed physical, spiritual, and political issues not discussed by the Earth Summit.

Although NGOs lack political power, they exert a moral authority which cannot be ignored lightly. Most importantly, they can transcend national boundaries, prejudices, and politics by engendering a spirit of "global constituency".

Indeed, the spread of NGOs throughout the world represents one of the most hopeful portents for the future. They not only act as a conscience and

Voice of the world

The giant bodies of state and transnational industries are powerful but often slow-moving and cumbersome. But the world is beginning to find its voice and give direction to these elephantine structures. Ordinary people talk to each other across political boundaries through non-governmental organizations (NGOs), and ally themselves with other groups to lobby in the corridors of power. A striking symptom of this global "clearing of the throat" is the proliferation of NGOs in the last 30 years, ranging from the Red Cross to anti-Vietnam War protest groups, and current front-runners like Amnesty International, OXFAM, Greenpeace, Friends of the Earth, and World Wildlife Fund. There are now over 12,000 such groups, an embryonic nervous system for our global society, affording a degree of sensitivity unmatched by the nation states and their 2,000 intergovernmental agencies, which are too often tongue-tied by diplomacy.

The rise of the NGOs
The number of international non-government organizations (INGOs) with activities spanning more than three countries, rocketed to 4,779 by 1983 (right). If we add to this total another 1,119 groups which are not fully autonomous, 1,111 which have a more formal structure, 607 religious bodies, and 4,514 internationally oriented NGOs based in a single country, we find a grand total of some 12,130 NGOs.

1910   1920   1930

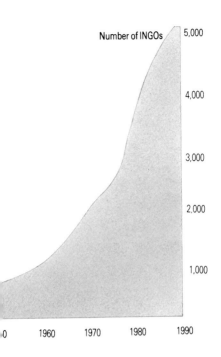

Number of INGOs

5,000

4,000

3,000

2,000

1,000

1960   1970   1980   1990

Planned parenthood
Fifty years ago, the dissemination of information about birth control was illegal in the US, and only 30 years ago, the population explosion was not considered to be a problem. Today, by contrast, birth control facilities are available in almost every country – and the majority of the world's governments have their own population programmes. Blazing the trail through this dramatic historical about-face has been the International Planned Parenthood Federation (IPPF), formed in 1952. After major clashes with opponents in government, the medical profession, and the Catholic church, the IPPF gained consultative status with the UN in 1964. The real turning-point came with the 1974 World Population Conference, the first to focus on birth control. Today, the IPPF has 114 member countries and a $55 million budget.

Voice of the corporates
The transnational corporations (TNCs) are enormous in both size and influence; just 500 TNCs control 70% of world trade. They also control whole industrial processes from shaping demand to resource extraction, from production to end use, and wield so much control through trade that they can influence governmental decisions. Their actions will dictate much of our future course to a sustainable society. Industry's response to the challenge has been to create the Business Council for Sustainable Development (BCSD) which spent two years compiling, for the 1992 Earth Summit, a report reconciling commercial objectives with the needs of the environment. It argues unequivocally for market forces, rather than domestic regulation, as the best means of promoting change. Environmentalists, not surprisingly, criticised this report for removing all positive references to government regulation, and are suspicious of industry-backed NGOs – such as Informed Citizens for the Environment, a US coal industry-funded organization with the aim of "repositioning global warming as a theory (not a fact)".

Environment-friendlier business
Traditionally, industry has been concerned about the expense of measures to prevent pollution. Now, however, consumers are becoming aware of the impact of their purchasing decisions, and avoid products that have an adverse effect. Faced with lost markets and increased competitiveness between corporations, industry is re-evaluating its processes and standards. One US-based company, 3M, has taken a lead in reducing the impact of some aspects of its operations. 3M produces a wide range of consumer goods and in 1975 introduced the Pollution Prevention Pays (3P) programme which is based on the concept that the best way to prevent pollution is not to generate it. By installing new machinery and implementing new processes, 3M have prevented more than 453,600 tonnes of waste emissions, and saved more than $500 million in the US alone.

A company which has protected and respected the environment from its beginning is The Body Shop. Anita Roddick opened her first shop in Brighton, UK, in 1976, and now has 726 outlets worldwide. The Body Shop sells cosmetics not tested on animals, uses only natural ingredients, and uses recycled, reusable packaging. Anita Roddick pays First World prices to Third World communities for goods, helping to set up a number of small businesses in developing countries and among rainforest peoples. In 1991 The Body Shop was valued at $900 million, with worldwide sales of $400 million.

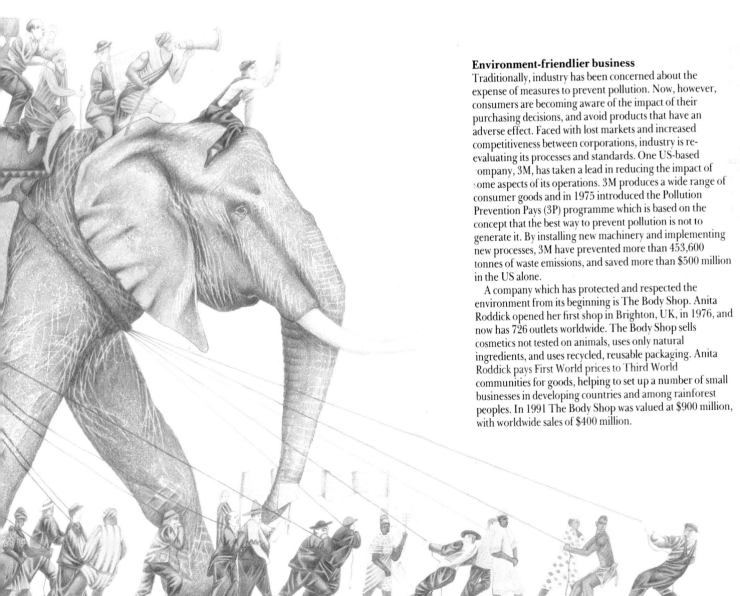

spur for the growing array of official agencies that act on behalf of the world community, but are the most visible expression of a growing global citizenry.

Broadening the base of power

Experts propose, people dispose. Without the spontaneous involvement of grass-roots communities, the best-laid plans of development technocrats often come to naught. By contrast, with active participation at "village level", whether it be in a true village, in a more extended farming community, or in an urban setting, modest means can achieve the most surprising results.

Such, as least, has been the experience of communities as far apart as Calcutta, Nairobi,

Glasgow, Watts (in Los Angeles), and the back-country areas of China, Sri Lanka, Costa Rica, and Burkina Faso. The traditional approach of "the expert knows best", sometimes called "top-down development", is giving way to a wholly new strategy known as "bottom-up development".

The growing enthusiasm for local self-help, with local people looking to outsiders for financial and practical support rather than for direction, represents one of the more striking advances of recent years. Local people, who have often lived with the problems at issue for decades, even generations, generally know where a problem's roots lie – and can differentiate between symptoms and root causes. Typically, too, they can see ways to solid,

Voice of the village

Why plant a new tree where an established, healthy tree already stands? Traditional communities, whether in villages or major cities, are like the established tree. Their roots are deeply anchored in the history and culture of the locality. Development proposals produced by remote government officials and development agencies often ignore these grass-roots management systems. Instead, they often try to transplant management systems developed in very different local circumstances. Now, as many of those transplants wilt in the heat of reality, there has been renewed interest in building on the base provided by some of the more traditional, local political and management systems. This is in no way a retrograde step – organizations like the UN must listen to what the people most affected say, and know, about the issues. But if this approach is to work, such systems must be given real responsibilities and backed with hard cash.

South Shore banks on the future
When the commercial banks pull out of a declining inner city area in the North, it is generally the last nail in the coffin. Chicago's South Shore community achieved an extraordinary victory when it refused to let the old South Shore National Bank move out. Local people took their case to the government, arguing that they needed the bank to help renew the neighbourhood. Federal regulators denied the bank's request to move, and the bank was bought by the Illinois Neighbourhood Development Corporation. All this was in the early 1970s: today the revitalized bank is serving as a powerhouse for the renewal of South Shore, boosting employment and funding housing.

workable solutions, which build on, rather than swamp, their own culture and experience. That a radical approach of this sort can be made to work is demonstrated by a host of recent initiatives, such as the Chipko and the People's Science movements in India, the Sarvodaya Shramadana or "Gift of Labour" organization in Sri Lanka, the Village Education Resource Centre in Bangladesh, and the Naam programme in Burkina Faso.

Some of this creative initiative is fostered by private activist bodies in the developed world, such as the Intermediate Technology Development Group in the UK and similar groups in Germany, the Netherlands, and the US. The essential element in these recent breakthroughs, however, lies in the willingness and capacity of local people to tackle their own problems.

Nor should we suppose that the "bottom-up" approach has been confined to the developing world. In the US and EC, for example, local citizens have taken matters into their own hands. Faced with the decay of the inner cities, of the areas and buildings in which they live, they have responded by setting up tenants associations and small-scale co-operatives. Local community groups have also scored remarkable successes in protecting habitats and opposing pollution from local factories and waste incinerators. In the US alone, 7,500 grass-roots groups are united in their call for "environmental justice".

Naam brings new confidence

The Naam movement began in the early 1970s in the Yatenga Province of Burkina Faso. The immediate cause was a prolonged drought, but the movement's roots are firmly entrenched in the local Mossi culture. The word "Naam" harks back to an old tradition of self-help and village- level co-operation. This was the weapon with which the movement's founder, Bernard Ledea Ouedraogo, aimed to defeat the crisis of confidence which he saw afflicting his people. This crisis had been brought on by years of environmental stress, in combination with a sense of impotence stemming from their dealings with apparently superior outsiders. Today, Mossi culture has been rekindled, and from about 100 Naam village groups in 1973, the number had grown to more than 2,500 by 1987. New dams have been built and new trees planted. Grain warehouses have been rebuilt, and villagers can now borrow grain in the difficult weeks before the harvest, repaying the loan in kind once the harvest is complete – saving them from predatory moneylenders.

Chipko protects the roots

When commercial loggers began large-scale felling of trees near to Reni, a village in the northern Indian district of Chamoli, their chainsaws were slicing through the very roots of local society and threatening the livelihoods of local communities. In an astonishing display of courage and determination, the Reni villagers wrapped their arms around the trees, to protect them from felling – sparking off the Chipko Andolan movement (the "movement to hug"). Eventually, following an inquiry, the government declared 12,000 sq km of the sensitive watershed region of the Alakananda basin "off limits" to loggers (see also p. 52). Today, the Chipko movement runs many reforestation programmes in other villages where livelihoods are threatened.

Clearly, political leaders, officialdom, and experts must "grow new ears" that pick up the voice of the village.

The basic unit of society

What do a community, a city, and a nation consist of? What makes them function, at the basic level? The answer: individuals and families – the most fundamental units of society. The nation state and the corporations may dictate the future course of development, but at a fundamental level it is the actions of individuals which drive them.

Over the last decade, the number of people living in liberal, multi-party democracies has increased. More citizens than ever before have the chance to support and join political parties, to stand for election, and to vote. In many emerging democracies, environmental concerns have been a strong motivator in people's desire for change. Expressed through the ballot box, these concerns gain power, even if the candidate many voters favour does not win. For example, the UK Green Party's capture of 15 percent of the vote in the 1989 European election was a spur to other parties to develop environmental policies. Political change or new legislation generally comes in response to public demand; the leaders follow where the people lead.

The desire for change is also expressed through consumer choice. Where people are fortunate enough to afford choice in their purchasing, a growing number expect products which are environment-friendly, in content and production. The "green consumer" movement has exercised considerable influence on industry. There has also been a positive response within the financial sector. Many investment companies offer ethical trust funds, which will invest your money in environment-friendly companies.

Many individuals have made changes in their own lives, realising that the accumulation of individual consumption places pressures on the planet. The consumer demand for bottle banks, paper skips, can bins, and so on, often outstrips supply. Consumer boycotts, such as that of South African produce while apartheid remained, are powerful tools, reminding producers that they can only be profitable if there is public demand.

For the profligate rich countries, a reduction in energy demand is essential. Many individuals have responded to this by using energy efficient products, and reducing overall consumption. More people travel by bicycle or public transport as they realise the full environmental impact of car use.

In many ways, it is actions by individuals which drive the political choices and the direction of industry. The next stage may well prove more difficult – to reduce the overall consumption of the rich countries. It is easier to care for the environment if all it involves is a change of brand. It is more difficult to stop an environmentally harmful process altogether, or to do without it. It will not be enough to consume better, we will have to learn to consume less.

Voice of the individual

The individual and the family represent the basic units of society, at the centre of a network of change. They find themselves with three roles: consumer, producer, and participator. Each of these roles is enmeshed in economic, ecological, and ethical spheres of influence. The last decade has seen the voice of the individual grow in power to become an important force for change. The spreading awareness of our relationship with the Earth has resulted in an unprecedented pressure for change on governments, producers, and the service industry. For example, many individuals are favouring "green" trust funds that avoid companies with links with the arms trade, repressive regimes, or the exploitation of animals, and instead invest in companies with positive policies towards the developing world and worker involvement. Just as the ethics and values of a society are transmitted through its education system, so we must ensure that young people learn how to implement change by individual action.

Ethical investment
In the US more than 10% of the money passing through Wall Street is ethically invested. In the UK, the successful organization EIRIS advises individuals, churches, trades unions, pension funds, and stockbrokers.

The four Rs
We have to move from a consumer to a conserver economy. That means learning the 4 Rs. *Refuse* produce if it is inefficient, will not last, or we can do without it. *Reuse* things as much as possible before disposing of them. *Repair* a product so that its life is prolonged and work may be generated. *Recycle* items to complete the resource loop.

The spiritual dimension
All the great religions support the concept of "enough" rather than ever-growing consumption and wealth. Multi-faith gatherings round an environmental agenda are becoming more common as people realise the need to re-establish their spiritual relationship with the living world.

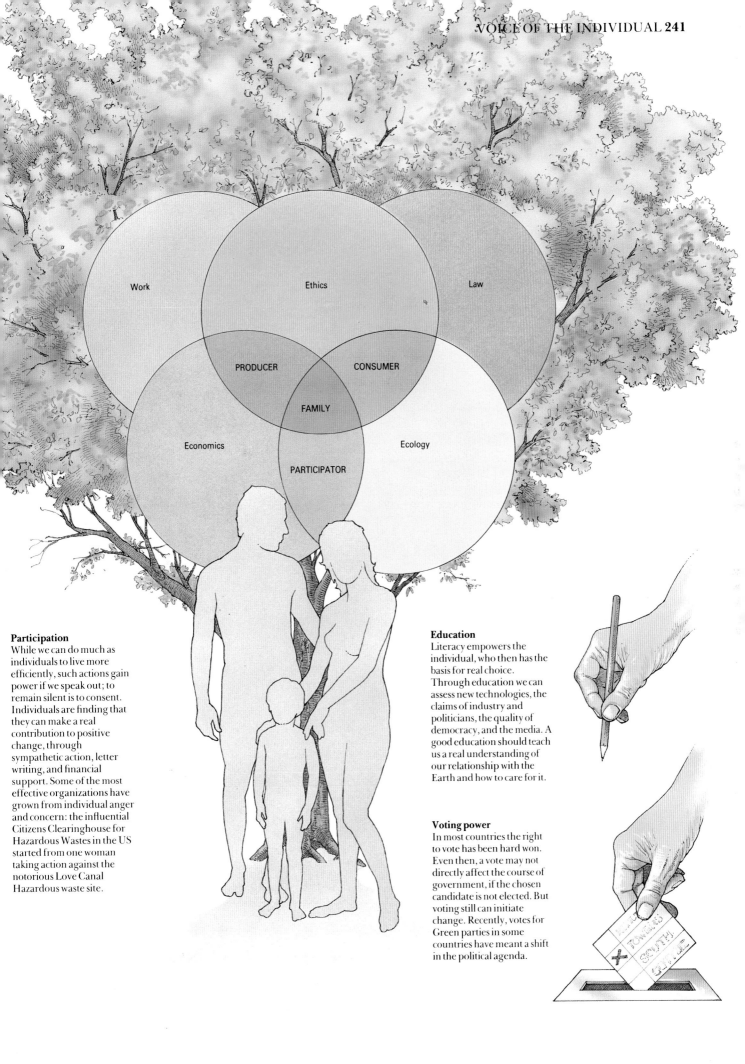

Work

Ethics

Law

PRODUCER

CONSUMER

Economics

FAMILY

Ecology

PARTICIPATOR

Participation
While we can do much as
individuals to live more
efficiently, such actions gain
power if we speak out; to
remain silent is to consent.
Individuals are finding that
they can make a real
contribution to positive
change, through
sympathetic action, letter
writing, and financial
support. Some of the most
effective organizations have
grown from individual anger
and concern: the influential
Citizens Clearinghouse for
Hazardous Wastes in the US
started from one woman
taking action against the
notorious Love Canal
Hazardous waste site.

Education
Literacy empowers the
individual, who then has the
basis for real choice.
Through education we can
assess new technologies, the
claims of industry and
politicians, the quality of
democracy, and the media. A
good education should teach
us a real understanding of
our relationship with the
Earth and how to care for it.

Voting power
In most countries the right
to vote has been hard won.
Even then, a vote may not
directly affect the course of
government, if the chosen
candidate is not elected. But
voting still can initiate
change. Recently, votes for
Green parties in some
countries have meant a shift
in the political agenda.

CRISIS: THE THREAT OF WAR

We are reaching the ultimate breaking point. If we do not "get our act together" fast, we shall suffer the consequences of management breakdown - confrontation rather than co-operation, conflict rather than harmony, and in the end, war of an intensity that we have not hitherto envisaged. What greater incentive could we possibly have to accomplish planet management?

The crisis of perception

As we over-stress the natural resource base of our planet, we set up strains and tensions that lead to fractures in our societies. Often, these environmental "breaking points" trigger conflicts, both within and between nations.

Human communities depend for their livelihood on a range of basic resources, on living resources such as grasslands, forests, and soils, and on nonliving resources such as water, fossil fuels, and minerals. If a community depletes its endowment of resources, and/or is denied fair access to resources elsewhere, its economy is undermined, its political structure becomes destabilized, and its social fabric starts to fray.

Such problems are greatly compounded by the nation-state system which has dangerous limitations when it comes to managing the emerging threats to our planetary life-support systems. National leaders, with honourable exceptions, have been slow to recognize the new sources of conflict: there is, in short, a crisis of perception which could prove the most critical of all. For without the political capacity to adapt to new threats, we face a continuous slide into chronic anarchy and endemic violence - as manageable problems deteriorate into unmanageable conflicts.

To focus on one of many problem areas, consider Central America. Whatever the ostensibly political nature of the upheavals that have afflicted this troubled region, the central problems can often be traced back to the maldistribution of resources. The grossly inequitable distribution of farmland, for example, applicable to most of the region's countries, means that conflict is perpetuating – and will continue to worsen as long as the problem remains unrecognized by political leaders. Even the 1983 Kissinger Commission, a major effort to come to grips with this inherent regional instability, failed to acknowledge this basic cause of recent upheavals.

This tragic story of misperception is being replicated in several other critical regions of the world,

Breaking points

"In the global context, true security cannot be achieved by a mounting build-up of weapons (defense in a narrow sense), but only by providing basic conditions for solving non-military problems which threaten them. Our survival depends not only on military balance, but on global cooperation to ensure a sustainable biological environment."
REPORT OF THE BRANDT COMMISSION

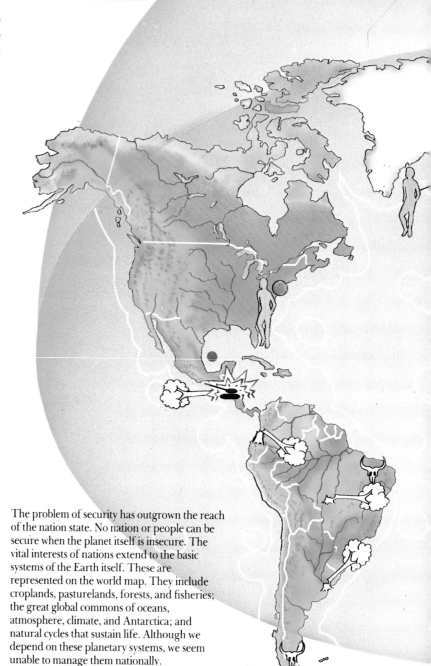

The problem of security has outgrown the reach of the nation state. No nation or people can be secure when the planet itself is insecure. The vital interests of nations extend to the basic systems of the Earth itself. These are represented on the world map. They include croplands, pasturelands, forests, and fisheries; the great global commons of oceans, atmosphere, climate, and Antarctica; and natural cycles that sustain life. Although we depend on these planetary systems, we seem unable to manage them nationally.

Since 1972, we have had a series of conferences covering planetary problems such as desertification, population, energy, food, and health. None of these problems is solely national or even international; none has anything to do with military strength; rather they are supranational or global in scope. In each case a World Plan of Action has been adopted – and promptly forgotten by governments immersed in geopolitical traditions. Problems of such dimensions, which are everybody's concern, tend to be nobody's business.

Environmental backlash
In the Third World, population pressures on non-renewable resources force more and more people out of rural areas into cities. Greater urbanization generates increasing violence, terrorism and political unrest. Dangerous cracks

Future shocks

Nuclear security, world economic instability, over-population, climatic dislocation, and pollution of air and ocean threaten even the most predictable parts of all our daily lives. And there is more to come. The projected increase of carbon dioxide in the global atmosphere may disrupt agriculture and the world economy (pp 110-13). Increased pressure for migration could swell the cities of the South, and lead to South to North population movements. North/South conflict will deal other future shocks. These two halves of the world are already locking horns in the battle over genetic resources which could represent a powerful weapon for developing nations. Mexico and Ethiopia, two major germplasm suppliers, among other developing world nations, are finding ways to exploit the dependence of the North on the genetic diversity of the South. If we make no attempt to manage these existing crises, we risk a slide into far deeper conflict.

Breaking point: Ethiopia

Gross degradation of the natural resource base of countries in the Horn of Africa precipitated the downfall of Haile Selassie in 1974, and conflict in the Ogaden desert in the late 1970s. Superpowers were alerted in response to the threatened security of the nearby oil-tanker route.

Ocean conflict

Nations conflict over the common resources of the oceans. Already national lines cut across Antarctica (below), while the new EEZs place 40% of the ocean under national control (pp 88-91). The UK and Iceland have come to the edge of hostilities over cod fisheries; 16 similar conflicts can be documented around the world. North has clashed with South over the Third Law of the Sea as developing countries attempt to secure their share of deep sea minerals.

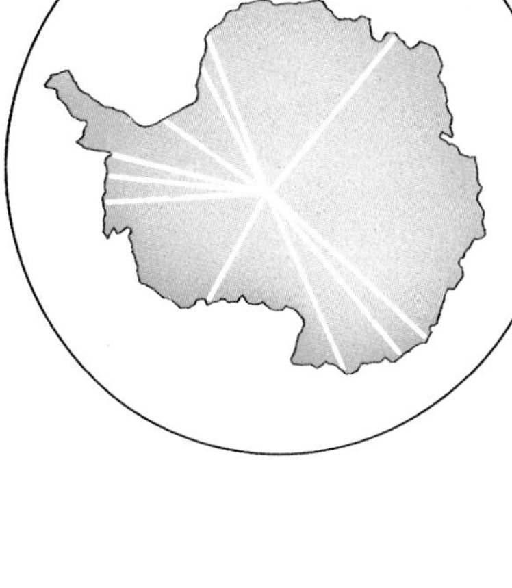

in our mismanaged system are appearing in the war zones, shown above. Obvious symptoms of stress include drought, pollution, deforestation, and the glaring divergence between the life expectancies of people living in the North and South.

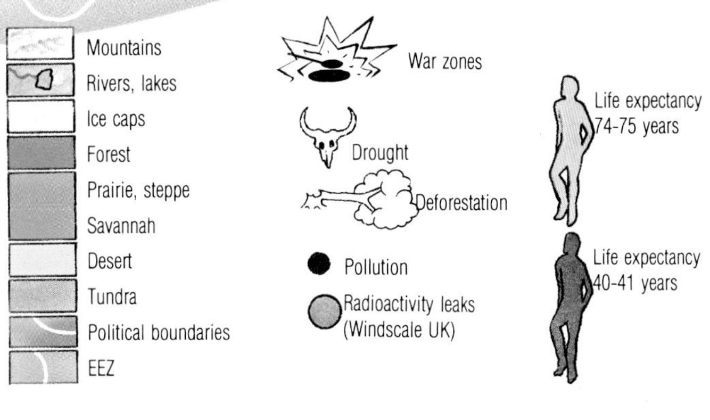

Mountains	
Rivers, lakes	
Ice caps	
Forest	
Prairie, steppe	
Savannah	
Desert	
Tundra	
Political boundaries	
EEZ	

War zones

Drought

Deforestation

Pollution

Radioactivity leaks (Windscale UK)

Life expectancy 74-75 years

Life expectancy 40-41 years

such as Southeast Asia, the Horn of Africa, and the Middle East. Israel's reluctance to share water supplies on the West Bank, water which it regards as vital to the development of its agriculture, probably constitutes a leading factor in determining the future security of the whole Middle East region. Competition for limited resources will spark many more such confrontations, unless we act promptly and imaginatively to defuse these issues before they finally explode.

The "Third World" war

Since World War II ground to its bloody conclusion, there have been at least 170 armed conflicts, almost all of them in the developing countries. As the world has grown increasingly resource-stressed and politically turbulent, the rate of conflict has risen. By the 1950s, the average number of outbreaks was nine a year, whereas during the 1980s it was 16. In 1987, no less than 27 wars were raging.

Of the wars in developing countries, at least half have been civil wars of one form or another, many fought over tribal or religious differences. But their real roots often lie in basic resource problems, reflecting the inequities in resource distribution both within and between countries.

Tragically, it is safer to be a soldier in battle than a civilian on the so-called sidelines. Of the estimated 22 million people killed in war since 1945 most have been civilians; around three-quarters of the dead during conflicts in the 1980s. War now involves and consumes entire communities: it has become total in a way we have not known before.

Nuclear armed country

Country with nuclear arms capability

Country with nuclear arms ambitions

Towards a violent planet

Our world has become an increasingly violent place to live. Wars now are shockingly more destructive and deadly than ever before. So far this century, there have been over four times as many war deaths as in the 400 years preceding. Since 1945 some 22 million lives have been lost in war, most of them civilian lives. Around half of this total has been in the Far East alone (see tombstones, below, showing war deaths 1945-1990).

But war deaths tell only part of the story. Millions more have been injured and maimed. In the last 20 years the number of war refugees has increased from 3 million to 15 million, mostly living miserable lives in refugee camps, unable to return to their homes. Many of the world's governments have turned on their own citizens. A number of others have turned to terrorism, and defiance of the law. Despite the end of the Cold War, the world is ridden by conflict, as it is proliferated by weapons of destruction, conventional or otherwise.

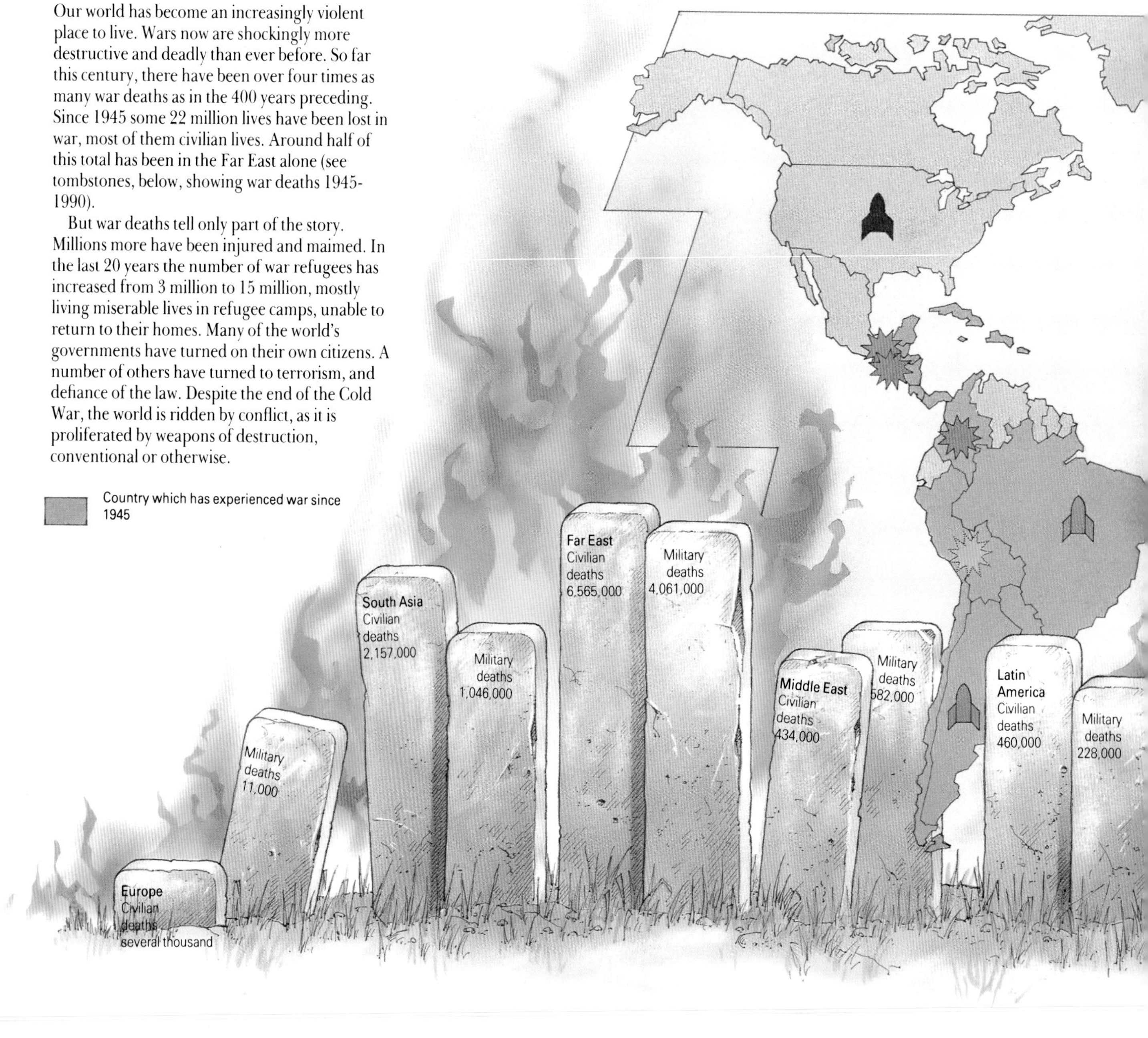

Country which has experienced war since 1945

South Asia
Civilian deaths
2,157,000

Military deaths
1,046,000

Far East
Civilian deaths
6,565,000

Military deaths
4,061,000

Middle East
Civilian deaths
434,000

Military deaths
582,000

Latin America
Civilian deaths
460,000

Military deaths
228,000

Military deaths
11,000

Europe
Civilian deaths
several thousand

Yet even though modern wars may be ostensibly local affairs, there tends to be a great deal of involvement by rich nations. In fact, a large number of developing-country wars have been "proxy wars", fought by locals on behalf of the big power blocs, and heavily supplied with arms and advisors from outside. Without this foreign support, most of these conflicts would have been shorter and less destructive.

Although the end of the Cold War has brought a thaw in relations between East and West, it has also brought ethnic and territorial rivalry between the newly-independent states, which threatens to draw others into conflict. And the old East-West axis of confrontation is itself shifting: to the North versus the South. Defence planners in the North increasingly see a very unstable world over the next few decades, foreseeing bitter divisions of wealth and poverty, coupled with the continued proliferation of advanced weapons, producing a dangerous world disorder. For example, nuclear planners operate up to three decades ahead, the lifetime of a nuclear weapons system, and so are planning the next system to respond to potential future threats, so as to maintain their nation's wealth and "security".

Ironically, wars consume resources which, used more wisely, could have addressed the root-causes of conflict much more cost-effectively. It is almost always more expensive to go to war than to choose an alternative course, if political leaders were to consider the needs of their people rather than the dictates of national and individual prestige. For example, rehabilitation costs in Iraq following

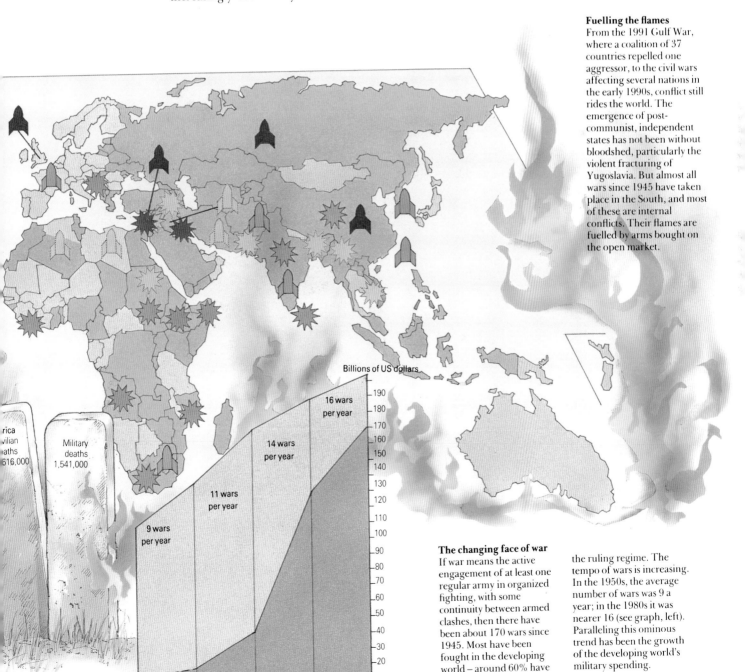

Inter-nation conflict

Civil/internal conflict

Potential conflict over rivers/water

Fuelling the flames
From the 1991 Gulf War, where a coalition of 37 countries repelled one aggressor, to the civil wars affecting several nations in the early 1990s, conflict still rides the world. The emergence of post-communist, independent states has not been without bloodshed, particularly the violent fracturing of Yugoslavia. But almost all wars since 1945 have taken place in the South, and most of these are internal conflicts. Their flames are fuelled by arms bought on the open market.

Billions of US dollars

16 wars per year

14 wars per year

11 wars per year

9 wars per year

190
180
170
160
150
140
130
120
110
100
90
80
70
60
50
40
30
20
10

rica ilian aths 616,000

Military deaths 1,541,000

Developing countries military expenditure

1950   1960   1970   1980   1990

The changing face of war
If war means the active engagement of at least one regular army in organized fighting, with some continuity between armed clashes, then there have been about 170 wars since 1945. Most have been fought in the developing world – around 60% have involved attempts to topple the ruling regime. The tempo of wars is increasing. In the 1950s, the average number of wars was 9 a year; in the 1980s it was nearer 16 (see graph, left). Paralleling this ominous trend has been the growth of the developing world's military spending.

the 1991 Gulf War, for instance, may cost as much as $200 billion. This is more than four times the country's GNP.

The price of "security"

Within the few minutes that it takes the reader to get through this portion of text, governments will devote several million dollars to military activities. Each year, we spend an average of around $180 per world citizen on the arms race. Add all this up and the total figure approaches $1,000 billion a year – or 6 percent of global GNP.

Suppose that we could divert a mere 10 percent of the world's military budget into constructive activities. Many of the problems we face on our over-burdened planet could be eliminated. The proposed UN Action Plan to halt Third World desertification could be funded for a whole year by the $5 billion that represents two days of global military expenditure. The money spent in just 24 hours on the 1991 war over Kuwait could have funded a child immunization programme for five years, and prevented the deaths of one million children annually. The money that UNEP has had to safeguard the global environment over the past decade is equal to only five hours of global military spending.

True, there has been a decline in military expenditure since the end of the Cold War, yet the decrease is relatively slight. At the same time, resources devoted to military research and development (R & D) have increased, reflecting the continuing efforts to achieve and maintain military technological superiority. Around $100 billion is ploughed into military R & D annually, and it employs one-fifth of the world's research scientists and engineers. The military absorbs scientific and technological capabilities that are ten times greater than those available to all developing countries.

Furthermore, the global consequence of decreasing military budgets in rich, arms- producing countries is increased pressure to sell weapons abroad. As domestic orders fall, defence industries look to exports to increase sales and recoup the huge R & D costs involved in weapons production. With the end of the Cold War, the US has overtaken the former USSR as the number one arms exporter. The arms trade is a major economic activity, second only to the oil industry: some $250 billion is given each year to the world's military to buy weapons. To be fair, the North often has willing customers in the South: the developing world continued to increase its military spending during the 1980s – and stands at 16 percent of the global total.

The supreme irony is that, while political leaders around the world proclaim that they will not surrender one square metre of territory to a foreign invader, they allow huge areas of once-fertile lands to wash or blow away each year. They could purchase more real security, in the proper sense

The cost of militarism

"I think that people want peace so much that one of these days governments had better get out of their way and let them have it."
PRESIDENT EISENHOWER

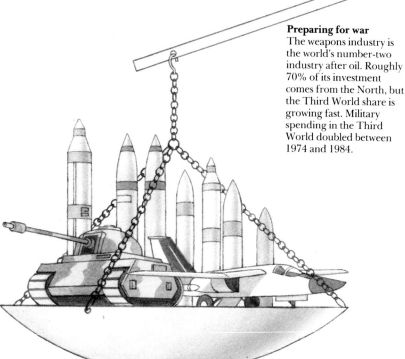

Preparing for war
The weapons industry is the world's number-two industry after oil. Roughly 70% of its investment comes from the North, but the Third World share is growing fast. Military spending in the Third World doubled between 1974 and 1984.

The cost of militarism
Count out 60 seconds and 27 of the world's children will have died for lack of food and adequate health care. Count out another 60 seconds: in this short space of time, the world will have spent $1.8 million on its military. Indeed, we are currently spending almost $1000 billion a year on the instruments of death – and, in the process, opening up a new battlefield of social neglect. Between 1945 and 1989, it is estimated that 22 million lives were lost in war. We now lose a similar number *every year* due to various forms of social neglect. We have got our priorities drastically wrong. The $140 billion now spent by developing countries on weapons is three times what it would have cost to provide essential health care, medicines, vaccinations, clean water, and sanitation for all.

The Romans were among the first to insist that the only way to secure peace was to prepare for war, regardless of the cost. But around the world the cost is mounting. Rehabilitation costs in Iraq following the 1991 Gulf War are estimated at $200 billion. In addition to the tens of thousands of civilian casualties, the allied bombing seriously damaged water supply, purification, and sewerage systems. The death rate of children under five years old increased almost fourfold. For the allies, the costs of the war run into billions. The US in particular, with its large budget deficit, has had to borrow even more. US research suggests that the medical and pension costs for wounded soldiers and other veterans tend to be almost three times the original outlays for a war.

Spending vs aid
There is a huge and growing gap between world military expenditure and world spending on development aid (see graph, right). Military expenditures of developed countries are now 30 times larger than their aid budgets for the developing countries. The tragedy is that sustainable economic development could remove many pre-war tensions.

☐ World military expenditure

☐ World foreign economic aid

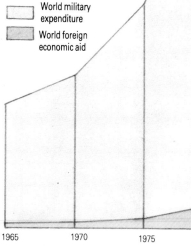

1965   1970   1975

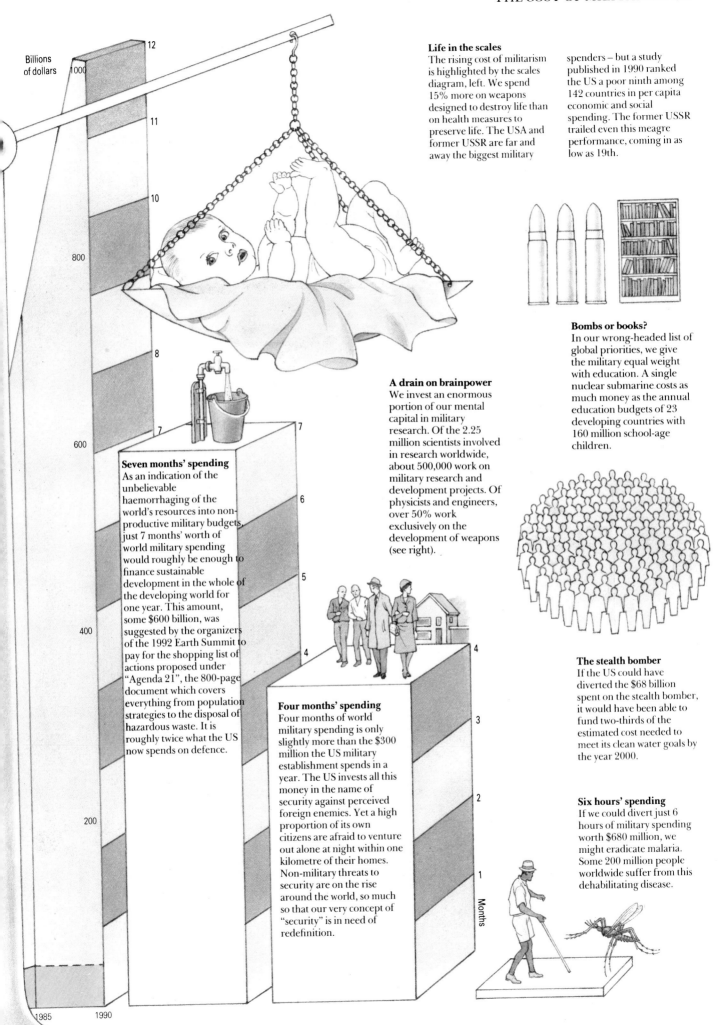

Billions
of dollars

Life in the scales
The rising cost of militarism is highlighted by the scales diagram, left. We spend 15% more on weapons designed to destroy life than on health measures to preserve life. The USA and former USSR are far and away the biggest military spenders – but a study published in 1990 ranked the US a poor ninth among 142 countries in per capita economic and social spending. The former USSR trailed even this meagre performance, coming in as low as 19th.

Bombs or books?
In our wrong-headed list of global priorities, we give the military equal weight with education. A single nuclear submarine costs as much money as the annual education budgets of 23 developing countries with 160 million school-age children.

A drain on brainpower
We invest an enormous portion of our mental capital in military research. Of the 2.25 million scientists involved in research worldwide, about 500,000 work on military research and development projects. Of physicists and engineers, over 50% work exclusively on the development of weapons (see right).

Seven months' spending
As an indication of the unbelievable haemorrhaging of the world's resources into non-productive military budgets, just 7 months' worth of world military spending would roughly be enough to finance sustainable development in the whole of the developing world for one year. This amount, some $600 billion, was suggested by the organizers of the 1992 Earth Summit to pay for the shopping list of actions proposed under "Agenda 21", the 800-page document which covers everything from population strategies to the disposal of hazardous waste. It is roughly twice what the US now spends on defence.

The stealth bomber
If the US could have diverted the $68 billion spent on the stealth bomber, it would have been able to fund two-thirds of the estimated cost needed to meet its clean water goals by the year 2000.

Four months' spending
Four months of world military spending is only slightly more than the $300 million the US military establishment spends in a year. The US invests all this money in the name of security against perceived foreign enemies. Yet a high proportion of its own citizens are afraid to venture out alone at night within one kilometre of their homes. Non-military threats to security are on the rise around the world, so much so that our very concept of "security" is in need of redefinition.

Six hours' spending
If we could divert just 6 hours of military spending worth $680 million, we might eradicate malaria. Some 200 million people worldwide suffer from this dehabilitating disease.

Months

1985   1990

of that term, if they used the funds to safeguard natural resources, and enhance human development, on which their country's future depends.

The ultimate foolishness

The ultimate horror of nuclear war is still a probability we face every day, despite the revolutionary changes of recent years. The existing nuclear states have arsenals now rated at 18,000 megatons. One megaton is equivalent to 80 Hiroshima-type bombs, so present arsenals contain explosive power equivalent to 1.4 million Hiroshimas.

With the end of the Cold War, it is accepted by the US and former USSR that the huge overkill of nuclear weapons can be curbed. At present, their arsenals contain about 47,000 weapons - both strategic and tactical. The Strategic Arms Reduction Treaty (START) and the Bush-Yeltsin agreement of June 1992 will, if carried out, cut the number of strategic weapons operationally deployed to around 3,500 each by the early years of next century. The US intends to retain 1,600 tactical weapons; the Russians may deploy none.

However, the destruction and storage of nuclear weapons creates major problems. Dismantling is a lengthy process; the capacity of Russian dismantling facilities is a rate of about 1,500 a year. So, it will take about 15 years for the Russian arsenal to be reduced. In the meantime, the weapons will have to be secure from theft and safely maintained - a difficult and highly- expensive business.

In spite of reductions by the US and former USSR, both appear to have every intention of keeping their nuclear forces. The day after the historic Bush-Yeltsin agreement in June 1992, the US detonated a nuclear device underground (one of more than 1,800 such tests since 1945). Several new nuclear weapons are under consideration by the US military, including a bomb designed to release an intense electromagnetic pulse to destroy electronic command and communications systems, and so-called "micro-nukes" to intercept Third World ballistic missiles. Military and nuclear planners operate up to decades ahead, and they take note of the development of potential threats.

More than a dozen developing countries have long-range missiles in service; about half of these states are believed to possess chemical weapons; two have nuclear weapons. Advanced military technology is spreading through the developing world, and there is an increasing effort by these countries to establish independent, indigenous capabilities.

Ironically, and tragically, advances in military technology, if allowed to continue unabated, could lead, détente notwithstanding, to perceptions of a nuclear first-strike capability and the possibility of limiting – and winning – a nuclear war. Such perceptions serve only to destabilize fragile international relations and increase the risk of war, particularly accidental or unintentional nuclear war. Furthermore, the military's monopoly

One million Hiroshimas

"The stone age may return on the gleaming wings of science, and what might now shower immeasurable blessings upon mankind may even bring about its total destruction. Beware, I say; time may be short."
WINSTON CHURCHILL

The nuclear payload carried by a single Trident submarine is equivalent to *eight* times the total firepower expended during World War II. Today's nuclear arsenals contain the combined potential firepower of more than one million Hiroshimas. The nuclear fireball, right, gives a striking impression of the degree of "overkill" now afforded by the world's nuclear weapons. The warheads carried by a single Trident submarine have the explosive capacity to destroy all the major cities of the northern hemisphere. Today, we command arsenals with the explosive energy yield of 18,000 megatons, while the 11 megatons that were released in three majors wars in this century (WW II, Korean, and Vietnam war) killed a total of 44 million people. Even though the start of the 1990s has brought promises of arms reductions from the US and former USSR, there is still a great proliferation of all types of weapons. Indeed, the ever-increasing accuracy and sophistication of the world's missiles make the degree of overkill achieved to date look even more ridiculous.

"The growing wealth of petro-nations and newly hegemonic powers is available to bullies and crazies, if they gain control, to wreak havoc on world tranquility."
Head of US Strategic Air Command, 1991

Hiroshima: 0.013 megatons
The first military use of the atomic bomb came with the dropping of the 13-kiloton "Little Boy" bomb in 1945, killing nearly 100,000 people.

World War II: 3 megatons
A single square, left, is equal to the total firepower used in WWII.

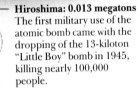

Poseidon: 9 megatons
The three squares shown, left, represent 9 megatons, equal to three WWIIs. They are also equivalent to the firepower of a single Poseidon submarine. The US has 33 strategic nuclear submarines.

Trident: 24 megatons
The eight squares, left, represent 24 megatons, the firepower equivalent of a single modern Trident submarine. Enough to destroy every major city in the North.

Goodbye to cities
A hundred squares equal 300 megatons, enough to destroy all the world's cities.

The nuclear club

The number of nations now moving towards full membership of the nuclear club is increasingly ominous. There is still no effective control over club membership, apart from the Nuclear Non-Proliferation Treaty. This was signed in 1970 and restricts the number of nuclear states and the sale of nuclear weapons to non-nuclear states, but has proved weak in practice. Although nuclear weapons are the first choice for many countries, they require advanced technology, long periods of time, and a great number of resources to produce. Six countries (US, UK, former USSR, France, China, and Israel) are considered fully fledged nuclear powers. Many developing countries have tried to join the nuclear club secretly, a few with some success. India and Pakistan are believed to have small arsenals. Argentina and Brazil, both with a nuclear capability, are showing no further intentions of extending it. North Korea, Iran, Libya, Syria, and Taiwan are suspected of having nuclear ambitions, but are still years away from producing weapons. Iraq was developing weapons, but the Gulf War and subsequent UN actions forcing inspection and dismantling of weapons, has considerably set back its progress.

Biological weapons

Throughout history, diseases have caused as many deaths during wars as weapons, and military strategy has often encouraged their spread among the enemy. Biological weapons use toxins, produced by bacteria or other organisms, or living germs that cause diseases, such as anthrax, cholera, and plague. These have the potential to kill humans and animals, and to damage plant crops. A 1972 treaty prohibits their production, possession, and use. US intelligence has frequently made allegations that the treaty has been violated by a number of countries, including the former USSR and Iraq.

Chemical weapons

The use of chemicals as weapons, such as asphyxiating or nerve gases, poisons and defoliants, was outlawed by international treaty 65 years ago, but there is no prohibition on production or possession. Chemical weapons are popular choices for developing countries because they are cheaper and easier to acquire or produce than nuclear weapons. Although they are not as effective as nuclear bombs, to those countries without such a capability chemical weapons can seem an appropriate and affordable deterrent. Iraq is the only developing country to admit to possessing chemical weapons, but around a dozen more are suspected to have undeclared stocks. Observers generally agree that the number of nations seeking to acquire chemical weapons has sharply increased. Substantial stocks are also known to exist in the developed world, some left over from WW II. In particular, the US and the former USSR have tens of thousands of tonnes each. In 1992 a draft chemical weapons convention was agreed following a quarter-century of negotiations. If ratified, participating nations will have ten years to destroy their stocks. This is likely to prove a major problem, in terms of costs, technology, and environmental damage.

Ballistic missiles

Ballistic missiles add another dimension to the proliferation of unconventional weapons since they can carry nuclear, chemical, or biological payloads. The race for advanced military technology began in the superpowers: now it is spreading rapidly in the developing world. The CIA has forecast that, by the year 2000, at least 15 developing countries will be producing their own missiles. Most have a range of 300 km, but Iraq successfully modernized a Scud missile to travel twice as far; India is developing the Aqui missile to travel 2,500 km; and Saudi Arabia has received from China the CSS-2 with a range of almost 3,000 km.

of millions of highly-trained workers, in the name of "security", diverts research talent from tackling urgent global problems. Without these additional skills, the likelihood of the rich-poor poverty gap widening is increased, together with North-South tension, and the attendant risks to world security.

Nuclear winter or environmental collapse?

The military absorbs more than money and brain-power. It is the world's single biggest polluter, according to research carried out by the Canadian organization Science for Peace. It calculates that the world's armed forces account for 10 percent of all emissions of greenhouse gases, and two-thirds of all emissions of the CFC formulation that does most damage to the ozone layer. Military accidents have left at least 50 nuclear warheads and 11 nuclear reactors on the ocean floor, and military activity consumes as much aluminium, copper, nickel, and platinum as the whole of the developing world.

These are just the peacetime consequences of military expenditure. In recent years, the environment has suffered greatly from war. Vast areas of Vietnam's forests and crops were destroyed by herbicides, which led to soil erosion. During the 1991 Gulf War, there was a massive oil spill off Kuwait, and 613 oil wells were ignited, burning between 4 and 8 million barrels of oil a day for months. The clouds of pollution were so large and dense that the sun was blacked out and surface temperatures lowered in many parts of the north-

Bang or whimper?

We are already engaged in World War III, to paraphrase Professor Raymond Dasmann of the University of California, a war against our Earth – and we are winning it.

In the South, we have successfully gutted large portions of tropical rainforest, and allowed good soil to be washed uselessly away. In the North, the Earth's river arteries and ocean life-tides carry untold poisons, while rain, that life-giver from the skies, sometimes proves to be as acid as vinegar.

And, just in case our straightforward ecological assaults should fail, we have recourse to the alternative strategy of nuclear war. We could trigger a "nuclear winter" with less than 1% of our current nuclear arsenal. We may convince ourselves that we are "secure", but globally we have never been more threatened.

ern Gulf.

These dark skies over Kuwait followed just a few years after studies that revealed a hidden threat behind the world's stockpile of almost 50,000 nuclear weapons. Only a fraction of these weapons needs to be detonated to lift so much smoke, soot, dust, and other debris into the atmosphere that the sun would be blotted out over the combatant nations – a nuclear winter. This would create Arctic conditions with sub-zero temperatures, and darkness would halt the process of photosynthesis. The biological consequences would be devastating, with massive destruction of agriculture, forests, and other ecosystems. Nor would the nuclear winter be confined to the latitudes of conflict; the cloud would extend across the globe, although the pall would thin out as it dispersed. Survivors would then also have to contend with the effects of huge damage to the ozone layer - another consequence of nuclear war – on top of the lack of drinking water, food, heat, medical care, and functioning social and economic systems.

So, our civilization could still go out with a multi-megaton bang. Meanwhile, our quest for "security" blinds us to a parallel threat already upon us. If we do not learn to live in harmony with our environment, we shall experience the laws of nature in full force. On this course, human society seems destined to go out with a whimper – or, rather, a prolonged wail.

UNDER NEW MANAGEMENT

We are a privileged generation, living at a great turning point in the story of Gaia and humanity. Faced by multiplying crises, we are challenged to a creative endeavour surpassing that required of any earlier society. To survive, Homo sapiens must now advance from a pioneer species, which is aggressive, prolific, and greedy for resources, into a climax species, which recognizes ecological limits, swaps assertiveness for co-operation, and expresses self-regulation as the golden rule. After setting ourselves apart from nature, we must become a part of nature again: truce, treaty, and reconciliation – and so become, truly, human kind. A challenge indeed, and one that requires a sweeping revolution in how we act, think, and feel; in how we use our technology; and above all in how we manage ourselves, and thus the impact on Gaia of each and every one of us and of the collective global human family and its institutions.

Global governance for a sustainable world

Today's international institutions and laws are compacts between governments, not people, and have failed to evolve in step with economic, political, and environmental realities. If we are to steer our way peaceably and sustainably through coming crises and into the next century, we will have to reshape these institutions to allow better "governance".

Better governance does not mean more, improved government. It means co-operative self-management by everyone. It means creating a global "net" of participatory mechanisms that involve people from all sectors of society and at all levels, from local to global, in decision-making. Whereas government involves states, governance involves us all.

The key organ of governance today is the United Nations. After a chequered history, it now plays a growing role in environment and security issues. Reform and democratization of the UN is thus of prime importance in broadening the global decision-making process - its aim: democratic international governance, representing self-reliant societies.

A broader-based UN could add to the existing General Assembly (nations) a new, non-governmental Assembly representing the "We, the peoples . . . " of the UN Charter – with delegates from NGOs, trade unions, the professions, women's groups, cities, villages, farms, corporations, financial bodies, the arts, and more. As a UN "upper house", with voting powers, such an Assembly would be pioneering and relatively non-aligned, more able to take a global

Global governance

The world community is evermore interdependent, with a growing commitment to internationalism – witness the transnational links between NGOs and professional bodies. But we need a political transformation, too, to meet the challenge of building global self-governance, with institutions assuring environmentally-sound and socially-equitable development. This new global governance will have its sovereignty rooted in empowered communities; manifested in national or regional governments with the accountability and legitimacy to govern responsibly; and expressed in international institutions with the strength to hold their members accountable. We are in an unusually dynamic era, with a historic opportunity for global governance. But if the UN now fails to evolve in radical directions and democratize, reaction will set in and the prospects will be bleak.

Phoenix of the new nations
Although political leaders pay lip service to the need to protect the sovereignty of the nation state, it is being steadily eroded as nations become ever more interdependent. Sovereignty has already been reduced by: supranational laws on human rights, the environment, and disarmament; international laws; and regional laws, such as those controlling the European Common Market. Although the nation state will not "wither away" in the forseeable future, its main role will generally be confined to enacting rules to manage global, regional, and local affairs.

The family and individual
An alert citizenry is the ultimate check on the activities of politicians and commercial and financial institutions. Effective governance will depend on individuals exerting their rights and responsibilities, so as to monitor the activities of governments and apply pressure to ensure that the rule of international law is not violated. Good "world citizens" will refuse to be influenced by the propaganda of governments or the media. They will be sensitive to the need to match consumerism with sustainable development, and use their voting power to ensure that economic and financial policies reflect proper care of the world's resources.

A new UN net

The structure of the UN has never been adequate to solve security, political, or ideological problems – although its record on humanitarian issues is good. Now that the East-West ideological conflict is over, the time is ripe for UN reform. Under the auspices of the international Conferences on a More Democratic United Nations (CAMDUN) there have been many reform proposals, from changing the patterns of voting power in the Security Council, to the establishment of a peoples' or parliamentary assembly as another UN organ – a counter-balance to the General Assembly. In the meantime, an advance would be to award NGOs participatory, rather than observer, status at all UN-sponsored conferences, and to include a non-governmental or parliamentary representative in all General Assembly delegations.

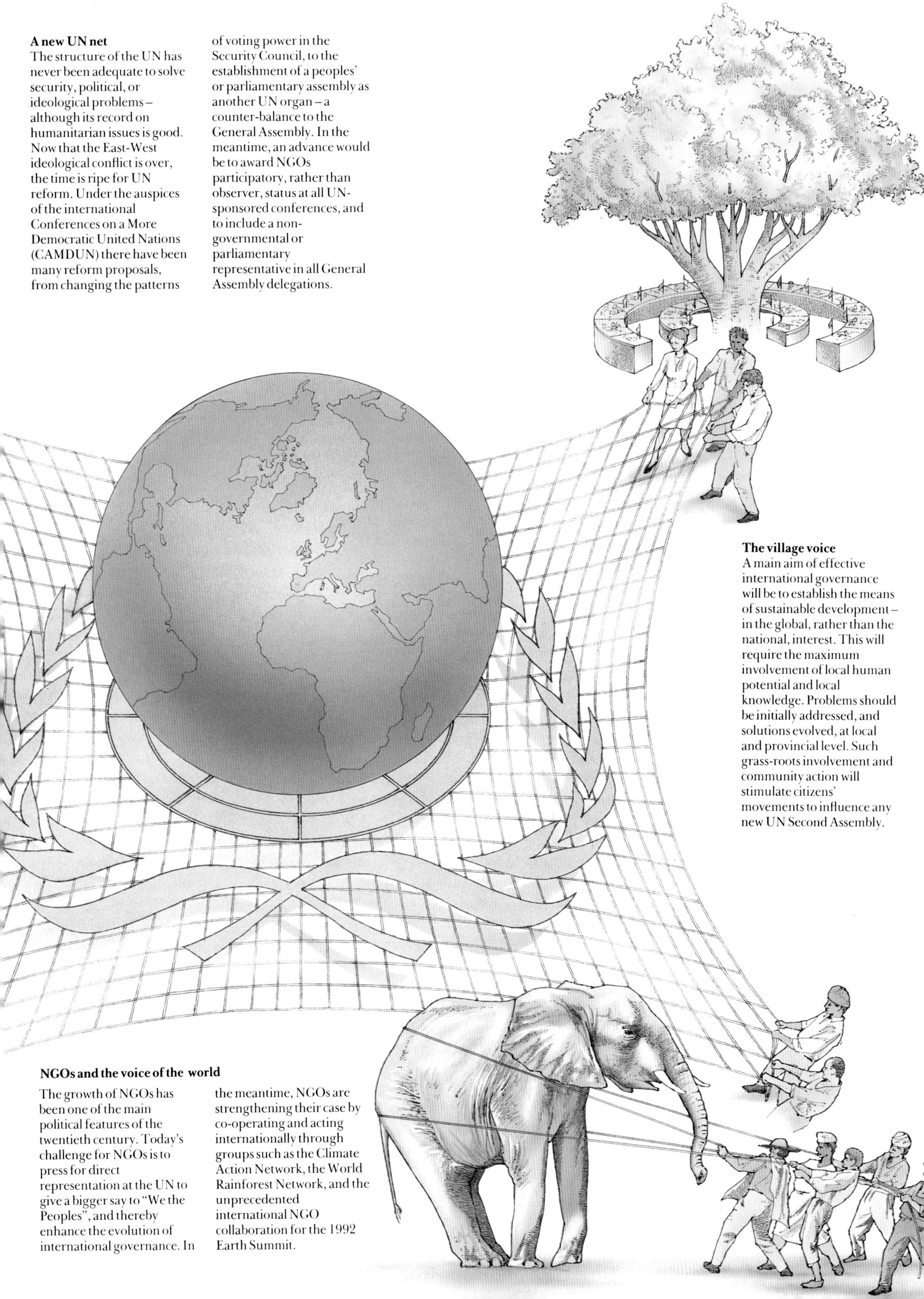

The village voice

A main aim of effective international governance will be to establish the means of sustainable development – in the global, rather than national, interest. This will require the maximum involvement of local human potential and local knowledge. Problems should be initially addressed, and solutions evolved, at local and provincial level. Such grass-roots involvement and community action will stimulate citizens' movements to influence any new UN Second Assembly.

NGOs and the voice of the world

The growth of NGOs has been one of the main political features of the twentieth century. Today's challenge for NGOs is to press for direct representation at the UN to give a bigger say to "We the Peoples", and thereby enhance the evolution of international governance. In the meantime, NGOs are strengthening their case by co-operating and acting internationally through groups such as the Climate Action Network, the World Rainforest Network, and the unprecedented international NGO collaboration for the 1992 Earth Summit.

view. The General Assembly would still decide much of UN policy. But the smallest village would also be able to influence world affairs.

Sustainable development may become the central thrust of ongoing UN reform. We need a UN system of collective security which can prevent conflicts and avert environmental threats. The Sustainable Development Commission, established by the 1992 Earth Summit, is one example of how the UN could develop a role as the supreme policy maker in this sphere.

Sustainable management

We are moving into a totally new phase of human development. Instead of ransacking the Earth without thought for tomorrow, we must guard and use it on a sustainable basis. This is the great transition we must make which will demand a management exercise on a scale unequalled in human history.

To be able to make the switch from runaway growth to sustainable management, it is vital that we calculate the productive capacities of our human and natural resources. At present, our observation, measurement, and monitoring of environmental change is grossly inadequate. We are now destroying crucial planetary resources without knowing the full consequences of our actions. Ironically, we do have the expertise to calculate these productive capacities, and, as we have seen, there already exists the communications power, and a diverse institutional framework, to bring effective planet management within our grasp.

Our transition to a sustainable society will encompass a new definition of "security"; one based on "the four Es" of Energy, Environment, Economics, and Equity. Attaining energy security means abandoning the supply-side addiction of exploiting ever more oil and gas fields, in favour of increased energy efficiency and the use of renewable sources. This will reduce our dependency on oil-rich regions, as well as cutting emissions of greenhouse gases, so contributing to environmental security. We must avoid the massive disruptions which climate change, continued population growth, and deforestation could unleash. Improving economic security will require the transfer of capital and human talent from the military to the civilian sectors. Building and maintaining the infrastructure for a sustainable, energy-efficient, and socially-equitable society would create far more jobs globally. Studies in the US suggest that spending $1 billion on guided-missile production creates only 9,000 jobs; the same money spent on education would create 63,000.

Perhaps our most turbulent transition will be the establishment of equitable security. Attaining a less imbalanced world is a prerequisite for sustainable management. Only by reducing the likelihood of North-South conflict over resources can we realize the potential for agreement over vital global issues. We must move beyond talk of North or South, us or them – the

The great transition

Although we live in the finite environment of Earth, we continue to exploit its resources as if they were infinite, or as if we had another planet parked out there in space, to use when Earth runs dry. We cannot continue in this way for long. Sooner, rather than later (when events may act for us in a manner not to our liking), we will have to bring under control the current dangerous trends of overpopulation, energy depletion, environmental abuse, and wasteful military spending shown by the exponential curves of the graphs, right (and also pp 16-17). This unsustainable, runaway growth is setting our civilization on a course for sudden and disastrous collapse. Before we are able to make the transition to sustainable and equitable global living in balance with Gaia and ourselves, we have to turn these growth rates into stable, "steady state" cycles of consumption and renewal. Any economic growth we achieve thereafter will have to come from greater efficiency and better use of our resources, not from a further rise in consumption. We have to learn to live on the income of our planet's bountiful natural productivity, not to burn up its capital as if there was no tomorrow.

1000
1100
1200
1300
1400
1500
1600
1700
1800
1900
2000 Population growth

Environmental abuse   Military spending   Primary-resource consumption

We already have a range of prophecies and prescriptions which encourage us to make the transition to sustainable planet management. Some extracts are reproduced below.

"If present trends continue, the world in 2000 will be more crowded, more polluted, less stable ecologically and more vulnerable to disruption.

"Serious stresses involving population, resources and environment are clearly visible ahead. Despite greater material output, the world's people will be poorer in many ways than they are today.

"For hundreds of millions of the desperately poor the outlook for food and other necessities will be no better. For many it will be worse. Barring revolutionary advances in technology, life for most people on earth will be more precarious in 2000 than it is now – unless the nations of the world act decisively to alter current trends."

THE GLOBAL 2000 REPORT TO THE PRESIDENT (1982)

"In the end, sustainable development is not a fixed state of harmony, but rather a process of change in which the exploitation of resources, the direction of investments, the orientation of technological development, and institutional change are made consistent with future as well as present needs."

OUR COMMON FUTURE (WORLD COMMISSION ON ENVIRONMENT AND DEVELOPMENT, 1987)

"The aim of Caring for the Earth is to help improve the condition of the world's people, by defining two requirements. One is to secure a widespread and deeply-held commitment to a new ethic, the ethic for sustainable living, and to translate its principles into practice. The other is to integrate conservation and development: conservation to keep our actions within the Earth's capacity, and development to enable people everywhere to enjoy long, healthy and, fulfilling lives."

CARING FOR THE EARTH (IUCN, UNEP, WWF, 1991)

"The future of our planet is in the balance. Sustainable development can be achieved, but only if irreversible degradation of the environment can be halted in time. The next 30 years may be crucial."

JOINT COMMUNIQUÉ BY THE UK ROYAL SOCIETY AND THE US NATIONAL ACADEMY OF SCIENCES (1992)

language of conflict - for the planet offers enough for all, present and future generations, provided they are equitably and sustainably shared.

Tools for transition

The transition to a sustainable society is far from impossible. We have the tools: we must find the will.

There are two fundamental needs: stabilizing population and consumption. Even the most likely predictions show population stabilizing by 2150 at the earliest. If we want to avoid having nature do the job for us earlier, in unpleasant fashion, we can and must achieve it earlier ourselves – by doubling and redoubling our efforts to provide income security and family planning for all.

The equally basic requirement for stable consumption looks more hopeful. Ever more energy and resource efficient processes and products are available,

we are making strides in developing technology for renewable energy, and the recycling industry is growing. Indeed, technology offers much: a quarter-tonne satellite can detect ecosystem damage with great accuracy. Another satellite now permits NGOs and scientists to keep in a network of constant communication and co-operation.

Information is one of the principal keys to transition: information flowing in new ways, from new sources to new recipients, carrying new ideas. Glasnost, for example, opened up information channels long closed in Eastern Europe, triggering the rapid transformation towards a new system, one more consistent with the new perceptions.

Such political change towards participatory democracies is another beacon of hope. A stronger commitment to popular participation in decision-making within communities, nations, and global fora, lies at the

Signs of hope

The rise of human numbers has cast a long shadow across the Earth. But now we have within our grasp the tools and the ability to make the transition to a stable, sustainable society – one not of shadows, but of light, and promise. A society where human civilization is integrated within Gaia, the managed lands and the wild synchronized in harmony, as this illustration suggests. Many signs of hope stem from the new ways of thinking about our common future. There are numerous paths towards sustainable development, and many groups and individuals have already started down them. The more people who join in that journey, the more surely a foundation for the future is put in place.

heart of the transition. New international political conditions also offer us the chance to redirect defence spending to meet environmental and social needs, and to finance sustainable development. We must take this chance: the peace dividend must become our urgently needed "sustainability investment".

The greatest sign of hope is the slow but irreversible shift in world opinion towards more sound and responsible care of resources. The principles of sustainable management, formulated in "Our Common Future", the 1987 report by the UN World Commission on Environment and Development, have been accepted and endorsed by a majority of governments, by key NGOs, and by all major international institutions. Since 1987 the twin issues of environment and development have risen to the top of our international, national, community, and individual agendas, in a surge of concern symbolized by the 1992 "Earth Summit".

Management or medicine

Many civilizations in the past have foundered through failure of their environment, leaving a ruined landscape to warn us of their fate. Where ancient Sumer once flourished with gardens and irrigated fields, now is arid desert. The Sahara itself was fertile and wooded before the Romans came there. Ecology today sees such past events in a new light – these cultures exploited their environment beyond its power of recovery, and so brought on their own end. The whole history of humanity begins to look like a similar tale of growing damage, brought about by an opportunist species which exploits its environment, blindly, for short-term gain.

Of course, all species exploit the environment; the changes caused by the human species are only dangerous because of our numbers, our appetite for resources, and our adaptable technology. But the danger in the end is to ourselves: it is a fundamental rule of Gaia that species which adversely affect their own environment are doomed. Their progeny have an increasingly selective disadvantage as they inherit the damage done by their forebears – until in the end, the race must become extinct.

Is Sumer's fate now to be ours but on a global scale? Or can we, by proper management, prevent it? "Planet management" is now an established term, even adopted by large governmental and corporate programmes for satellite communications and long-term environmental monitoring. The risk is that these powerful interest groups may begin to use the term to imply control or rule of the planet by humans (and by elite groups of humans at that). Once again, they are falling into the trap of setting humans "above" nature. And this hubris has rightly drawn protests from thoughtful ecologists.

We are a part of nature, one member of Gaia, that "very democratic entity" described by James Lovelock. Of course we cannot "manage Gaia" nor rule the biosphere. We are ourselves ruled by the

biosphere. No worse fate could befall the Earth, or us, than for humans to take on forever the task of running the living world, with its infinite web of connections between biota, climate, soils, water, and air. Should we so damage Gaia that we are obliged to attempt this role, then we should truly be aboard "Spaceship Earth" - and about to crash!

But this is not the proper meaning of planet management. It means *managing ourselves, with the planet in mind*. Our civilization is global, with planet-scale impacts on Gaia's life systems. Yet, whether as political leader or corporate executive or individual citizen, we rarely think or act with planetary awareness. Instead, we each act for our own, personal, local, or national interest and ignore the cumulative result. We are practising planet *mismanagement* on a staggering scale.

The art of planet management is to do just the opposite: to manage our human affairs so that the well-being of Gaia (including habitats, species, and people, present and future) comes first – because we all depend on the whole for our survival. Such a system of self-management cannot be imposed from above by governments. To be effective it needs to draw on all the diversity of human culture, just as evolution draws on the biological diversity of Gaia.

The debate over what should be the human role in relationship to planet Earth will sharpen in years to come. Rather than "managers" some have preferred the idea of "stewards" or "caretakers". Lovelock himself, noting that on past performance humans make pretty poor managers, suggests the role of "shop stewards" - active representatives of all species. He has also proposed that we need a new practice of "planetary medicine" - a pragmatic approach to the health of Gaia, like that of the family doctor who, after cleaning up any cuts and bruises, prescribes rest, fresh air, and good food, and allows nature to do the healing.

Thus as planetary doctors, we need to clean up the damage we have done; stop polluting the air and water and depleting the nutrients of the soil; and allow forests and other ecosystems of the Earth some rest from our attack – so that Gaia's systems can recover their naturally robust good health.

Whether management or medicine, stewardship or shop stewardship, in practice all agree one thing is plain: we must engage in an active "damage limitation" exercise, and we must begin now, before it is altogether too late for Homo sapiens.

In one area, at least, we have already begun. We are coming to grips with global ozone depletion, via the Montreal Protocol – a beacon signalling that scientists, politicians, corporates, and consumers can react quickly when the need is understood. But the result cannot be immediate. Although it will be only a handful of years until the Montreal Protocol is fully implemented, the ozone-depleting chemicals will not completely leave the atmosphere for more than a century.

"The unleashed power of the atom has changed everything except our way of thinking . . . we need an essentially new way of thinking if mankind is to survive."
ALBERT EINSTEIN

"I have a high regard for our species, for all its newness and immaturity as a member of the biosphere. As evolutionary time is measured, we arrived here only a few moments ago and we have a lot of growing up to do. If we succeed, we could become a sort of collective mind for the Earth. At the moment, for all our juvenility as a species, we are surely the brightest and brainiest of the Earth's working parts. I trust us to have the will to keep going, and to maintain as best we can the life of the planet."
LEWIS THOMAS

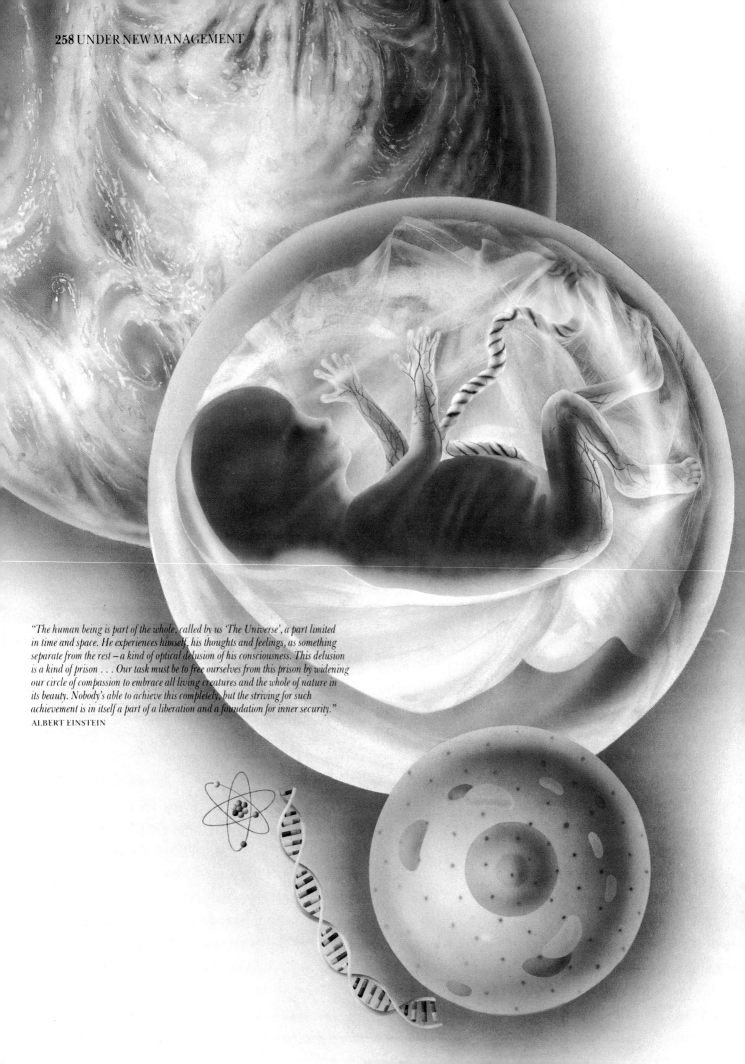

"The human being is part of the whole, called by us 'The Universe', a part limited in time and space. He experiences himself, his thoughts and feelings, as something separate from the rest – a kind of optical delusion of his consciousness. This delusion is a kind of prison . . . Our task must be to free ourselves from this prison by widening our circle of compassion to embrace all living creatures and the whole of nature in its beauty. Nobody's able to achieve this completely, but the striving for such achievement is in itself a part of a liberation and a foundation for inner security."
ALBERT EINSTEIN

A new ethic

Alone among species, humankind has been able to leave the biosphere and study it from the outside. We are now able to look in on the fragile miracle, Gaia, and on our own behaviour, with a sharpened perception. And as our vision clears, and our danger becomes apparent, we increasingly question our role in an otherwise harmonious world.

We have already acquired considerable power; that much is obvious. We are witnessing the consequences of our economic and technological penetration of the planetary cycles of energy and materials – cycles that have co-evolved with and sustained life for the past four billion years.

The challenge we face is to bring our activities within the carrying capacity of the Earth's supporting ecosystems, while still improving the quality of life for people today, yet without reducing the prospects for future generations. This is the challenge of sustainability. Meeting it will require changes in how we all evaluate our needs, how we behave, and how we perceive each other, other life, and the Earth. At base, it is a matter of ethics.

The issue of ethics is fundamental, because ethics form the principles and values upon which society is built. It is ethics which determine how we pursue economic, technological, and political goals. What we do depends on what we believe; a widely shared belief is far more compelling than any government edict.

Sustainability was the original mode of our species. Pre-industrial peoples lived sustainably because it was all too apparent that they had to. If, for example, they expanded their populations beyond the available resource base, then sooner or later they would starve or be forced to migrate. The sustainability of their way of life was empowered by a special consciousness of nature: the people were spiritually connected to the land, the animals, and plants on which they subsisted; they were part of the natural world, respecting and honouring it, not set apart as if masters. They belonged to their landscape, tied to it spiritually and economically in a way we cannot now conceive.

Eventually, and slowly, this era came to an end. A growing urban population demanded intensive agriculture, yielding a surplus, and then an expansion of production, met by conquest, or colonization, or improved technology. A different consciousness, a different ethic, sustains this "civilized" mode of life. The Earth and its creatures are increasingly considered the property of humankind: for domination, control, and manipulation, until ultimately, the whole of nature is seen like some giant machine, designed for human benefit.

The belief that we are "apart from" the rest of nature is an intrinsic part of the dominant world view — a human-centred or anthropocentric philosophy. Yet on what grounds can we support that view – our higher intelligence? Our temporary dominance?

We are too many now to return to the sustainable economy of our distant ancestors. But we can create a new ethic, suitable to the new era we are entering. An ethic that embraces care for species, and all humanity - a new humanism, a new world view, a new planetary concern. Creation of such an ethic is no side issue: it is recognized as an imperative by world institutions. It features prominently, for example, in Caring for the Earth — a Strategy for Sustainable Living launched by IUCN, UNEP, and WWF in 1991 which proposes "a world ethic for living sustainably" as a foundation for its principles. No longer apart from nature, "Every human being is a part of the community of life, made up of all living creatures. This community links all human societies, present and future generations, and humanity and the rest of nature. It embraces both cultural and natural diversity."

Many would go further, and stress that because of the interdependence of all living things, the natural world has certain rights, including the right to existence, which are quite independent of its utilitarian value to humankind.

No more powerful contribution to the debate over anthropocentrism has been made than that generated by James Lovelock's Gaia hypothesis – and the concept of the Earth as "superorgansim", where life and the environment have "co-evolved" in such a way as to ensure the maintenance of life. This "organic" concept of the Earth – one which has a long tradition in human civilization – has great emotional force. Perhaps its re-emergence signals the awakening of a new, mature vision: a culture and a technology that seeks not to control the world but to participate with it, not to operate upon nature, but to co-operate with nature. Some have even taken this further, to an ethic of planetary healing – healing our troubled relationship with Gaia, the living Earth.

We now have the skills to work with natural systems to derive benefit from them, without destroying their integrity and ability to renew themselves. We can live off the biosphere's income, rather than ever-depleting its capital. With this skill in our hands, we could assist forests to spring up on bare lands, safeguard species from the myriad pressures driving them to extinction, renew the soil, stabilize and reverse the pollution of oceans and atmosphere, harness the renewable power of the sun, and minimize our consumption of resources, and creation of wastes. And just as we have long controlled the population of many other species, we could control our own, at a sustainable level.

It is time for humanity to embrace a new ethic. We must have the vision and the courage to face ourselves, admit our power of life and death, and bring it under permanent, watchful control. We must use our skills to take sides with life, for the sake of ourselves as well as other life on the planet. In the words of Albert Schweitzer: "Until he extends the circle of his compassion to all living things, man will not himself find peace".

Epilogue

"The most beautiful object I have ever seen in a photograph, in all my life, is the planet Earth seen from the distance of the moon, hanging there in space, obviously alive. Although it seems at first glance to be made up of innumerable separate species of living things, on closer examination every one of its working parts, including us, is interdependently connected to all the other working parts. It is, to put it one way, the only truly closed ecosystem any of us knows about. To put it another way, it is an organism. It came alive, I shall guess, 3.8 billion years ago today, and I wish it a happy birthday and a long life ahead, for our children and their grandchildren and theirs and theirs."
LEWIS THOMAS

We hardly know the planet. There are countless gaps – in our understanding and in our ability to compute both the cost of environmental damage and the cost of inaction. The world's environmental database is incomplete and of variable quality. For instance, scientists do not even agree, within 50 percent either way, on how much vegetation there is on Earth, critically important as that is to our understanding of where we can grow crops, support livestock, or harvest wood. They know even less about our wild habitats, upon which we make greater and greater demands. We are extinguishing species not yet even discovered – indeed, we have only conflicting estimates of the Earth's total complement of life-forms. As for knowledge of the effects of human actions, very few nations establish the true rates of soil loss, deforestation, or pollution; some cannot accurately count their hungry, poverty-stricken, or workless populations. In the oceans, we do not know the size of the fish stocks we are depleting, nor the rate of spread of the toxins we pour out. Our ignorance is so vast, in fact, that we are not aware of it.

Our understanding of our planet is also sadly deficient. True, we have grown familiar with spacecraft photos of that lonely-looking globe hanging in the void, covered with a life-sustaining biosphere that makes our planet uniquely beautiful and leaves it uniquely fragile. Yet we know next to nothing about the workings of the Earth's ecosystem. While we are now unlocking the secrets of DNA and life's molecules, we have only begun to grasp the nature of planetary life as an organic whole. Scientists are only just formulating questions such as "To what extent does the network of all living things actively control the make-up and temperature of the atmosphere, to maintain our biosphere as a single life-support system?" What, in other words, is the nature of the Gaian functioning of our planet? Far from supplying essential answers, we are not even sure what are the right questions to ask. And despite our primitive understanding, we pursue unwitting experiments of global scope, altering the Earth's climate in a manner we barely recognize and with results that will irreversibly affect every citizen of the planet.

It is difficult to conceive of the speed at which we are overloading the Earth's ecosystems. Our numbers have almost doubled in 35 years, and are still growing fast. We are "consuming" entire regions – forests, grasslands, arable zones – at a pace that should remind us of the image of lily pads extending across a pond. If the pads double in area each day, and if the pond is to be entirely covered in 30 days, on which day is the pond half-covered? Answer: the 29th day.

Still more worrying, the degradative processes under way have built up so much momentum that they cannot be halted overnight. Our situation is like that of the captain of a supertanker. If he decides he wants to turn around, he will need several miles to slow the ship, let alone to head in a different direction. Time is not on our side.

The consequences of our impact on the Earth's ecosystem are already so pervasive and profound that they will surely persist for a long stretch into the future. The toxic chemicals we pour into the environment could well leave their poisoning impact until long after our children are gone. The desert-lands we are creating will still be apparent in several centuries' time. The green band of forest around the equator is being transformed into a bald ring that will probably endure for millennia. The extinction spasm we are imposing on wildlife species may not be made good for millions of years.

Against this backdrop of accelerating global deterioration, there are sure signs of hope. The environment agenda has undergone remarkable progress and built up an unstoppable momentum, thanks largely to the lead taken by "ordinary people" – in community groups and NGOs - whether working in conservation, recycling, and clean-up schemes or actively raising awareness and lobbying the policy-makers. Environment and development have moved from the periphery into the forefront of numerous concerns: political, economic, and security issues, from local to global. With gaining momentum, governments have begun to enact tougher environmental standards, and are steadily building ecological considerations into their policies. Underlying all this activity is the growing realization by every country that no nation can quarantine itself from ecological destruction. As the contours of our economic activity have assumed global dimensions, so our understanding that the environment knows no boundaries has become clearer.

This was well illustrated by the 1992 UN Conference on Environment and Development - the Earth Summit - where the largest number of world leaders so far assembled acknowledged many of the common threats to our common future. Sadly, the conference showed that the world's governments were not yet ready to make the leap to a collaborative sharing of the challenge of safeguarding the planet. But there were positive steps forward on a journey begun two decades before at the Stockholm Conference – a journey whose destination is sustainability.

Whether or not we hold to this course, the world now faces inevitable, convulsive change. Will it be from strong global initiatives that reverse degradation and lead to a sustainable society – or will it come from environmental deterioration, economic decline, and social disruption? Decisions taken now mean either development or decline in the future – there is no in-between. The task of building a sustainable society is truly massive: restructuring the economy, reshaping human reproductive behaviour, lifestyles, and values. Time is not on our side for such a "revolution", comparable in scale only to the two other great revolutions which have transformed human civilization.

The Agricultural Revolution began about 10,000 years ago; the Industrial Revolution little more than 200 years ago. Both required centuries to fully develop. By contrast, the revolution we now need - the Environmental Revolution - must be compressed into decades.

The Revolution has already begun, although we are hardly aware that it has. We need to force the pace, to build up the collective momentum to speed its flowering – and that force is slowly coming from all levels of society. The proliferation and influence of citizen's groups and NGOs has been a driving force, stoking the momentum for change to a sustainable society. Nation states, with a pivotal role to play, are finding that it is possible to reach a consensus and collaborate on many issues. Some corporations, too, have acted – reducing air and water pollution, waste generation, and energy use; looking within to their own production processes, as well as out to the success of their product in the marketplace. The public, who buy the products of corporations, are also playing a key role; the waves of ethical and green consumerism, and the appearance of product labelling, testify to this. Change starts with individuals, and individuals can and do make lifestyle decisions - using energy more efficiently and reducing overall levels of consumption – to promote a sustainable society.

The two previous revolutions brought changes quite unforeseen – or unimaginable – to those who lived through their dawn. Those people who, around 10,000 years ago, began to domesticate some animals and plants, could not know that their descendants would as a result abandon a nomadic hunter-gatherer lifestyle, and stay in one place – a totally new concept that began to alter the face of the planet. The advent of the Industrial Revolution changed everything again – machines, railways, factories, even capitalism were born. In each revolution, human thought was also profoundly changed. In what new ways will our thinking and our world appear after the full flowering of the Environmental Revolution? The prospect of change is daunting, but the promise – of living in accord with our planet Gaia – is too compelling for us to remain unmoved.

Norman Myers with the Gaia Editors

Appendices

GENERAL EDITOR
NORMAN MYERS

Dr Norman Myers is a best-selling author and journalist, and an international consultant on environment and development, especially in Third World regions of the tropics. He travels widely, his consultancy having taken him to over 70 countries, and has worked with numerous organizations notably the World Bank, US AID, the International Union for Conservation of Nature, the World Wide Fund for Nature, FAO, UNEP, Rockefeller Brothers Fund, and the US State Department. In 1983 he won the WWF Gold Medal for his work on conservation. His articles have been published in popular and scientific journals worldwide, from *National Geographic* and the *New York Times* to *Science*, and he has broadcast frequently on British and American television. His published books include *The Long African Day, The Sinking Ark,* and *The Gaia Atlas of Future Worlds.*

with
John Elkington, Ken Laidlaw, Joss Pearson and Philip Parker, Timothy O'Riordan, Frank Barnaby (revised edition)

PRINCIPAL CONSULTANTS
Erik Eckholm David Hall, Tom Burke, Sidney Holt, Dorothy Myers, Uma Ram Nath, David Pimentel, Melvin Westlake

CONSULTANTS

Erik Eckholm Overall consultant and reader. Editor of *Natural History,* New York City. Formerly, on the State Department's Policy Planning Staff, Senior Researcher with Worldwatch, and Visiting Fellow at the IIED.
David O. Hall BSc PhD Overall consultant and reader. Professor of Biology, King's College, University of London. Chairman of the School of Biological Sciences, London.
Project Leader in the EEC Energy Programme.
Tom Burke General consultant, and special advisor on ELEMENTS. Director of Green Alliance. Press Officer, European Environment Bureau, London.
Sidney Holt DrSc Consultant on OCEAN. Consultant on marine mammals to international organizations. Former Director of Fisheries Resources, FAO. Holder of the World Wildlife Fund Gold Medal.
Dorothy Myers Consultant on HUMANKIND, CIVILIZATION, MANAGEMENT. BA Research Fellow University of Nairobi. Consultant to the UN Centre for Human Settlements (Habitat). Closely involved with UN Environment and Human Settlements programmes, she is widely published on those and related issues.
Uma Ram Nath MA Consultant on HUMANKIND. Lecturer, University Of Delhi. Co-editor of *Third World: EEC,* London. Writer, editor, broadcaster on women, health, art and Third World issues.
David Pimentel Reader and consultant for LAND. Professor with the Department of Entomology and Section of Ecology and Systematics, New York State College of Agriculture and Life Sciences, Cornell University.
Melvin Westlake Consultant on CIVILIZATION. Business and Economics editor of *South.* Formerly, economics writer on *The Times.* Writer and broadcaster on all aspects of finance and economics including Third World policy.
Timothy O'Riordan Overall reader for the revised edition and contributor to ELEMENTS. Professor in the School of Environmental Sciences, University of East Anglia. European Executive Editor for *Environment* magazine.
Frank Barnaby Consultant on MANAGEMENT and contributor to original and revised editions. Widely published on disarmament and peace-related issues. Formerly, Director of Stockholm International peace Research Institute. Co-chair of World Disarmament Movement.

CONTRIBUTORS

Peter Adamson MA Writer, journalist and broadcaster for the UN and the BBC on development issues. Co-founder and former Editor of *New Internationalist.* Main contributor to the UNICEF State of the World's Children report.
Stewart Ainsworth BA Lecturer in Education, Jordan Hill College of Education, Glasgow.
Raúl Hernan Ampuero BA MA Course Manager, The Open University, UK. Producer, BBC External Services.
Brian Anson Dipl Arch National Planning Aid Officer, Town and Country Planning Association. Consultant for the Irish Government on environmental planning.
David Baldock BA Director of Earth Resources Research Ltd, London. Widely published on environmental aspects of food and agricultural policy.
Barry Barclay New Zealand environmentalist and film-maker, specializing in genetic conservation and Third World issues.
Pamela M. Berry BSc PhD Lecturer, Mansfield College and St Edmund's Hall, Oxford. Demonstrator, School of Geography, Oxford University.
Patricia W. Birnie BA PhD Lecturer in Law at the London School of Economics. Widely published on the Law of the Sea and environmental laws.
John R. Bowers BA Arch MSc Senior consultant with international development organizations on urban design and environmental land planning.
Julian Caldecott BSc PhD Researcher, World Wildlife Fund, Malaysia.
William D. Clark MA President, International Institute for Environment and Development, London. Formerly, Vice-President, External Relations, World Bank; PR Adviser to the Prime Minister, UK; Commonwealth Fellow and Lecturer in Humanities, University of Chicago.
Trevor D. Davies BSc PhD Lecturer, School of Environmental Sciences, University of East Anglia. Meteorologist, specializing in pollutants.
Paul R. Ehrlich Bing Professor of Population Studies at Stanford University. Expert on ecology and evolution and widely published on environmental crisis, population control and racial justice.
John Elkington contributed to the shaping and polishing of the manuscript of the first edition of this book. He is a Director of Earthlife UK and of Bioresources Ltd. His published work concentrates on the environmental implications of changing technologies.
T. Scarlett Epstein PhD Research Professor, African Studies, University of Sussex. Writer, journalist, documentary film-maker.
Peter G. H. Evans BSc BPhil Research Population Geneticist at the Edward Grey Institute of Field Ornithology, Department of Zoology, Oxford University.
Graham Farmer BSc PhD Senior Research Associate, Climatic Research Unit, University of East Anglia. Climatologist, specialist in crop-weather modelling and agricultural drought.
Lois Marie Gibbs Founder and Executive Director of the Citizen's Clearinghouse for Hazardous Waste (Center for Environmental Justice), National Office in Virginia, USA. Campaigner and networker on grass-roots environmental issues.
A. John R. Groom BSc MA PhD Reader in International Relations, University of Kent. Widely published in the field of conflict research.
Jeremy Harrison BSc Head of the Protected Areas Data Unit of the IUCN Conservation Monitoring Centre, based at Kew, UK.
Paul Harrison MA Writer and journalist specializing in the Third World. Regular contributor to *The Guardian, New Society, New Scientist* and to major development agencies such as WHO and ILO.
Tony Hill BSc Development Desk Programme Officer, Catholic Institute for International Relations.
Colin Hines BA Environmentalist and author. Researcher, Earth Resources Research Ltd, London. Coordinator, London Energy and Employment Network.
P. Mick Kelly BSc PhD Senior Research Associate, Climatic Research Unit, University of East Anglia.
Gillian Kerby BSc Researcher, Animal Behaviour Research Group, Department of Zoology, Oxford University.
Derrick Knight MA Researcher and journalist Christian Aid. Writer, documentary film-maker with special interest in Central America.
Ken Laidlaw was involved in conceptual planning and research for the first edition of this book. He is a journalist specializing in environment and development issues, and author of *Crisis Decade in the Eighties,* ICDA.
Roy Laishley MA Journalist, specialist in agricultural development and commodity issues in developing countries.
Jean Lambert Principal speaker of the Green Party, UK. Council member of Charter 88 and a founder of the Ecological Building Society.
Alan Leather MPhil Research Officer, Trade Union International Research and Education group, Ruskin College, Oxford.
Stephanie Leland Feminist, active campaigner in the Green movement. Founder of Women for Life on Earth, UK.
James E. Lovelock Fellow of the Royal Society. Contributor to the NASA space programme since 1974. Author of *Gaia: a new look at life on Earth,*

OUP, and *Healing Gaia*, Harmony Books.
Simon Lyster MA Solicitor, Member of the New York Bar. Specialist in international wildlife law.
Donald J. Macintosh BSc PhD Research Fellow, Lecturer, Institute of Aquaculture, Stirling University. Adviser in aquaculture to the ODA.
Peter Marsh BSc Reporter on the *Financial Times* and formerly on the *New Scientist.*
Simon Maxwell BA MA Rural development and agricultural economist. Consultant for ODA, EEC, FAO.
Hugh Miall BA PhD Historian. Energy Consultant, Earth Resources Research Ltd, London. Gave evidence at the Sizewell B Inquiry.
Stephen Mills MA Writer, film-maker, specializing in scientific and wildlife subjects.
Barbara Mitchell Research Associate, International Institute for Environment and Development, London.
Adrian R. D. Norman MA MBA Fellow BCS. Senior Management Consultant, Arthur D Little Ltd. Adviser to governments on information technology.
David Olivier BSc Energy Consultant, Earth Resources Research Ltd, London.
Peter O'Neill BA MPhil Editor of Third World: EEC, London. Consultant to WHO. Writer, journalist, broadcaster specializing in the Third World and related issues.
Philip Parker BSc Author and editor of books and articles on environmental issues. Former principal Information Officer of Greenpeace UK, and editor of the Greenpeace magazine. Co-ordinated the revised edition of this book.
David C. Pearson MCP Berkeley, California, RIBA Principal Architect, Technical Services, London local authority. Writer and lecturer. Author of *The Natural House Book*, Fireside.
Brian Price Honorary Degree, The Open University, UK. Consultant on pollution. Adviser to Friends of the Earth.
John Rowley BSc Publications editor, IPPF, London. Journalist, editor, writer on population issues.
Peter Russell Honorary Scholar, Cambridge. Consultant to large corporations on the development of the human mind. Author of *The Awakening Earth*, Routledge and Kegan Paul.
Dr Nafis Sadik Executive Director of the United Nations Population Fund (UNFPA). Author of the annual *State of World Population* report.
David Satterthwaite BA Working Associate, Development Planning Centre, University College, London. Researcher, Human Settlements programme, IIED.
Steve Sawyer Executive Director of Greenpeace International, Amsterdam.
Viktor Sebek PhD Secretary of the Advisory Committee on Pollution of the Sea.
Dr Paul Shears MB MSc Refugee Health Coordinator, OXFAM.
Jonathan I. R. Simnett BSc MSc Coordinator of PREST, University of Manchester.
Robin Stainer Journalist. Specialist in commodities.
Kaye Stearman BEcon MA Deputy Director, Minority Rights Group, London.
Hugh Synge BSc Head of the Protected Areas Data Unit of the IUCN Conservation Monitoring Centre, based at Kew, UK.
Jorge Terena Founder of the Nucleus for Indigenous Rights and a representative of the Union of Indian Nations. Special Adviser to the Brazilian Secretary of the Environment.
Harford Thomas BA Journalist, writer, columnist. Formerly, Deputy editor and Financial editor of *The Guardian.*
Jane Thornback BSc MSc Senior Research Officer on the Mammals Red Data Book, IUCN. Widely published on population genetics, sea birds, conservation biology.
John Valentine Writer. Specialist in nuclear affairs. Participated in the Sizewell B Inquiry.
Karl Van Orsdol BA PhD Worked for the New York Zoological Society and the World Wildlife Fund, on natural resource conservation.
János Vargha President of ISTER (East European Environmental Research Institute) in Budapest. Founder and former leader of the influential "Danube Circle" group.
Peter Willets BA MSc PhD Lecturer in International Relations, the City University, London. Writer specializing in the Non-aligned Movement.
Thomas Wilson Graduate of Princeton University. Writer and lecturer on environment, world security, growth policy, human rights. Consultant, UN Environment Programme.

Help and information were provided by the very willing staff of many organizations, including the following: *Acid News.* Amnesty International, London. Art in Action. BBC. *BBC Wildlife.* British Oil Spill Coastal Association. BP. British Library. Canadian High Commission, London. Cancer Research Campaign. Centre for Our Common Future. Channel 4. Climate Action Network. Commonwealth Secretariat. Conservation Monitoring Centres, UK. Cornucopia Project. Council for the Preservation of Rural England. Department of the Environment, UK. Earthlife. Dfax Associates. Earthscan. Earth Resources Research Ltd, London. Eco. Ecological Parks Trust, London. Future Studies Group, University of Berkeley. FAO. Friends of the Earth UK. Gaia Foundation. Green Alliance. Greenpeace (UK and International). *The Guardian.* IBPGR. ICDA. IFDA. IIED. ILO. IMMP. IMO. Institute of International Relations. Institute of Oceanographic Sciences. Intermediate Technology Development Group, London. International Coffee Organization. International Co-operative Alliance. Intergovernmental Panel on Climate Change. IPPF. IUCN Environmental Law Centre, Bonn. IUCN, London and Gland. London School of Economics. The Minority Rights Group Ltd. National Maritime Institute. Natural Environment Research Council, UK. *New Internationalist.* The Open University, London. OXFAM. Panos Institute. Reading University. Rocky Mountain Institute. The Royal Society. Scott Polar Research Institute, Cambridge. *South.* Standard Telephone and Cables. Survival. Swedish Embassy. Traffic, UK. Tropical Products Institute, London. United Nations Conference on Environment and Development office, Switzerland. UNECE. UNEP. UNESCO. UNFPA. UNHCR, London. UNICEF. UN Information Centre, London. UN University. UN Water Decade. US Arms Control and Disarmament Agency. Warren Spring Laboratory, UK. World Conservation Centre, Gland. World Development Movement. World Meterological Organization. World Priorities Inc. Worldwatch Institute, Washington. *WWF.*

Publisher's Acknowledgements
First edition Gaia Books extends warmest thanks and appreciation to all the many, many individuals who gave their time unstintingly to the research and preparation of this book, and first and foremost to Norman and Dorothy Myers, without whom the project could never have been carried through. We also give special thanks to David Hall, for his patient guidance and advice, to Erik Eckholm, for his work as US reader and consultant, to Ken Laidlaw, for his two years of work and loyalty to the task, to John Elkington, for setting aside all his many commitments to join the team in the final months, to James Lovelock and Peter Russell for their inspiration, to Peter Brent, for his help on Evolution, to Barry Barclay, Bryan Poole and Yvette van Giap for their early work on the concept and approach; to Uma Ram Nath and Melvin Westlake for their work on the later chapters; to Sidney Holt for his long work on Ocean, to Colin Hines for his research on Land; to Tom Burke, Jon Tinker, and Paul Wachtel for their support and advice on the content; also to Maureen Keaveney, Maureen McNeil, Derek Pryce, Graham Henley, Benno Glauser, Joss Lemmers, Roy Laishley, David Kardish, and Susan Leigh for invaluable assistance in the first months' preparations; to Kyle Cathie and Phil Pochoda for their faith in the book. We would also like to thank the following for their assistance and support: Jonathan Hilton, Christopher Pick, Caroline Simpton, Sue Burt, Lesley Gilbert, Imogen Bright and Lucy Lidell for help and support; Giovanni Caselli for early illustration work; Keith Holroyd, Chris Schumacher, and Dr. Sandra Wallman; Sara Mathews, Viv Pover, Mat Cook, Brenda Breslan, Jean Stevens, Sasha Devas, and Fred Ford for assistance on design and paste up; David Thomas for research assistance; John Brewer, Alan Woodcock, and Barry Smith of Marlin Graphics; Michael Burman and Bob Gray of F.E.Burman Ltd; Anita Porcari of Mondadori Ltd; On Yer Bike; the staff of many picture agencies, in particular Aspect, Susan Griggs, and the Alan Hutchinson Library; and the host of unknown people who answered our telephone queries on data around the world. Most of all, we should like to thank our artists, who worked long and cheerfully, checking and rechecking all the minutiae of maps and diagrams, and creating the body of this Atlas.

Revised edition Gaia Books extends very deepest thanks to Timothy O'Riordan, for his constructive comments on the revised text, and to Frank Barnaby, for his assistance on the final chapter. We are extremely grateful to Nina Behrman for the high quality of her research, and to Jan McHarry, Guy Arnold, and Paul Vodden for additional research. The co-operation of the authors of the new chapter introductions is also greatly appreciated, and the invaluable assistance of Helena Paul (Gaia Foundation) and Willem Beekman and Joan Guitart (Greenpeace Marine Division) is noted. Deep thanks are also due to Ann Chandler for good-humoured typing, to Pamela Mainwaring for proof reading, to Michelle Atkinson for tireless assistance, research, and final proof reading; Bill Donohoe for illustrations; Susan Mennel for picture research; Patrick Nugent for design; and Susan Walby for production.

ESSENTIAL READING
There is an enormous amount of reference material now published in the field of environment and development. Many strongly argued polemic or philosophical popular books are also available, and more come out every month as the world's consciousness is aroused. Out of this mass of information, we recommend:

Annual or biannual publications
Brown, Lester R; *State of the World*, an annual Worldwatch Institute Report on Progress Toward a Sustainable Society, W W Norton.
Sivard, Ruth Leger; *World Military and Social Expenditures*, World Priorities Inc.
UNICEF; *State of the World's Children*, OUP
United Nations Environment Programme; *Environmental Data Report*, Blackwell.
World Resources Institute; *World Resources*, OUP

Books
Barnaby, Frank (gen ed); *The Gaia Peace Atlas*, Doubleday (1988).
Benedick, Richard Eliot; *Ozone Diplomacy*, Harvard University Press (1991).
Berry, Wendell; *Home Economics*, North Point Press (1987).
Burger, Julian; *The Gaia Atlas of First Peoples*, Doubleday/Anchor Books (1990).
Capra, Fritjof; *The Turning Point*, Simon and Schuster (1982).
Carson, Rachel; *Silent Spring*, Penguin (1965).
Caufield, Catherine; *In the Rainforest*, Heinemann (1985).
Caufield, Catherine; *Multiple Exposures*, Secker and Warburg (1989).
Clark, Mary E; *Ariadne's Thread*, St Martin's Press (1989).
Commoner, Barry; *Making Peace with the Planet*, Gollancz, UK (1989).
Conroy, Czech; *The Greening of Aid*, Earthscan, UK (1988).
Cowell, Adrian; *The Decade of Destruction*, Hodder and Stoughton, UK (1990).
Daly, Herman; *Steady-State Economics*, Island Press (1991).
Daly, Herman and Cobb, J; *For the Common Good*, Beacon Press (1989).
Dauncey, Guy; *After the Crash*, Greenprint, UK (1989).
Davidson, J. Myers, D and Chakraborty, M; *No Time To Waste*, Oxfam, UK (1992).
Durning, Alan; *How Much is Enough?*, W W Norton (1992).
Durrell, Lee; *Gaia State of the Ark Atlas*, Doubleday (1986).
Ehrlich, Paul R and Ehrlich, Anne H; *Healing the Planet*, Addison-Wesley (1991).
Ekins, Paul; *The Gaia Atlas of Green Economics*, Doubleday/Anchor (1992).
Ekins, Paul (ed); *The Living Economy*, Routledge and Kegan Paul, UK (1986).
Elkington, J and Hailes, J; *Green Consumer Guide*, Victor Gollancz, UK (1988).
Elsworth, Steve; *A Dictionary of the Environment*, Paladin, UK (1990).
Fukuyama, Francis; *The End of History and the Last Man*, The Free Press (1992).

George, Susan; *A Fate Worse Than Debt*, Penguin (1988).
George, Susan; *The Debt Boomerang*, Penguin (1992).
Gever, John, Kaufmann, David Skole and Vorosmarty, Charles; *Beyond Oil: The Threat to Food and Fuel in the Coming Decades*, Ballinger (1986).
Girardet, H; *Earthrise*, Paladin (1992).
Girardet, H; *The Gaia Atlas of Cities*, Doubleday (1993).
The Global Tomorrow Coalition; *The Global Ecology Handbook*, Beacon Press (1990).
Goldsmith, E and Hildyard, N; *The Earth Report 3*, Mitchell Beazley, UK (1992).
Goodland, Daly, Serafy, and Drost; *Environmentally Sustainable Economic Development: Building on Brundtland*, UNESCO, Geneva (1991).
Gribben, John; *The Hole in the Sky*, Corgi, UK (1988).
Gribbin, John; *Hothouse Earth: The Greenhouse Effect and Gaia*, Bantam Press (1990).
Gribbin, J and Kelly, M; *Winds of Change: Living with the Greenhouse Effect*, Headway, UK (1989).
Harrison, Paul; *The Greening of Africa*, Grafton Books, UK (1990).
Huxley, Anthony; *Green Inheritance*, Doubleday (1986).
IUCN-UNEP-WWF; *Caring for the Earth: A Strategy for Sustainable Living*, Earthscan, UK (1991).
Joseph, Lawrence; *Gaia: The Growth of an Idea*, Arkana, UK (1991).
King, Alexander and Schneider, B; *The First Global Revolution*, Pantheon Books (1991).
Lappe and Schurman; *Taking Population Seriously*, Earthscan, UK (1989).
Lean, Hinrichsen, and Markham; *WWF Atlas of the Environment*, Hutchinson, UK (1989).
Leggett, Jeremy (ed); *Global Warming: The Greenpeace Report*, OUP (1990).
Leonard, George; *The Transformation: A Guide to the Inevitable Changes in Humankind*, J P Tarcher (1981).
Litvinoff, Miles; *The Earthscan Action Handbook*, Earthscan, UK (1990).
Lovelock, J E; *Gaia: A New Look at Life on Earth*, Oxford University Press (1979).
Lovelock, J E; *Healing Gaia: Practical Medicine for the Planet*, Harmony Books (1991).
Lovelock, J E; *The Ages of Gaia*, W W Norton (1988).
Lovins, Amory B; *Soft Energy Paths*, Ballinger (1977).
Margulis, L and Olendzenski, L; *Environmental Evolution*, MIT (1992).
Margulis, L and Sagan, D; *A Garden of Microbial Delights*, Harcourt Brace Jovanovich Inc (1988).
Margulis, L, and Sagan, D; *Microcosmos*, Allen and Unwin (1987).
Mathews, Jessica Tuchman (ed); *Preserving the Global Environment: The Challenge of Shared Leadership*, W W Norton (1991).
May, John; *The Greenpeace Book of Antarctica*, Dorling Kindersley, UK (1988).
McCormick, John; *Acid Rain: The Global Threat of Acid Pollution*, Earthscan/WWF, UK (1988).
McKibben, Bill; *The End of Nature*, Viking (1990).

Meadows, Donella H; *The Global Citizen*, Island Press (1991).
Meadows, D H, Meadows, D L and Randers, J; *Beyond the Limits*, Earthscan, UK (1992).
Mendes, C and Gross, T; *Fight for the Forest: Chico Mendes in His Own Words*, Latin American Bureau, UK (1989).
Milbrath, Lester W; *Envisioning a Sustainable Society*, State University of New York Press (1989).
Myers, Norman; *The Gaia Atlas of Future Worlds*, Doubleday/Anchor Books (1990).
Myers, Norman; *The Primary Source*, W W Norton (revised 1992).
Nichol, John; *The Animal Smugglers*, Christopher Helm, UK (1988).
North, Richard; *The Real Cost*, Chatto and Windus, UK (1986).
Orr, David W; *Ecological Literacy*, State University of New York Press (1992).
Patterson, Walter; *The Energy Alternative: Changing the Way the World Works*, Boxtree, UK (1990).
Pearce, D (ed); *Blueprint for a Green Economy*, Earthscan, UK (1989).
Pearce, Fred; *Green Warriors*, The Bodley Head, UK (1991).
Pearson, David; *The Natural House Book*, Simon and Schuster (1989).
Peters, Arno; *Peters Atlas of the World*, Longman, UK (1989).
Porritt, Jonathon: *Seeing Green*, Blackwell (1984).
Porritt, Jonathon; *The Coming of the Greens*, Fontana, UK (1988).
Robertson, James; *Future Wealth*, Cassell, UK (1989).
Russell, Peter; *The Awakening Earth*, Arkana, UK (1984).
Sagoff, Mark; *The Economy of the Earth*, Cambridge University Press (1988).
Schneider, Stephen H; *Global Warming: Are We Entering the Greenhouse Century?*, Lutterworth Press, UK (1989).
Schneider, S H and Boston, P I; *Scientists on Gaia*, MIT (1991).
Schumacher, E F; *Small is Beautiful*, Harper & Row (1973).
Seymour, J and Girardet, H; *Blueprint for a Green Planet*, Dorling Kindersley, UK (1987).
Shiva, Vandana; *Staying Alive: Women, Ecology, and Development*, Zed Books (1989).
Shackleton, K and Snyder, J; *Ship in the Wilderness*, Dent, UK (1986).
de Silva, Donatus; *Against All Odds: Breaking the Poverty Trap*, The Panos Institute (1989).
Spretnak, C and Capra, F; *Green Politics: the Global Promise*, Grafton Books, UK (1984).
Stark, Linda; *Signs of Hope: Working Towards Our Common Future*, OUP (1990).
Timberlake, Lloyd; *Only One Earth*, BBC Books/Earthscan, UK (1987).
Trainer, Ted; *Developed to Death*, Greenprint, UK (1989).
Wilson, E O (ed); *Biodiversity*, National Academy Press (1988).
World Commission on Environment and Development; *Energy 2000: A Global Strategy for Sustainable Development*, Zed Books (1987).
World Commssion on Environment and Development; *Our Common Future*, OUP (1987).

MAPS AND DATA

In our familiar view of the world, we have learned to think of the North Pole as the "top" and the South as the "bottom". This, in itself, conditions much of our thinking. Worse still, since it is impossible to convey both accurate *areas* and accurate *shape* outlines of our curved continents on a flat map, every projection sacrifices one or the other. The most well known projection is the Mercator, which tends to show land masses increasingly larger as it approaches the Poles. The result is to create a Northern view of the world, with the Southern continents (which actually straddle the centre of the globe, around the Equator) reduced in significance and size.

In this book, we have often used a projection devised by Gaia, based mainly on Gall's cylindrical projection, but viewed in a different perspective, bringing the South into prominence, and countering the Northern view of the world. We have also used numerous other projections to alert the reader to different views of the planet, and to convey data in the clearest graphic form. The more recent Peters projection, used in the later chapters of this book, represents the true surface areas of the continents, thus emphasizing Latin America and Africa, but distorting their familiar shapes and patterns.

The "broken-up" maps, where the continental outlines are rearranged under tables of visual data, along graphs, or raised to different heights, are all based on statistical groupings.

Just as maps can never be entirely free of distortion, so global statistics always mask wide variations. The "average" calorie intake of the poorer South is much lower than that of the North, for instance; but this level, itself adequate, conceals the plight of around half a billion starving people. Statistics rely, too, on nations' capacity to collect and willingness to admit to, reliable data about themselves. FAO figures, for example, though "authoritative", can only be as good as the contributing national statistical services. Satellite remote sensing will, eventually, give us clearer data on global resources, but its work is only beginning. To offset global uncertainties, we have used case studies, informed analysis, and research data as a countercheck wherever possible.

The base year for current data in this book is 1990/2 unless otherwise stated. Future projections are mostly carried through to 2025. Past studies are more rare, as the correlation of world data is relatively recent. Even 15 years ago, categories used in UN statistics were very different from those in use today.

Despite all such caveats, however, the data are overwhelmingly clear in their import. Most devastating are those which show rates of soil erosion, desertification, deforestation, species loss, pollution, as well as the very fully documented facts on militarization, increasing violence, income division, human suffering, and wasted potential. Even if some estimates vary – for instance, between 17 and 25 million people are counted as refugees – they are still dreadful. And most of them are more likely to be under, rather than over, estimates.

ABBREVIATIONS

ACP African Caribbean and Pacific Countries
AFP Agence France Presse
AP Associated Press (US)
ASEAN Association of South East Asian Nations
BCSD Business Council for Sustainable Development
BODA British Overseas Development Administration
CAMDUN Conference on a More Democratic United Nations
CARICOM Caribbean Common Market
CCAMLR Convention on the Conservation of Antarctic Marine Living Resources
CCHW Citizens Clearinghouse for Hazardous Waste
CFCs chlorofluorocarbons
CHP Combined Heat and Power
CITES Convention on International Trade in Endangered Species
CLRTAP Convention on Long-Range Transboundary Air Pollution
COMECON (or Comecon) Council for Mutual Economic Aid, or Assistance (Communist Nations)
COPES Convention on the Protection of Endangered Species
CRAMRA Convention on the Regulation of Antarctic Minerals
DNA deoxyribonuclei acid
ECOWAS Economic Community of West African States
EC European Community
EEZ Extended Economic Zones
EIA Environmental Impact Assessment
ERR Earth Resources Research
ERS European Research Satellite
FAO Food and Agricultural Organization (specialized Agency of the United Nations)
FoE Friends of the Earth
GATT General Agreement on Tariffs and Trade
GDP Gross Domestic Product
GGP Gross Global Product
GNP Gross National Product
GRP Gross Regional Product
IAEA International Atomic Energy Authority
IBPGR International Board for Plant Genetic Resources
ICA International Co-operative Alliance
IIASA International Institute for Applied Systems Analysis
IIED International Institute for Environment and Development
ILO International Labour Organization (specialized Agency of the United Nations)
IMF International Monetary Fund (specialized Agency of the United Nations)
IMMP International Multi Media Promotions
IMO International Maritime Organization
INFACT International Formula Action Coalition
INGOs International Non-Governmental Organizations
IPCC Intergovernmental Panel on Climate Change
IPPF International Planned Parenthood Federation
ISA International Seabed Authority
IT information technology

ITDG Intermediate Technology Development Group
IUCN International Union for the Conservation of Nature
IWC International Whaling Commission
LDCs Less Developed Countries
MAB Man and Biosphere programme
MARPOL International Conventions for the Prevention of Pollution from Ships
MTOE Million Tonnes Oil Equivalent
NASA National Aeronautics and Space Administration (US)
NATO North Atlantic Treaty Organization
NCS National Conservation Strategy
NGO Non-Governmental Organization
NICs Newly Industrialized Countries
NIEO New International Economic Order
NOx nitrogen oxides
OAS Organization of American States
OAU Organization of African Unity
ODA Overseas Development Administration
OECD Organization for Economic Co-operation and Development
OPEC Organization of Petroleum Exporting Countries
OTEC Ocean Thermal Energy Conversion
OXFAM Formerly, Oxford Committee for Famine Relief. International charity formed in World War II, to bring aid to poverty and famine-stricken countries.
PADIS Pan-African Documentation and Information System
PHC Primary Health Care
PPMV Parts Per Million by Volume
PQLI Physical Quality of Life Index
PREST programme of Policy Research in Engineering, Science & Technology (University of Manchester)
SCAR Scientific Committee on Antarctic Research
SEWA Self-Employed Women's Association (India)
SITE Satellite Instructional Television Experiment (India)
START Strategic Arms Reduction Treaty
3P Pollution Prevention Pays
TNCs Transnational Corporations
UN United Nations
UNCLOS United Nations Conference on the Law of the Sea
UNDP United Nations Development Programme
UNEP United Nations Environment Programme
UNESCO United Nations Educational, Scientific and Cultural Organization
UNFPA United Nations Population Fund
UNHCR office of the United Nations High Commissioner for Refugees
UNICEF United Nations Children's Fund
UPI United Press International (US)
UV ultraviolet
VOC Volatile Organic Compound
WCED World Commission on Environment and Development
WCS World Conservation Strategy
WHO World Health Organization (specialized Agency of the United Nations)
WIPO World Intellectual Property Organization
WWF World Wide Fund for Nature

CREDITS

INTRODUCTION
Overall consultants: David O. Hall, Norman Myers, Joss Pearson.
pp10-11 The fragile miracle
Artist Chris Forsey
pp12-13 Accelerating evolution
Artist Bill Donohoe
pp14-15 Latecomers to evolution
Artist Gary Marsh
pp16-17 The long shadow
Artist John Shipperbottom
pp18-19 Crisis or challenge? **Photograph** Daily Telegraph Colour Library and NASA

LAND
Overall consultants: Norman Myers, David O. Hall, Erik Eckholm. **Research co-ordination:** Ken Laidlaw, Colin Hines. **Contributors:** Roy Laishley, Barry Barclay, Joss Pearson.
Principal map projection: Gaia projection.
pp22-3 The fertile soil
Artist Bill Donohoe
pp24-5 The green potential
Artist Bill Donohoe
pp26-7 The global forest
Artist Bill Donohoe
pp28-9 Tropical forests
Artist Bill Donohoe
Map Nordic projection
pp30-1 The world croplands
Artist Bill Donohoe
pp32-3 The world's grazing herd
Artist Chris Forsey
pp34-5 The global larder
Artist Bill Donohoe
pp36-7 The disappearing soil
Artist Chris Forsey
Photographs left Arthur Rothstein/Peter Newark's Western Americana; right Ivan Strasburg/Alan Hutchinson Library
pp38-9 The shrinking forest
Artist Chris Forsey
Photograph Victor Englebert/Susan Griggs Agency
pp40-1 Destroying the protector
Artist David Ashby
Photographs left Alan Hutchinson Library; right, top Jean Pierre Dutilleux/Observer Magazine; bottom OXFAM
pp42-3 The encroaching desert
Artist Eugene Fleury
Photograph Horst Munzig/Susan Griggs Agency
pp44-5 Hunger and glut
Artist Eugene Fleury
pp46-7 The cash-crop factor
Artist Eugene Fleury
Photograph Horst Munzig/Susan Griggs Agency
pp48-9 The global supermarket
Artist Chris Forsey
pp50-1 Harvesting the forest
Artist Chris Forsey
Photographs top Ed Parker/Still Pictures; bottom Duncan Poole/WWF
pp52-3 Forests of the future
Artist Bill Donohoe
Photographs top Kotoh/Zefa; bottom BBC
pp54-5 Managing the soil
Artist Chris Forsey
pp56-7 Green revolution?
Artist Bill Donohoe

pp58-9 Agriculture in the balance
Artist Chris Forsey
pp60-1 Towards a new agriculture
Artist Chris Forsey

OCEAN
Overall consultants: Sidney Holt, Norman Myers, Erik Eckholm. **Contributors:** Patricia W. Birnie, Sidney Holt, Donald J. Macintosh, Barbara Mitchell, Viktor Sebek, Jonathan I. R. Simnett, with the assistance of PREST, University of Manchester.
Principal map projection: Lambert zenithal equal-area.
pp64-5 The world ocean
Artist Chris Forsey
pp66-7 The living ocean
Artist David Mallott
pp68-9 The vital margins
Artist David Mallott
Photographs top Erica Coleman/Des Barratt; bottom left Fredrik Ehzenstrom; bottom right Donald J. Macintosh
pp70-1 The global shoal
Artist Chris Forsey
pp72-3 Ocean technology
Artist Chris Forsey
pp74-5 The polar zones
Artist Eugene Fleury
Map Azimuthal equal area projection
pp76-7 The empty nets
Artist Eugene Fleury
pp78-9 Polluting the oceans
Artist Chris Forsey
pp80-1 Destruction of habitat
Artist Eugene Fleury
Photographs left R. Eugene Turner/University of Louisiana; right Field Studies Council
Map Nordic projection
pp82-3 Whose ocean?
Artist David Mallott
pp84-5 Harvesting the sea
Artist Chris Forsey
pp86-7 Clean-up for the ocean
Artist Eugene Fleury
pp88-9 Managing Antarctica
Artist Eugene Fleury
Map Azimuthal equal-area projection
pp90-1 The laws of the sea
Artist David Mallot
pp92-3 Future ocean
Artist Aziz Khan
Photographs top National Maritime Institute; bottom J. Roessler, Planet Earth Pictures/Seaphot
Map Adapted from Athelstan Spilhaus whole ocean map

ELEMENTS
Overall consultants: Norman Myers, David O. Hall, John Elkington. **Contributors:** Hugh Miall, P. Mick Kelly, Graham Farmer, Robin Stainer, Trevor Davies, David Baldock, John Valentine, David Olivier, Brian Price, Timothy O'Riordan, Philip Parker.
Principal map projection: Gaia projection.
pp96-7 The global powerhouse
Artist David Mallott
pp98-9 The energy store
Artist Bill Donohoe

pp100-1 The climate asset
Artist David Mallott
Map Interrupted Mollweide's homolographic projection by George Philip & Son Ltd
pp102-3 The freshwater reservoir
Artist David Mallott
pp104-5 The mineral reserve
Artist Bill Donohoe
pp106-7 The oil crisis
Artist Aziz Khan
pp108-9 The fuelwood crisis
Artist Ann Savage
Map Nordic projection
pp110-1 The greenhouse effect
Artist Alan Suttie
pp112-3 Climate chaos
Artist Bill Donohoe
pp114-5 Holes in the ozone map
Artist Bill Donohoe
pp116-7 The invisible threat
Artist Bill Donohoe
Map Oblique azimuthal equidistant projection by George Philip & Son Ltd
pp118-9 Water that kills
Artist Chris Forsey
pp120-1 Widening circle of poison
Artist Chris Forsey
pp122-3 The nuclear dilemma
Artist Bill Donohoe
pp124-5 The new energy path
Artist Bill Donohoe
pp126-7 An energy-efficient future
Artist Bill Donohoe
pp128-9 Managing energy in the South
Artist David Cook
Photograph Erik Eckholm
pp130-1 The laws of the air
Artist Bill Donohoe
pp132-3 Managing water
Artist Alan Suttie
pp134-5 Clean water for all
Artist Alan Suttie
Photograph Anne Charnock
Map Nordic projection
pp136-7 Waste into wealth
Artist Alan Suttie

EVOLUTION
Overall consultants: Erik Eckholm, David O. Hall. **Contributors:** Pamela M. Berry, Julian Caldecott, Peter Evans, Stephen Mills, Jane Thornback, Gillian Kerby, Simon Lyster, Karl G. Van Orsdol, Joss Pearson.
Principal map projection: Gaia projection.
pp140-1 The life pool
Artist Bill Donohoe
pp142-3 The web of Gaia
Artist Bill Donohoe
pp144-5 Life strategies
Artist Shirley Willis
pp146-7 Partners in evolution
Artist John Potter
pp148-9 The genetic resource
Artist David Salariya
pp150-1 Keys to the wild
Artist Norman Barker of Linden Artists
pp152-3 The irreplaceable heritage
Artist George Thompson
pp154-5 The destruction of diversity
Artist Ann Savage

pp156-7 Genetic erosion
Artist John Potter
pp158-9 Towards a lonely planet
Artist Roger Stewart of Virgil Pomfret Agency
pp160-1 Conserving the wild
Artist Eugene Fleury
Map Equal area homolosine projection
pp162-3 Preserving the genetic resource
Artist Ann Savage
pp164-5 Laws and conventions
Artist Alan Suttie
pp166-7 Towards a new conservation
Artist Martin Camm of Linden Artists

HUMANKIND
Consultants Norman Myers, David O. Hall, John
Elkington, Ken Laidlaw. **Contributors:** John
Rowley, Stewart Ainsworth, Harford Thomas,
Uma Ram Nath, Peter O'Neill, Raúl Hernan
Ampuero, Derrick Knight, Stephanie Leland,
Kaye Stearman.
Principal map projection: Peters projection.
pp170-1 People potential
Artist Alan Suttie
pp172-3 The world at work
Artist Eugene Fleury
Maps Gaia projection
pp174-5 Homo sapiens
Artist George Thompson
pp176-7 The numbers game
Artist Eugene Fleury
pp178-9 The work famine
Artist Chris Forsey
pp180-1 Sickness and stress
Artist George Thompson
pp182-3 The literacy chasm
Artist John Potter
pp184-5 Outcasts and refugees
Artist John Potter
Photograph Adam Woolfitt/Susan Griggs Agency
pp186-7 Managing numbers
Artist John Potter
pp188-9 The voice of women
Artist John Potter
Photographs *top* Howard Smith; *centre* Amnesty
International; *bottom* Ed Baker
pp190-1 Health for all
Artist George Thompson
pp192-3 Tools and ideas
Artist Norman Barber of Linden Artists

CIVILIZATION
Overall consultants: Norman Myers, David O.
Hall, John Elkington, John Bowers.
Contributors: David Satterthwaite, Peter Marsh,
Adrian Norman, Melvin Westlake, Simon
Maxwell, Jonathon Hilton, Philip Parker.
Principal map projection: Peters projection.
pp196-7 The world city
Artist Ann Savage
pp198-9 The world factory
Artist Roger Stewart of Virgil Pomfret Agency
pp200-1 The power of communications
Artist Chris Forsey
Map Mollweide's interrupted homolographic
projection
pp202-3 The world market
Artist Chris Forsey

pp204-5 The world's wealth
Artist Clive Spong of Linden Artists
pp206-7 Chaos in the cities
Artist Aziz Khan
Photographs *top* Brian Anson; *bottom* UNICEF
pp208-9 Monopolies in media control
Artist Roger Stewart of Virgil Pomfret Agency
Map Oblique azimuthal equidistant projection
pp210-1 Market tremors
Artist Clive Spong of Linden Artists
pp212-3 Haves and have-nots
Artist Aziz Khan
pp214-5 Environmental trade-off?
Artist Bill Donohoe
pp216-7 The poverty bomb
Artist Ann Savage
Photograph FAO Photo Library
pp218-9 Urban regeneration
Artist David Cook
pp220-1 The will to communicate
Artist Bill Donohoe
pp222-3 Technology transfer
Artist Nicky Snell of Virgil Pomfret Agency
pp224-5 Growing interdependence
Artist Clive Spong of Linden Artists
pp226-7 Closing the gap
Artist Ann Savage
pp228-9 New rules of the game
Artist Bill Donohoe

MANAGEMENT
Overall consultants: Frank Barnaby, Erik
Eckholm, David O. Hall, John Elkington.
Contributors: Frank Barnaby, Dr A. John R.
Groom, Peter Russell, V. Tarzie Vittachi, Peter
Willetts, Thomas Wilson, Philip Parker.
Principal map projection: Peters projection.
pp232-3 Phoenix nation state
Artist Bill Donohoe
pp234-5 Reluctant internationalists
Artist Alan Suttie
pp236-7 Voice of the world
Artist Francesca Pelizzoli
pp238-9 Voice of the village
Artist Francesca Pelizzoli
pp240-1 Voice of the individual
Artist Bill Donohoe
pp242-3 Breaking points
Artist Aziz Khan
pp244-5 Towards a violent planet
Artist Ann Savage
pp246-7 The cost of militarism
Artist Norman Barber of Linden Artists
pp248-9 One million Hiroshimas
Artist Eugene Fleury
pp250-1 Bang or whimper?
Artist Roger Stewart of Virgil Pomfret
Agency
pp252-3 Global governance
Artist Bill Donohoe
pp254-5 The great transition
Artist Gary Marsh
pp256-7 Signs of hope
Artist John Shipperbottom of Virgil Pomfret
Agency
pp258-9 A new ethic
Artist Jim Channel of Linden Artists

ADDITIONAL PHOTOCREDITS
Contents pp6-7 *Peru, pilgrims at shrine, Quyllur,*
Victor Englebert/Susan Griggs Agency; **Foreword**
The Maldive Island, aerial photograph by Adam
Woolfitt/Susan Griggs Agency; **Land** p.21
*Philippines – agriculture, Tom Spiegel/Susan Griggs
Agency;* **Ocean** p.62 *Rainbow Warrior at Sea,*
Ferrero/Greenpeace; p.63 *Walruses, Alaska,* S.
Krasemann/Natural History Photographic
Agency; **Elements** p.95 *San Raphael Falls, Ecuador,*
Tony Morrison; **Evolution** p.139 *Planes minutus on
Pegea Socia,* G. R. Dietzmann/Seaphot Ltd: Planet
Earth Picture; **Humankind** p.169 *Ethiopia, roofing
village hut,* Dr. Georg Gerster/The John Hillelson
Agency; **Civilization** p.195 *Leningrad and the Gulf
of Finland,* NASA and Daily Telegraph Colour
Library; **Management** p.231 *Ethiopia, Lalibela
clerics dancing,* Dr. Georg Gerster/The John
Hillelson Agency.

INDEX